Clinical Microbiology for Diagnostic
Laboratory Scientists

Clinical Microbiology for Diagnostic Laboratory Scientists

Sarah J. Pitt

School of Pharmacy and Biomolecular Sciences
University of Brighton, UK

Registered Office(s)
John Wiley & Sons, Inc., 111 River Street, Hoboken, NJ 07030, USA
John Wiley & Sons Ltd, The Atrium, Southern Gate, Chichester, West Sussex, PO19 8SQ, UK

Editorial Office
The Atrium, Southern Gate, Chichester, West Sussex, PO19 8SQ, UK

For details of our global editorial offices, customer services, and more information about Wiley products visit us at www.wiley.com.

Wiley also publishes its books in a variety of electronic formats and by print-on-demand. Some content that appears in standard print versions of this book may not be available in other formats.

Library of Congress Cataloging-in-Publication Data

Names: Pitt, Sarah J.
Title: Clinical microbiology for diagnostic laboratory scientists / by Sarah J. Pitt.
Description: Hoboken, NJ: Wiley-Blackwell, 2018. | Includes bibliographical references and index. |
Identifiers: LCCN 2017033966 (print) | LCCN 2017035046 (ebook) | ISBN 9781118745830 (pdf) |
   ISBN 9781118745823 (epub) | ISBN 9781118745854 (paperback)
Subjects: LCSH: Diagnostic microbiology. | BISAC: SCIENCE / Life Sciences /
   Biology / Microbiology.
Classification: LCC QR67 (ebook) | LCC QR67 .P58 2018 (print) | DDC 616.07–dc23
LC record available at https://lccn.loc.gov/2017033966
Highlight Quote

Cover Design: Wiley
Cover Image: © Fredrick Kippe/Alamy Stock Photo

Set in 10/12pt Warnock by SPi Global, Pondicherry, India

Printed in Singapore by C.O.S. Printers Pte Ltd

10 9 8 7 6 5 4 3 2 1

For Alan

Love appears to be an infection which cannot be cured

And for my parents

Margaret and John

Contents

Preface

This book is intended for post-registration and post-graduate level scientists who are developing careers in diagnostic clinical microbiology. It is suitable for those who are working in hospital laboratories while studying for advanced qualifications, as well as full-time MSc students. The aim is to prompt readers to make connections between the clinical symptoms, pathogenesis of infections and the approaches used in laboratory diagnosis. This is not a comprehensive account of all aspects of clinical microbiology, but a consideration of a range of infections caused by selected pathogenic bacteria, viruses, fungi, protozoa and helminths. The idea is to use these examples to illustrate clinical and diagnostic issues, to stimulate critical appraisal of published evidence and to encourage problem-solving in the clinical laboratory context.

There is an introductory chapter, which outlines the scope of clinical diagnostic microbiology and the key areas for the laboratory scientist to be aware of. In the subsequent six chapters, a type of infection is reviewed in depth, using particular pathogenic microorganisms to illustrate salient points. As well as journal articles, there are references to publically available epidemiological data and professional guidelines throughout the book. This includes links to specialist websites. I hope that the reader will find the mixture of sources of information useful and a helpful place to start their own exploration of topics which interest them. At the end of each chapter there are three exercises related to management of a diagnostic service and assessing the suitability of test methods to specific contexts. There are no right or wrong answers to these, but the reader could discuss them with their laboratory colleagues or university tutor. Chapters 2 to 6 also include clinical case studies based on the content of the chapter. Application of appraisal and problem-solving skills should lead to the solutions in each case, but the outline answers are provided in Appendix 1, so the reader can check their interpretation of the information.

Pathology laboratory services seem to be continually in the process of reorganisation and reconfiguration. The ways in which laboratory scientists are expected to work is changing, due to the possibilities afforded by new techniques. This often includes requirements for multi-site working patterns and a more multidisciplinary outlook. However, the laboratory scientist should never lose sight of the point of their work, which is to help the patient. Microorganisms and the diseases that they are associated with are endlessly fascinating – and new pathogens are being discovered all the time! The graduate microbiologist can make a valuable contribution to patient care through questioning received wisdom, investigating different laboratory methods and evaluating research.

Sarah J. Pitt

Acknowledgements

I would like to thank my colleagues Rowena (Bertie) Berterelli, Heather Catty, Cinzia Dedi, Dr Annamaria Gal, Joseph Hawthorne and Simonne Weeks at the University of Brighton and Lindsey Dixon, Judith Feeney, Cassandra Malone, Nicholas O'Flanagan, Clare McKeon and Clare Reynolds at Brighton and Sussex University Hospitals NHS Trust/Frontier Pathology for their help and support. It is thanks to all of them that I was able to collect the pictures of slides and photographs of test strips and cultures used for many of the figures in this book. I would also like to thank Ian Phillips who recently retired from working at Public Health Wales, Cardiff, but not before allowing me to use some of his pictures.

I would especially like to thank Dr Alan Gunn of Liverpool John Moores University for his unfailing encouragement throughout the process of writing this book, for proof-reading and commenting on drafts and also for providing material for some of the figures.

Thanks are also due to the staff at Wiley-Blackwell for their assistance at the various stages of the project.

1

Overview of Clinical Diagnostic Microbiology

1.1 Introduction

The scope of diagnostic microbiology has developed along with technological advances in laboratory science. In the nineteenth century, detection of organisms relied on light microscopy and a limited range of *in vitro* culture methods. Some of these techniques are still used to detect and identify bacteria and parasites, in some cases with little or no modification (e.g., Gram's stain). The discovery of proteins and then later nucleic acids during the twentieth century, along with the advent of the electron microscope, opened up the microbial world. The ability to determine and characterise the exact cause of an infectious disease, and thus to devise control measures, treatments, and prophylaxes has reduced morbidity and premature mortality throughout the world. Indeed, some diseases – smallpox in humans and rinderpest in cattle – have been eliminated altogether.

The ever-expanding range of known microorganisms has led to the separation of diagnostic microbiology into distinct specialist areas – bacteriology, virology, mycology and parasitology. The diversity of microorganisms necessitates the training of individuals to be experts in one particular type of organism and they are often considered separately during study. It is important to remember that microorganisms do not operate in isolation and that a patient can be infected with more than one pathogen simultaneously. Advances in technology, in particular automation, make it possible to test for markers across pathology disciplines (not just within microbiology) on one single specimen in relatively short turnaround times. This means that the diagnostic microbiologist must be aware of the implications of an eclectic mixture of results and should certainly be able to solve problems and make decisions about patients with a wide range of infections.

N.B: This book assumes microbiological knowledge to at least graduate level. There are some very good general textbooks available for those who would like to revise any areas. For example, Greenwood *et al.* (2012) provides a brief overview of a comprehensive range of organisms, while clinical features are addressed clearly and simply in Murray *et al.* (2013) and Goering *et al.* (2013). Laboratory aspects are covered by Ford *et al.* (2014) and Wilkinson (2011), which were both written by practising diagnostic microbiologists.

Point to consider 1.1: What do you think is the most important scientific discovery in microbiology during the last 100 years and why?

1.2 Organisation and Management of Diagnostic Microbiology Services

The configuration of pathology services seems to be constantly changing in response to management trends, as well as financial constraints. Technological advances mean that it is easier to identify microorganisms to the level of a strain within a species, in a much shorter amount of time than it was 30 years ago. The techniques available have also been developed such that many aspects of the laboratory work can be done in automated systems (e.g., Fournier *et al.*, 2013). The graduate and postgraduate microbiology scientist must therefore be able to interpret ever more complicated data, while retaining a working knowledge of the principles behind the tests used to generate that information. As laboratory microbiologists develop expertise and are given more responsibility, they sometimes find themselves working in larger teams, while at other times they are concentrating on a specialised area with a small number of colleagues. Cooperation with staff across pathology and in other departments in the hospital and primary care is becoming increasingly important in the daily running of the diagnostic microbiology service. Also, responding to the requirements of external bodies such as those involved in public health surveillance, quality assessment and service monitoring can create enough work for a full-time post. Employing organisations can come and go, while the people who staff the laboratories often stay the same. This can be disconcerting, but it is important to keep sight of the facts. Firstly, each sample received in the laboratory has been collected from a patient with a clinical problem which needs addressing such as, 'Has there been an adequate immunological response to this course of a vaccine?', 'What is causing this rash?' and 'Why does this antimicrobial treatment appear to be ineffective?' Secondly, microbiology is endlessly fascinating; organisms can change very quickly – rendering diagnostic methods and treatments out of date – and infectious diseases are always in the news!

Point to consider 1.2: How many news stories involving microorganisms have you noticed during the last seven days? Find one article and think carefully about how well it was reported. Which points of information do you think were communicated well? Where there were inaccuracies, can you think of reason for the misunderstandings?

1.3 Techniques

Some techniques used to isolate and identify microorganisms have not changed substantially since they were first introduced – which in some cases is well over 100 years ago. This is usually because they are still useful and cost effective in helping to isolate and identify microorganisms from clinical samples. Others have been superseded by methods which give a more accurate or rapid result. It is therefore useful to critically evaluate techniques and to challenge assumptions about them. Does a traditional technique such as the Gram's stain still have a place in twenty-first century routine diagnostic microbiology? Why did virology services phase out virus isolation in monolayer cell culture? Is 'PCR' really the answer to every search for the optimal microbiology test?

The principles behind each technique will not be considered in any detail within this book. Readers are referred to laboratory protocols or standard operating procedures as appropriate, along with Wilkinson (2011) and Ford *et al.* (2014) for details.

Table 1.1 outlines some of the discussion points to consider about a selection of the techniques commonly used in diagnostic microbiology settings. The suite of tests used to investigate a particular clinical problem depends on a range of factors, which may include workload, available space, skill mix of workforce and service user requirements, as well as cost.

Point to consider 1.3: Select one of the techniques listed in Table 1.1 and think about how it is currently used in routine diagnostic microbiology (it might be helpful to focus on a type of infection, such as urinary tract infections, or a group of organisms, such as viruses). Do you think this method is used in the best way? Could the technique be improved? Would another type of test give better results?

1.4 Point-of-Care Testing (POCT)

There is a wide and expanding range of point-of-care tests (POCTs) for detection of microbial antigens or specific antibodies (Pitt, 2012; Moore, 2013; St John and Price, 2014). Many POCTs (also referred to as 'near patient tests' [NPTs] and 'rapid diagnostic tests' [RDTs]) utilise the lateral flow/immunochromatographic assay format. However, some use quite simple technology such as latex agglutination and others more complex principles including polymerase chain reaction (PCR) or other molecular methods (Moore, 2013). All are designed to be used outside of the main laboratory, by personnel who might not be laboratory staff – both healthcare professionals and in some cases, patients themselves. While it can be very useful clinically to get a rapid test result, there are a number of limitations of POCTs which scientists must be aware of (see Text Box 1.1).

It is important that senior laboratory scientific staff are involved in discussions about possible implementation of a POCT assay outside of the laboratory, such as on a hospital ward or in primary care. The main scientific issue with currently available POCT antigen detection tests is that most have relatively low sensitivities, although specificities are generally very good (Text Box 1.1; Moore, 2013). This means that the positive results can be trusted, but when the POCT does not detect the organism, follow-up laboratory tests are required to confirm the result. Maintenance of quality assurance is also a concern (Pitt, 2012; St John and Price, 2014) and the purpose of required procedures and documentation must be highlighted to those performing the test (which might include the patient). How the assay results are to be recorded must also be considered, in order that POCT test results are included in the patient's pathology profile (St John and Price, 2014). Some bench top analysers (e.g., PCR devices) keep an electronic record of the result, but this is not useful unless it can be connected to the main laboratory database. The other key factor in the decision of whether and how to use a POCT in microbiology is the cost per test (Pitt, 2012; St John and Price, 2014). While the price of kits themselves is usually relatively high, this has to be balanced against the staff costs (a senior scientist in the laboratory is more expensive per hour than a healthcare assistant on the ward) and time taken to make the diagnosis, implement treatment and discharge the patient. The proposed

Table 1.1 Comparison of selected techniques used in diagnostic microbiology departments to isolate and characterise microorganisms.

Technique	Use in diagnostic microbiology	Advantages	Disadvantages
Light microscopy (with appropriate stains to aid visualisation)	Detection of bacteria, fungi and parasites	Low cost Sample preparation straightforward and relatively quick Equipment easy to obtain and maintain 'Catch all' technique	Relatively low sensitivity Limit of resolution around 0.2 μm Recognition of organisms requires training and skill Slides examined individually so labour-intensive and time-consuming Follow-up tests usually necessary for full identification.
Light microscopy with fluorescent-labelled antibodies and UV microscope	Detection of bacteria, fungi and parasites	Very specific due to immunological detection method Good sensitivity as fluorescence easy to detect.	Expensive compared to use of general stain Detects pre-determined organisms only Slides examined individually so labour-intensive and time-consuming If no organism detected, further investigations may be indicated Requires ultraviolet (UV) microscope.
Culture on agar including selective, differential and chromogenic media	Isolation and initial identification of bacteria and fungi	'Catch-all' technique Selective media or specialised culture conditions exclude all but organism of interest to grow Chromogenic media allow presumption identification on basis of colonial appearance and colour change Live viable organism obtained to do further investigations –e.g., antimicrobial sensitivities, analysis of strain variation	Preparation of media time-consuming (most laboratories buy pre-prepared plates) Follow-up tests usually necessary for full identification.

Technique	Application	Advantages	Disadvantages
Electron microscopy	Detection of viruses and small protozoa in clinical samples Detection/identification of viruses isolated in monolayer culture.	Limit of resolution is around 2 nm 'Catch all' technique Organisms identified due to morphological appearance.	Relatively low sensitivity Grids examined individually so labour-intensive and time-consuming Equipment very expensive to purchase and maintain Identification of structures requires specialised training and skill.
Monolayer cell culture	Detection of viruses (and some other obligate intracellular organisms) in clinical samples	'Catch-all' technique Detects viable viruses Live organism obtained to do further investigations – e.g., antiviral sensitivities, analysis of strain variation.	Takes days or weeks for viruses to grow and cells to show cytopathic effect (CPE) Requires specialised training and skill to interpret CPE Expensive Ethically questionable to use animal cells – especially primary cells.
Detection of antigen by enzyme immunoassay	Detection of bacterial, viral and protozoal antigens in clinical samples particularly blood and faeces	Good sensitivity and specificity Majority of assays provide result within a few hours Generic training sufficient to perform assay Can be automated to allow large batches of samples to be tested simultaneously.	Detects pre-determined organisms only If no organism detected, further investigations may be indicated Detects antigen rather than live viable organism Antigen may be genus-specific only Species-specific assays may not detect strain variation.
Detection of antibodies by enzyme immunoassay	Detection of specific antibody response to bacteria, viruses, fungi, protozoa and helminths in blood and other fluids	Good sensitivity and specificity Majority of assays provide result within a few hours Generic training sufficient to perform assay Can be automated to allow large batches of samples to be tested simultaneously Assays available to detect different classes of antibody which helps to indicate stage of infection.	Detects antibody to pre-determined organisms only If no specific antibody detected, further investigations may be indicated Detects evidence of presence of organism rather than live viable organism.

(Continued)

Table 1.1 (Continued)

Technique	Use in diagnostic microbiology	Advantages	Disadvantages
Polymerase chain reaction	Detection of specific DNA or RNA sequence characteristic of virus, bacterium, fungus, protozoan or helminth in clinical samples	Very specific and sensitive Majority of assays provide result within one working day Generic, albeit specialised, training sufficient to perform assay Can be automated to allow batches of samples to be tested simultaneously Multiplex assays allow assay for suite of organisms in single test PCR product can be sequenced to allow further characterisation.	Detects genome of pre-determined organisms or set of organisms only If no genome detected, further investigations may be indicated Detects genome of organism rather than live viable organism Relatively expensive Requires specialised equipment, several preparation and analytic steps and separate rooms to minimise contamination.
Matrix-assisted laser desorption/ ionisation – time of flight (MALDI-TOF)	Identification of bacteria and fungi from isolates	Specific and good sensitivity Provides result quickly and easily in comparison to a full set of biochemical tests Relatively cheap Protein sequence data can indicate antimicrobial sensitivities.	Requires comprehensive database – may not recognise new or unusual organisms.

Text Box 1.1 Advantages and Limitations of using Point-of-Care Tests for Infectious Diseases

Advantages

Provide rapid results; allows quick decision about management.
Require small amount of relatively easy to collect sample.
Easy to use; can be performed by anyone who has been given suitable training (and by following SOP instructions).
Kits often contain all required reagents and equipment.
Specificity usually very good (up to 100%).

Limitations

Expensive compared to standard laboratory assays.
Any excess sample discarded; follow-up tests require fresh collection.
Quality assurance can be overridden/overlooked.
Sensitivity usually relatively low (50% in some cases).

---

role of the POCT test in the patient care pathway must therefore be considered carefully (Moore, 2013; St John and Price, 2014).

Because they are quick and easy to carry out, microbiology POCTs are also widely used as part of main laboratories' repertoires, as initial or confirmatory tests – in spite of their limitations. In this situation, the information from the POCT can be put with other results. Outside of the laboratory, the most obvious role for POCT tests is where a wait for the results might compromise patient care. Examples of this type of situation include an outbreak of an infection (e.g., norovirus) where diagnosis is important for the individual, but is also necessary to minimise spread by isolation and cohort nursing of affected patients (Moore, 2013). Genitourinary medicine clinics have found POCT tests useful in situations where patients might not be prepared to wait or return at a later time for a result, for instance for a syphilis test (Pitt, 2012). They also have a role where the main laboratory is not easily accessible, such as the bench top device for measuring blood CD4+ counts in HIV patents living in remote areas (St John and Price, 2014). Use of a *Plasmodium* RDT in rural clinics in endemic countries (instead of relying solely on a clinical picture of pyrexia and rigors), should reduce misdiagnosis of malaria. Erroneous presumption of *Plasmodium* infections can lead to over use of anti-malarial drugs and the potential lack of appropriate treatment for other conditions (Pitt, 2012). Another interesting way in which microbiology POCTs have been used is in 'satellite' laboratories, within a multi-site healthcare organisation (e.g., Cohen-Bacrie *et al.*, 2011). The idea is to provide results at each hospital site where possible, to avoid the costs, logistical difficulties and risk of loss of integrity of the specimen, inherent in transport between locations. In this case laboratory staff are performing the tests, so they should have good insight into the limitations and also immediate access to support from the main laboratory (in terms of advice and the option of sending samples for further tests) if necessary. For more in depth consideration of the range and potential applications of POCT within diagnostic services, the reader is referred to other sources, such as Price and St John (2012).

Point to consider 1.4: How do you think microbiology POCTs could be used to best effect?

1.5   Antimicrobials

The advent of antimicrobial agents during the twentieth century drastically changed people's attitudes to infection. The list of feared, commonly fatal diseases changed from including infections including scarlet fever, gangrene from infected wounds, tuberculosis and syphilis, to various type of cancer. The course of non–life-threatening infections can also be shortened, which reduces the number of days the individual loses to illness, as well as decreasing the risk of secondary infections (Davey *et al.*, 2015). Since antibiotics are generally considered as safe drugs to take and courses of treatment are time-limited, the expectation by patients that they can be taken to treat even mild infections has arisen. Widespread overuse and misuse of the drugs have created the niches that microorganisms needed to develop resistance and this is now considered a major threat to public health (e.g., Shallcross and Davies, 2014). In response, guidelines are being issued for prescribing, which encourage more careful deliberation of the clinical need (e.g., NICE NG15, 2015). The microbiology laboratory scientist should have an understanding of the range of antimicrobial agents, their mechanisms of action (including selective activity) and the nature of mutations leading to resistance. This is to help inform decisions about antimicrobial testing protocols within the laboratory and also to understand how to contribute to surveillance for reduced susceptibilities. An overview of antibiotics, antivirals and antifungals is given below, while the treatment and management of the infections included in this book are addressed in the relevant chapters. Detailed discussion of antimicrobial activity, resistance mechanisms and treatment regimes is beyond the scope of this book and the reader is referred to Davey *et al.* (2015) – or other suitable texts – for in depth consideration of this subject.

1.5.1   Antibacterial Agents (Antibiotics)

As Figure 1.1 indicates, there is a range of potential weak points in the bacterial life cycle which could be targets for antibacterial agents. The differences in structure and metabolism between prokaryotic and eukaryotic cells mean that the effects can be quite specific, with low toxicity for the mammalian host. Table 1.2 summarises the mode of action and mechanism(s) of resistance for selected antibacterial drugs. Some antibacterial agents also work against other organisms. Examples include co-trimoxazole (trimethoprim plus sulphamethoxazole) which is used to treat *Pneumocystis jirovecii* pneumonia and metronidazole is often the drug of choice for infections with the gut protozoa *Entamoeba histolytica* and *Giardia duodenalis*.

1.5.2   Antiviral Agents

There are relatively few effective antiviral agents and they usually have limited and very specific uses. This is partly due to the lack of opportunities for selective toxicity – since viruses are reproducing inside host cells it can be hard to find a target peculiar to the virus. The other problem is the speed at which resistant virus strains can arise, particularly in RNA viruses. The first safe, effective and widely used antiviral was aciclovir (Zovirax®). It was a discarded candidate anti-cancer drug which was found to have activity against herpes simplex and varicella zoster viruses (Elion, 1983). It transpired

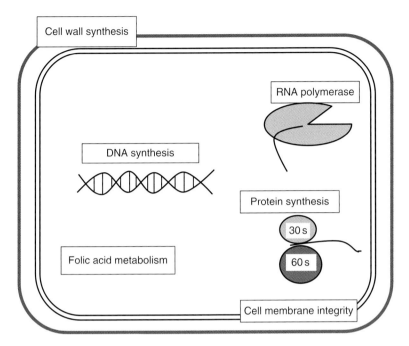

Figure 1.1 Outline of generic bacterial cell indicating possible targets for antibacterial agents.

that aciclovir is a substrate readily taken up by the viral thymidine kinase in purine metabolism (as outlined in Figure 1.2), but which does not have affinity for the equivalent host cell thymidine kinase. Subsequent progress in finding similarly specific 'viral enzyme-substrate' reactions in other virus species has been slow. Table 1.3 outlines some details about the main antivirals currently available.

1.5.3 Antifungal Agents

As fungi are eukaryotes, there is a restricted range of metabolic pathways which are not shared with mammalian host cells. Thus, most anti-fungal agents are rather toxic to patients and topical preparations are used wherever possible. The first antifungal to be introduced was griseofulvin (Davey *et al.*, 2015). It is thought to work by affecting mitosis in the dividing fungal cell. However, it is only effective when given orally and only works against dermatophytes. A key target area for anti-fungals is the sterols in the fungal cell membrane. Polyenes (e.g., amphotericin, nystatin) bind to sterols, thus disrupting the structure of the membrane. These are unfortunately not selectively toxic and only amphotericin is safe enough to give systemically (with careful monitoring!). Ergosterols are an essential membrane component found in fungal, but not mammalian membranes. Azoles (imidazoles and triazoles) inhibit the activity of lanosterol demethylase, which catalyses the conversion of lanosterol to ergosterol within the fungal cell. Examples include clotrimazole (an imidazole) and fluconazole (a triazole). Azoles are hepatotoxic in humans, so external treatment is used wherever feasible. Triazoles are used in preference to imidazoles in oral treatment regimens as they have more suitable pharmacological properties, but their use is reserved for severely debilitating or

Table 1.2 Site and mode of action and mechanism of resistance for selected antibiotics.*

Site of action	Antibiotic class (examples)	Mode of action	Mechanism of resistance	Susceptible organisms[1]
Cell wall synthesis	β-lactams (penicillins, cephalosporins)	Disruption of peptidoglycan cross linking; Inhibition of cell wall formation by attaching to penicillin binding proteins.	Production of β-lactamase; Modification of penicillin-binding proteins; Alteration of membrane permeability.	
	Glycopeptides (vancomycin, teicoplanin)	Inhibition of cell wall building through binding to acyl–D-alanyl-D-alanine	Substitution of alanine for lactate or serine	
Protein synthesis	Aminoglycosides (streptomycin, gentamycin, tobramycin)	Binding to ribosomes to interrupt formation of mRNA	Productions of enzymes which alter aminoglycoside structure or modifications to membrane proteins thus inhibiting transport into bacterial cell	
	Chloramphenicol	Obstruction of peptidyl-transferase reaction on ribosome	Production of chloramphenicol acetyltransferase to modify molecule and alteration in protein sequence of 50S subunit to prevent binding to ribosome; Change in membrane structure to inhibit transport into bacterial cell	
	Tetracyclines (tetracycline, doxycyline, tigecycline)	Binding to 30S subunit, preventing attachment of t-RNAs	Production of membrane protein associated with rapid removal of antibiotic from bacterial cell; Production of protein in cytoplasm which prevents binding to ribosome	
	Macrolides (erythromycin, clarithromycin)	Removal of nascent peptide chain from ribosome	Production of enzyme which alters structure of 23S subunit by adding methyl group	

Nucleic acid synthesis	Sulphonamides	Competitive substrate against PABA for dihydropteroate synthetase in folic acid pathways thus preventing production of purines	Production of alternative dihydropteroate synthetase which sulphonamides cannot attach to
	Trimethorprims	Competitive substrate against dihydrofolic acid for dihydrofolic acid reductase in folic acid pathways thus preventing production of pyrimidines	Production of alternative dihydrofolic acid reductase which sulphonamides cannot attach to
	5-Nitroimidazoles (metronidazole)	Not fully understood; possibly integration into DNA	Not fully understood
	Rifampycins (rifampicin)	Binding to β-subunit of DNA–dependent RNA polymerase	Modification to β-subunit
Membrane function	Polymixins (colistin)	Disruption of cell membrane integrity	Limited information since drug not widely used (has same effect on mammalian cell membrane)

*Information taken from Davey et al. (2015); see this reference for further details particularly Chapters 2,3,4 and 11.

[1] The final column has been left blank for the reader to fill in. See Exercise 1.1 below.

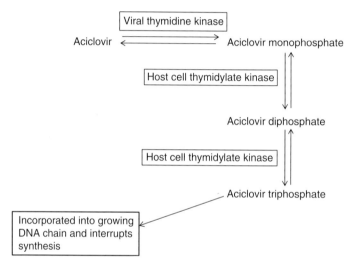

Figure 1.2 Mode of action of aciclovir.

life threatening infections (Davey *et al.*, 2015). Flucytosine is an example of a nucleotide analogue; it is metabolised within the fungal cell to 5-fluorouracil which is taken up by RNA polymerase, thus interrupting production of RNA. Unfortunately, this reaction only occurs in yeasts.

1.5.4 Antiparasitic Agents

Protozoa and helminths are also eukaryotic and they are relatively complex compared to other microorganisms. They tend to have multiple life cycle stages and most species use sexual as well as asexual replication mechanisms. This means that the similarities with mammalian cells are greater than for bacteria and viruses. Therefore the range and scope of safe, efficacious anti-parasitic agents is limited (Davey *et al.*, 2015). Long-standing treatments, such as quinine/chloroquine for malaria and arsenical compounds for trypa-nosomiasis are still used in spite of associated problems. Chloroquine resistant strains of

Table 1.3 Mode of action and selective toxicity for selected antiviral agents.*

Antiviral drug	Virus(es) it is effective against	Mode of action	Selective toxicity
Aciclovir (ACV)	Herpes simplex; varicella zoster	Analogue of guanosine; phosphorylated inside cell to be substrate for thymidine kinase in nucleoside-TP production; ACV-TP interrupts DNA replication by causing chain termination and inhibiting DNA polymerase	Phosphorylation only virus-infected cells ACV is a substrate only for viral thymidine kinase; cannot be activated by host enzyme
Ganciclovir (GCV)	Cytomegalovirus	Analogue of guanosine; phosphorylated inside cell by phosphotransferase enzyme; GCV-TP interrupts DNA replication by causing chain termination and inhibiting DNA polymerase	Not selectively toxic – GCV also substrate for host phosphorylation enzymes
Ribavirin	Respiratory syncytial virus Hepatitis C virus	Nucleoside analogue; phosphorylated inside cell. Exact effects unclear, presumed to be different in RNA and DNA viruses	Substrate for viral enzymes; however can affect host cells (e.g., causing haemolytic anaemia)
Foscarnet	Cytomegalovirus	Pyrophosphate analogue; direct DNA polymerase inhibitor	Substrate for viral enzymes; however can affect host cells (e.g., kidney cells causing acute renal failure)
Zidovudine (AZT)	Human immunodeficiency virus	Nucleoside analogue; activated to triphosphate form by host enzymes; AZT-TP causes chain termination in DNA replication and inhibits reverse transcriptase	Not selectively toxic – AZT activated by host enzymes; toxic to fast metabolising cells such as in bone marrow
Lamivudine	Hepatitis B virus	Nucleoside analogue; activated to triphosphate form by host enzymes; Lamivudine-TP causes chain termination in DNA replication and inhibits DNA polymerase	Substrate for viral enzymes
Tenofovir	Hepatitis B virus	Analogue of adenine mono-phosphate; inhibits DNA replication	Substrate for viral enzymes
Efavirenz	Human immuno-deficiency virus	Bind to reverse transcriptase and inhibits its activity	Substrate for viral enzymes
Oseltamivir	Influenza virus A and B	Blocks virus entry into host cell by inhibiting neuraminidase	Binds to specific viral protein
Maraviroc	Human Immuno-deficiency virus	Blocks virus entry by inhibiting binding to CCR5 receptor on host cell	Blocks virus-specific activity
Amantidine	Influenza A virus	Blocking uncoating of virus particle within host cell	Binds to specific viral protein; can affect neurone function causing insomnia and confusion

*Information taken from Davey *et al.* (2015); see this reference for further details, particularly Chapters 5, 6 and 7.

Plasmodium spp., particularly *P. falciparum* are widespread. While melarsoprol, which is an arsenic-based drug kills *Trypanosoma brucei* spp., it can also kill the host! Amphotericin is effective against *Leishmania* spp., but as mentioned above it is toxic to mammalian cell membranes as well. There are a number of anti-*Plasmodium* drugs (Davey *et al.*, 2015), although resistance is constant problem in malaria. As mentioned above, metronidazole is effective treatment for the gut protozoa *E. histolytica* and *G. duodenalis*. It works by disrupting DNA synthesis, although the mode of action has not yet been fully elucidated. Although there are only a few anthelmintic agents (note the spelling – a shortened version of 'anti-helminthic', which is much easier to say), they each tend to have a broad spectrum. Albendazole and mebendazole (benzimidazoles) are used to treat nematodes such as hookworm and *Ascaris lumbricoides*. They are thought to interrupt several points in the worm's metabolism including formation of microtubules in the cytoskeleton and production of ATP. The worm is weakened and eventually unable to move. Praziquantel is an effective treatment for flukes, such as *Schistosoma* spp. (but not *Fasciola hepatica*) and tapeworm infections (Davey *et al.*, 2015). It is thought to work by altering the calcium balance across the cells, similarly resulting in paralysis. It also disrupts the tegument (outer coat) which exposes the helminth's antigens to the host's immune system. Ivermectin also affects the neuromuscular system and can be used against ectoparasites such as the scabies mite (*Sarcoptes scabei*) as well as some parasitic worms (Gunn and Pitt, 2012). All these anthelmintic drugs can be toxic to humans, but often only one or two doses are required to fully clear the infection.

Point to consider 1.5: How could more detailed knowledge of the genetics and metabolism of viruses and funguses help in the search for more selectively toxic agents?

1.5.5 Antimicrobial Sensitivity Testing

Testing for antimicrobial sensitivities accounts for a sizeable amount of the workload in diagnostic microbiology departments. The importance of doing this is one of the arguments in favour of retaining laboratory techniques which allow isolation of viable organisms from patient samples. For bacterial and fungal susceptibility testing, laboratories in the UK should follow the guidelines issued by the British Society for Antimicrobial Chemotherapy (BSAC). This organisation sets standards for laboratory methods and conducts surveillance of resistance for clinically important bacteria and fungi. It uses the information collected to provide tables of breakpoints for each organism against all routinely prescribed antibiotics (www.bsac.org.uk). Since the issues related to antimicrobial resistance are international, BSAC works closely with similar organisations in other parts of the world, notably in Europe. In order to coordinate resistance data more effectively, UK clinical microbiology laboratories phased in the change to the European Committee on Antimicrobial Susceptibility Testing (EUCAST) protocols and breakpoint data, from January 2016 (www.bsac.org.uk). This means using a standardised disc diffusion assay (based on the Kirby-Bauer method and similar to the BSAC method) for antibacterial susceptibilities and a broth microdilution method to determine minimum inhibitory concentrations (www.eucast.org). A microdilution method is also recommended for susceptibility testing of isolates of yeasts, with a tube dilution method promoted for filamentous fungi (www.eucast.org). The EUCAST testing guidelines and surveillance remit also extends to organisms of veterinary importance (www.eucast.org). For a review of EUCAST and decision-making systems to determine and monitor antibiotic resistance see Winstanley and Courvalin (2014).

As matrix-assisted laser desorption/ionisation-time of flight (MALDI-TOF) becomes more widely used for routine characterisation of microorganisms, the power of the database in distinguishing between strains within species is increasing. This means that the MALDI-TOF profile of a clinical isolate has the potential to be used not only in full identification, but also in predicting its antimicrobial susceptibilities (Weiser *et al.*, 2012). There are clear advantages to this from a patient management point of view, particularly in situations where waiting an additional 24 hours for the antibiotic results on a new isolate may be unacceptably long. Although the MALDI-TOF result is based on a protein profile and not biological activity, it does appear to allow the identification of some antibiotic resistance profiles (Weiser *et al.*, 2012). Nevertheless, conventional testing methods will continue to have an important role in monitoring newly arising resistance (i.e., resistance gene spread to new species or new mutations affording mechanisms of resistance).

In contrast, as routine virus detection has moved to molecular methods, viral susceptibility testing has largely followed suit and is usually based on genome sequencing (Jeffery and Aarons, 2009). Resistance to a particular antiviral is associated with one or more specific genetic mutations, which can be ascertained by comparing the sequence to a database. Phenotypic assays for antivirals and monitoring of (live) isolates is still undertaken in reference laboratories and this helps to keep the reference databases current. Since the available treatment options for parasite infections are limited, susceptibility testing is not usually undertaken except in reference laboratories and research centres.

Point to consider 1.6: Surveillance of antimicrobial resistance is conducted at national and international level. Chose a specific example and think about how effective has this been in reducing the spread of resistance in the last five years.

1.6 Selection and Evaluation of Diagnostic Tests

The logistical and financial factors which are currently driving the configuration of pathology services can pose interesting challenges. One of these is combining the available equipment and expertise across two or more geographically separated sites, which had hitherto operated independently. The laboratory staff in this situation often find themselves navigating their way through a series of apparently conflicting issues and interests, but scientific and clinical considerations must be central in decision making (Price, 2012). Other points to consider include financial and logistical – for example, whether a complex assay in the main laboratory could be complemented (or replaced entirely!) by a point-of-care test (Cohen-Bacrie *et al.*, 2011) – and also which tasks can appropriately be assigned to which grade of staff. Dialogue with the service users is also important to ensure that the results provided by an assay are appropriate to the clinical need (Price, 2012; Price and Christensen, 2013). Expenditure can be maintained at manageable levels by reducing repetition of tests (e.g., the same test requested by the GP and then at the hospital a few hours later) and ensuring that the assay still detects the most relevant marker (Price and Christensen, 2013).

The postgraduate level laboratory scientist should be aware of the context in which the pathology service is operating in order to contribute effectively to decisions made by service managers. The concept of 'evidence-based laboratory medicine' (EBLM) as developed by Price and Christensen (Price *et al.*, 2009) is a useful approach. They describe a five-point cycle:

1) Ask the right question
2) Acquire suitable evidence
3) Appraise the information
4) Apply the outcome
5) Audit the service provided (Price, 2012).

Whether the laboratory team is thinking about changing the configuration of a service, replacing one assay with another for a particular marker or introducing a novel test, Price and colleagues suggest that applying EBLM can help. For detailed consideration of the topic, the reader is referred to Price *et al.* (2009) or Price (2012), but it is worth noting that the emphasis is on applying scientific understanding and ways of thinking to improving the overall outcome for the patient (Price, 2012). The complexity of the situation is highlighted by their discussion of the first step in the cycle – asking the right question. There are a range of people with different per-spectives all asking questions relating to the service and they each need to be included in the analyses (Price and Christenson, 2013). To illustrate this, take the example of the possible introduction of a POCT test for norovirus to be used on the wards to enhance rapid diagnosis in patients with gastroenteritis. The patient will be wondering what type of sample is to be used, whether it will be painful and uncomfortable to collect, how long will it take for the result to be ready and what the result could mean for them. Is it a simple stool sample? Will the result be ready on the same day? The staff on the wards will be thinking about how easy it is to perform the test, how much training might be required before being able to carry it out, where they could perform the assay, as well as turnaround time and interpreta-tion of the results. Does the test come in a kit with straightforward instructions? If the result is positive does that mean that the patient definitely does have norovirus and should be nursed separately? The laboratory scientist will be thinking about technical considerations including comparative sensitivity and specificity of the POCT method, effective training of ward staff to carry out the test and quality con-trol procedures. Is the test sensitive enough to pick up all infections even in patients presenting quite early (or late)? Is it specific enough that negative results can be trusted? How easy is it to ignore some part of the instructions and still obtain a result with this particular kit format? There will also be management issues such as overall cost, implications for the relevant laboratory and ward staff workloads, planning and commissioning of services, infection control and reliability of data used for surveillance. The laboratory manager will want to know whether the cost of the POCT kits will come out of the pathology or the ward budget, but also how many samples might have to be re-tested in the main laboratory (if it is above a certain level, the POCT test might not be suitable). If there is a regional or national outbreak of norovirus, the local commissioners and or government depart-ments might want to put more resources into diagnosis. Again, they will need to know the overall cost per test (which includes staff time, as well as the price of

---

Text Box 1.2 Requirements of an Assay Used in Diagnostic Pathology

Tests for a clinically useful marker (whole organism, genome, antigen, antibody or other marker).

Gives reliable results – that is, scientifically accurate and clinically credible.

Results are reproducible – that is, testing aliquots of the same sample repeatedly will give the same results each time.

Assay has high sensitivity and specificity (see Text Box 1.3).

The turnaround time for the result is appropriate to the clinical condition.

The method is straightforward for the operator after suitable training has been undertaken.

The assay uses equipment and kits (or reagents) which are readily available from a reliable supplier.

The test procedure is cost-effective overall.

---

equipment and consumables) and time taken to obtain a reliable result – which affects patient management. It is important to appreciate all these points of view to inform decision making (Price and Christenson, 2013). Evidence should be collected from consultation with service users, systematic review of the literature and discussions with colleagues in other laboratories; the outcome of the appraisal can then be used in a report or business case for presentation to managers or funding bodies (Price, 2012).

When the evidence suggests that a particular test should be evaluated, the laboratory-based scientist usually begins with an overview of its suitability. Text Box 1.2 outlines the characteristics required of a diagnostic test in pathology. It should be noted that it can be a false economy (and unhelpful to the patient) to prioritise cost per test and turnaround time at the cost of accuracy of the result. A test with a low specificity will produce false positives, which will need repeat analysis, and probably re-testing in a confirmatory assay. If fresh samples are required for this follow-up work, then the patient may also be needlessly distressed.

For the technical part of the assessment, the 'gold standard' method for that marker has to be identified. This is usually the method that is generally accepted to be the most sensitive and specific, reliable and accurate. Sometimes it is not feasible to use this (e.g., where the gold standard is only carried out in reference centres) and in that case, the assay currently used by the laboratory can be designated as the gold standard for the calculations. A well-designed study should be implemented involving a predetermined number of samples, with appropriate inclusion criteria and suitable controls. The samples should be run in the new assay and the gold standard in parallel and the results compared. As a reminder, the equations used to calculate the sensitivity, specificity and positive and negative predictive value (PPV and NPV) are shown in Text Box 1.3. It is important to note that in a population where the prevalence of a diagnostic marker is low, the PPV can be significantly reduced. In a diagnostic situation this may not be a prime consideration, but it should be borne in mind when implementing screening programmes (e.g., testing of asymptomatic patients for HIV antigen/antibody in a low-risk population in Western Europe).

Text Box 1.3 Comparison of an Assay With the 'Gold Standard'

1) Set out the data in the following format:

	Result in gold standard	assay
Result in new assay	Positive	Negative
Positive	True positive	False positive
Negative	False negative	True negative

2) Calculate sensitivity and specificity:

$$\text{Sensitivity} = \frac{\text{true positives}}{\text{true positives} + \text{false negatives}} \times 100 = \% \text{ sensitivity}$$

$$\text{Specificity} = \frac{\text{true negatives}}{\text{true negatives} + \text{false positives}} \times 100 = \% \text{ specificity}$$

3) Calculate positive and negative predictive values:

$$\text{PPV} = \frac{\text{true positives}}{\text{true positives} + \text{false positives}} \times 100 = \% \text{ PPV}$$

$$\text{NPV} = \frac{\text{true negatives}}{\text{true negatives} + \text{false negatives}} \times 100 = \% \text{ NPV}$$

1.7 Quality Management

It is clear that the overall quality of a clinical microbiology service is dependent on a range of scientific, technical, managerial and human factors (Pitt and Sands, 2002). The laboratory scientist must be aware of and fully engaged with all the recommended internal procedures for quality assurance and all the external quality assessment and accreditation requirements. It is assumed that each reader will be introduced to this as appropriate for their job role, but sources of useful information include: Microbiology UKNEQAS for Microbiology (www.ukneqasmicro.org.uk) Quality Control for Molecular Diagnostics (www.qcmd.org), United Kingdom Accreditation Service (www.ukas.com) and the European Society of Clinical Microbiology and Infectious Diseases (www.escmid.org). There are also textbooks available to provide guidance on preparing for accreditation, such as Burnett (2013).

1.8 Infection Control, Monitoring and Surveillance

It is useful for the scientist to appreciate the laboratory's role in infection control and outbreak investigation at a local level. It contributes to identifying cases and possible sources of infection, allowing measures to be put in place to manage individuals and reduce spread of disease. Recognising and reporting notifiable diseases is also imperative, as that information feeds into surveillance mechanisms at regional, national and international level. Monitoring of genetic variation within species is also a key role (although this is often carried out in specialist centres and reference laboratories) as this can lead to recognition of factors which might account for increased virulence or antimicrobial resistance. A number of techniques are proving valuable in this respect for monitoring outbreaks of microbial infections (e.g., Sabat *et al.*, 2013) and these include 'whole genome sequencing' (e.g., Halachev *et al.*, 2014) and the high throughput 'next-generation (or 'deep') sequencing' (e.g., Sherry *et al.*, 2013; Quiñones-Mateu *et al.*, 2014).

An 'outbreak' of infection is an epidemiological concept which is defined by precise criteria:

1) A situation where at least two people who can be connected – due to contact with each other or visiting a particular place at the same time – are exhibiting similar symptoms; or
2) The incidence of a particular infection is higher than usually recorded in that geographical area, at that time of year and in that population; or
3) One reported case of an uncommon infection such as diphtheria, rabies or poliomyelitis; or
4) Known or expected contamination of a food or water source.

When these criteria are met, this allows for an outbreak control team to be convened to manage the situation (McAuslane *et al.*, 2014). Laboratory diagnosis of suspected cases is not necessarily required to trigger the outbreak procedure, but it is needed to move the individual case definition to 'confirmed'.

Investigations of outbreaks of gastrointestinal disease such as norovirus infection or food poisoning can be challenging, since in healthy adults, they are usually mild and self-limiting. People often disperse after being in contact with the source and may not make the connection between being in a particular building or eating a certain meal and their symptoms of diarrhoea and vomiting. If they do not seek medical help, their illness will not be recorded. It may be days or weeks before there is sufficient indication of an outbreak. Identifying cases can be difficult, because people tend to shed viable organisms for very limited time periods after recovery. In addition, examination of food items is rarely possible, as they have usually been discarded. Therefore reviews of such events rely heavily on gathering of epidemiological evidence and deciding the most likely source and probable scenario of spread. For example Edwards *et al.* (2014) reported an outbreak of *Campylobacter* food poisoning, which was associated with the consumption of chicken liver pâté at a wedding reception. Someone reported that a number of people had become ill after being at the wedding. Once a pattern had been noticed, enquiries into potential contacts (i.e., wedding guests) yielded a list of 118 people. These were sent a questionnaire to ascertain various details including how long they were present at the reception and what they ate (this was important as two separate meals were provided by caterers during the course of the day). 'Cases' were defined as those who had eaten at the reception and who experienced diarrhoea or vomiting within 5 days (Edwards *et al.*, 2014). Those who had been abroad immediately before the wedding and anyone whose household contacts had diarrhoea symptoms, but had not been at the event, were excluded from the study. One hundred and eight of those approached returned a questionnaire and of those 49 met the criteria to be categorised as a case. Interestingly, 31 people had sought medical advice for gastrointestinal symptoms, but faecal samples were only obtained for 26 of them. *Campylobacter* infection was confirmed in 22 people, but it was not possible to obtain any food for testing. Examination of the data collected suggested that the common factor amongst all those affected was consumption of the chicken liver pâté (Edwards *et al.*, 2014). Discussions with the food preparation staff indicated that they had not followed the Food Standards Agency's guidelines about cooking chicken liver designed to kill *Campylobacter* spp. and thus it was concluded that this bacteria was the causative agent. It is worth noting that the reason that the investigation was begun was because one person knew several of the wedding guests; when people at an event do not know each other very well, that initial connection is often not made. Also, in this case a reasonable number of people were sufficiently worried about their symptoms to see a doctor and stool samples were collected from most of those at the time of presentation. This may not happen and it is therefore sometimes hard to identify those who were infected but whose symptoms did not trouble them much at the time; where few people submit samples and there is no food or fomite from the presumed source, it may not be possible to find the cause at all.

Surveillance is important as part of general public health measures, thus obtaining data to fulfil the 'outbreak' criteria number 2 and number 3 in the list above. Laboratory support, to confirm suspected cases, is a very valuable part of this. Data reported to central agencies at national level can be collated to investigate actual or possible links between outbreaks. For example, outbreaks of measles can occur in areas where MMR coverage is sub-optimal. In the UK, laboratory confirmed cases of measles are notifiable, which allows connections between individuals to be noticed and for healthcare professionals to be alerted. There was an unexpectedly high incidence of measles in England and Wales in 2012. The majority of cases were recorded in either the South East (Sussex) or the northwest (Merseyside) of England, so the infection appeared to be contained within limited geographical pockets. Then at the end of 2012, a very significant outbreak covering the whole of Wales began (Pitt, 2013); this highlighted the value of surveillance data and the importance of vigilance in control of infectious diseases, as well as providing a lot of work for the virologists! The incidence of measles in Scotland and Northern Ireland did not increase during this time, which is probably attributable to better MMR vaccination rates in those two countries (Pitt, 2013). Similarly, while the spread of West Nile Virus across the United States between 1999 and 2008 was rapid and alarming (https://www.cdc.gov/westnile), surveillance data did contribute to understanding the epidemiology of the infection and thus to bringing it under control (Lindsey *et al.*, 2014).

Member states in the European Union (EU) and European Economic Area (EEA) must report confirmed cases of certain infectious diseases to the European Centre for Disease Prevention and Control (ECDC: http://ecdc.europa.eu); this allows trends across the continent to be noted. For instance, in the summer and early autumn of 2012, Germany, the Netherlands and the UK all reported increased incidence of *Cryptosporidium* infection. An investigation into possible common factors and an assessment of Europe-wide risks (ECDC, 2012) concluded that there was no common source of the parasite even within each country and that the likelihood of similar occurrences across a wider area of the continent was low (ECDC, 2012). However, it was useful to highlight the issue, to alert microbiologists and public health professionals in other countries to consider cryptosporidiosis as a diagnosis for people who had recently travelled to any of the three countries. From a wider perspective, the World Health Organisation (www.who.int) also collects and analyses data from 194 countries, using its six regional offices to coordinate responses to epidemics and pandemics and work to improve control and prevention of infection. An example is the activity by the WHO European Region to understand and contain Middle Eastern Respiratory Syndrome Coronavirus, MERS-CoV (http://www.euro.who.int/en/health-topics/communicable-diseases/influenza/middle-east-respiratory-syndrome-coronavirus-mers-cov/mers-cov-in-the-who-european-region).

1.9 Exercises

1.1 Find examples for the final column in Table 1.2 – labelled 'susceptible organisms'.

In each case think about:
A How many examples of susceptible organisms are available for antibiotics in that group and the implications of this in terms of spread of resistance
B The clinical application and continued usefulness of each group of antibiotics

1.2 Chose a protocol used for either the investigation of a clinical condition (e.g., UTI) or diagnosis of a specific pathogen (e.g., herpes simplex virus) in your laboratory or from the published literature.
A Think about which techniques are used and why they might have been selected.
B Do you think that this protocol is the optimal method?
C Are there any ways in which it could be improved to enhance microbiology service delivery?

1.3 Select a microbiology POCT assay currently on the market.
A Search the literature to investigate research its technical characteristics, how well it has been assessed and how it is used in diagnostic microbiology.
B Design an evaluation study which would help to decide whether to use this test in a *specific* clinical context (e.g., paediatric ward, primary care).
C Prepare the outline of a short oral presentation or written report in which you would present your findings to a group of clinicians and managers who would decide whether this POCT should be implemented.

References

Burnett D (2013). *A Practical Guide to ISO 15189 in Laboratory Medicine.* London: ACB Venture Publications.

Cohen-Bacrie S *et al.* (2011). Revolutionizing clinical microbiology laboratory organization in hospitals with *in situ* point-of-care. *PLOS One,* **6**(7): e22403. doi: 10.1371/journal.pone.0022403

Davey P, Wilcox MH, Irving W and Thwaites G (2015). *Antimicrobial Chemotherapy*, 7th edn. Oxford: OUP.

ECDC (2012). Rapid risk assessment: Increased *Cryptosporidium* infections in the Netherlands, United Kingdom and Germany, 2012. Stockholm: European Centre for Disease Surveillance and Control.

Elion GB (1983). The biochemistry and mechanism of action of acyclovir. *Journal of Antimicrobial Chemotherapy*, **12(suppl B):** 9–17.

Ford M *et al*. (2014). *Medical Microbiology*, 2nd edn. Oxford: OUP.

Fournier P-E *et al*. (2013). Modern clinical microbiology: New challenges and solutions. *Nature Reviews Microbiology*, **11**: 574–585. doi: 10.1038/nrmicro3068

Goering *et al*. (2013). *Mims' Medical Microbiology*, 5th edn. Philadephia: Elsevier-Saunders.

Greenwood D, Barer M, Slack R and Irving W, eds. (2012). *Medical Microbiology: A Guide to Microbial Infections*, 18th edn. London: Churchill Livingstone-Elsevier.

Gunn A and Pitt SJ (2012). Parasitology: An integrated approach. Chichester: Wiley-Blackwell.

Halachev M R *et al*. (2014). Genomic epidemiology of a protracted hospital outbreak caused by multidrug-resistant *Acinetobacter baumannii* in Birmingham, England. *Genome Medicine*, **6**: 70. doi: 10.1186/s13073-014-0070-x

Jeffery K and Aarons E (2009). Diagnostic approaches. In: Zuckerman *et al*., eds., *Principles and Practice of Clinical Virology*, 6th edn. Chichester: Wiley-Blackwell.

Lindsey NP *et al*. (2014). West Nile virus and other arboviral diseases — United States, 2013. *Morbidity and Mortality Weekly Report*, **63(24):** 521–526.

McAuslane H *et al*. (2014). *Communicable Disease Outbreak Management: Operational guidance*. London: Public Health England Publications.

Moore C (2013). Point-of-care tests for infection control: Should rapid testing be in the laboratory or at the front line? *Journal of Hospital Infection*, **85**: 1–7. doi: 10.1016/j.jhin.2013.06.005

Murray PR *et al*. (2013). *Medical Microbiology*, 7th edn. Philadephia: Elsevier-Saunders.

NICE NG 15 (2015). http://www.nice.org.uk/guidance/ng15

Pitt SJ (2012). Point of care testing for infectious tropical diseases. *Biomedical Scientist*, **56**: 33–35.

Pitt SJ (2013). Measles: potentially fatal but eminently preventable. *Biomedical Scientist*, **57**: 444–448.

Pitt SJ and Sands RL (2002). Effect of staff attitudes on quality in clinical microbiology services. *British Journal of Biomedical Science*, **59**: 69–75.

Price CP (2012). Evidence-based laboratory medicine: Is it working in practice? *The Clinical Biochemist Reviews*, **33**: 13–19.

Price CP and Christenson RH (2013). Ask the right question: A critical step for practicing evidence-based laboratory medicine. *Annals of Clinical Biochemistry: An international journal of biochemistry and laboratory medicine*, **50**: 306–314. doi: 10.1177/0004563213476486

Price CP and St John A (2012). *Point of Care Testing: Making Innovation Work for Patient Care*. Washington DC: American Association for Clinical Chemistry Press.

Price CP, Lozar Glenn J and Christenson R (2009). *Applying Evidence-Based Laboratory Medicine: A Step-By-Step Guide*. Washington DC: American Association for Clinical Chemistry Press.

Quiñones-Mateu ME, Avila S, Reyes-Teran G *et al.* (2014). Deep sequencing: Becoming a critical tool in clinical virology. *Journal of Clinical Virology*, **61**: 9–19.

Sabat AJ *et al.* (2013). Overview of molecular typing methods for outbreak detection and epidemiological surveillance. *Eurosurveillance*, **18**(4): 20380.

Shallcross LJ and Davies SC (2014).The World Health Assembly resolution on antimicrobial resistance *Journal of Antimicrobial Chemotherapy*, **69**: 2883–2285. doi: 10.1093/jac/dku346.

Sherry NL *et al.* (2013). Outbreak investigation using high-throughput genome sequencing within a diagnostic microbiology laboratory. *Journal of Clinical Microbiology*, **51**: 1396–1401. doi: 10.1128/JCM.03332-12

St John A and Price CP (2014). Existing and emerging technologies for point-of-care testing. *Clinical Biochemistry Reviews*, **35**: 155–167.

Wieser A *et al.* (2012) MALDI-TOF MS in microbiological diagnostics—Identification of microorganisms and beyond (mini review) *Applied Microbiology and Biotechnology*, **93**: 965–974. doi: 10.1007/s00253-011-3783-4

Wilkinson M (2011). *Medical Microbiology*, 2nd edn. Banbury: Scion Publishing Ltd.

Winstanley T and Courvalin P (2011). Expert systems in clinical microbiology. *Clinical Microbiology Reviews*, **24**: 515–556. doi: 10.1128/CMR.00061-10

2

Infections of the Blood

2.1 Introduction

The persistent presence of detectable levels of microorganisms in the blood is rarely good news for the patient. They could be growing and dividing in the blood stream opportunistically (e.g., sepsis), passing through the blood extracellularly as part of their usual life cycle (e.g., hepatitis B virus) or infecting host red (e.g., *Plasmodium* spp.) or white (e.g., human immunodeficiency virus) blood cells. The occurrence of the organism could be associated with an acute and potentially life-threatening condition (e.g., sepsis), while in other cases it can be a chronic infection, causing long-term damage to the individual and raising public health issues (e.g., hepatitis B). The appearance in the blood can be a prelude to or a consequence of infection elsewhere in the body. For example, bacteraemia can lead to meningitis, while hepatitis B virus virions enter the blood because of active replication in the liver.

In clinical and laboratory practice, blood stream infections and blood-borne diseases are usually considered separately, since the symptoms and patient history would generally indicate one or the other. For clarity, this chapter will similarly address them under separate sections. Nevertheless, it is always important to consider how one type of blood infection might predispose a patient to another. The blood-borne virus infection acquired immune deficiency syndrome (AIDS) adversely affects the patient's immune response; repeated healthcare interventions to control the viral activity increase the risk of nosocomial blood stream infection, from which a poor outcome is likely, due to the viral-induced changes in immunity (Bearman and Wenzel, 2005). Also, co-infection is always a possibility due to common transmission routes. While the blood-borne viruses hepatitis B virus, hepatitis C virus and human immunodeficiency virus are often cited examples of this, it is also interesting to consider vector-borne diseases. Invertebrate vectors include mosquitoes in the genus *Anopheles*. These mosquitoes are 2–5 mm in length and may have patterned wings. As Figure 2.1 shows, the maxillary palps in this species are almost as long as the proboscis, which is a characteristic, distinguishing feature. Also when at rest, they hold their bodies at a 45-degree angle to the surface they are sitting on, whereas other types of mosquito tend to lie in parallel to it. *Anopheles* spp. can transmit both *Plasmodium falciparum* and *Wuchereria bancrofti* and in areas where both parasites are endemic, co-infection has been reported in both the vector and humans (Manguin *et al.*, 2010).

Clinical Microbiology for Diagnostic Laboratory Scientists, First Edition. Sarah J. Pitt.
© 2018 John Wiley & Sons Ltd. Published 2018 by John Wiley & Sons Ltd.

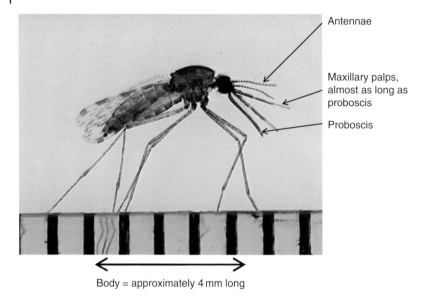

Body = approximately 4 mm long

Figure 2.1 Female *Anopheles* spp. mosquito. Source: Courtesy of Dr A Gunn. (*See insert for colour representation of the figure.*)

Point to consider 2.1: Before reading this chapter, make a list of organisms you know which are associated with blood stream or blood-borne infections. From your own experience and using the information in the introductory chapter, which laboratory techniques do you think you be most appropriate for diagnosis for each of the organisms you have listed and why?

2.2   Blood Stream Infections (Bacteraemia/Fungaemia)

Microorganisms are continually entering the blood stream via internal and external sites. They are usually removed swiftly by the innate (or, in some cases, specific) immune response (SMI B37, 2014). Occasionally a surge of organisms occurs following a disturbance of their habitat, such bacteraemia after brushing teeth (Lockhart, 2012) or insertion of a needle, but this is also usually transient, unremarkable and does not require intervention. Some dentists advise patients at particular risk of infectious endocarditis (IE, see below) to take prophylactic antibiotics before dental procedures (Lockhart, 2012). Current recommendations in the UK and Europe are that this is not necessary except when the patient is at high risk of IE and is undergoing major dental work (Habib *et al.*, 2009; Watkin and Elliot, 2009).

   Recurrent or intermittent infection can arise due to an internal localised infection (e.g., abscess), in the early stages of some diseases (e.g., pneumococcal bacteraemia prior to pneumonia) (SMI B37, 2014) or as part of the course of the infection (e.g., undulating fever in brucellosis, cyclical release of parasites ['schizogony'] in malaria). The immune system may be effective against the organisms to some extent, but antimicrobial treatment is likely to be needed. Information about the patient's symptoms might be important in such cases, in order to ensure that blood samples are collected at appropriate times (or time intervals), to maximise the chance of detection of the organism (SMI B37, 2014). In some cases a bacteraemia or fungaemia can overwhelm the body's immune system to cause sepsis ('septicaemia'). An apparently innocuous infection in a previously healthy adult can progress to septicaemia quite rapidly, while the risk of more severe disease and a worse outcome is clearly increased where the patient is immunocompromised in the first place. Another obvious point to note is that patients with compromised immune systems are more likely to be undergoing medical interventions which could lead to iatrogenic infections with opportunistic organisms. The clinical consequences of high concentrations of active bacteria or fungi in the blood can also include meningitis or infective endocarditis (see below).

2.2.1 Sepsis

The definition of sepsis is the systemic inflammatory response syndrome (SIRS) which has been produced due to an infection (SMI B37, 2014). A patient experiences SIRS as part of the early immune reaction to any kind of tissue injury and it is defined by the occurrence of at least two of the following: body temperature markedly above or below normal ($>38.0\,°C$ or $<36.0\,°C$), raised heart rate (>90 beats minute^{-1}), increased ventilation rate (>20 breaths minute^{-1}), white blood cell count either high ($>12,000$ cells μL^{-1}) or low ($<4,000$ cells μL^{-1}) (SMI B37, 2014). If untreated, this sepsis can develop into severe sepsis (where at least one organ has been incapacitated by the infection) or septic shock (where severe sepsis has progressed to hypotension). All three of these are associated with high mortality (Stearns-Kurosawa *et al.*, 2011), so accurate and timely laboratory diagnosis is invaluable in such cases.

2.2.1.1 Pathology

The pathogenesis of sepsis is complex and not completely understood, but it is currently conceived as an imbalance between the pro-inflammatory immune response produced in response to the presence of pathogenic microorganisms (Delves *et al.*, 2011) and the anti- inflammatory mechanisms which usually keep inflammation processes from being excessive (Anas *et al.*, 2010). Innate immunity to microorganisms is achieved via humoral and cellular factors. Pattern recognition receptors (PRRs) cells involved in the immune response bind to corresponding pathogen-associated molecular patterns (PAMPs) (Bochud and Calandra, 2003; Anas *et al.*, 2010). Examples of PRRs on human cells include toll-like receptors (TLRs), while the lipopolysaccharide (LPS), peptidoglycan and flagellin of Gram negative bacteria are all PAMPs (Bochud and Calandra, 2003; Anas *et al.*, 2010). Research has found the interactions to be quite specific. For instance TLR-4 recognises LPS, while lipoteichoic acid from Gram positive bacteria is bound by TLR-2 and flagellin is identified by TLR-5 (Anas *et al.*, 2010). These specific interactions on the surface of the immune cells trigger a series of pathways leading the production of pro-inflammatory cytokines (Bochud and Calandra, 2003) such as tumour necrosis factor-α (TNF-α) and interleukin-1 (IL-1). In sepsis, the response to PAMPs is disproportionate (possibly due to the repeated stimulation by high concentrations of bacterial antigens) and becomes damaging to the patient (Anas *et al.*, 2010). The body also produces danger-associated molecular patterns (DAMPS), such as heat shock proteins, in response to the inflammation; these can increase the production of cytokines, also mediated through TLR-4 (Anas *et al.*, 2010). A number of other proteins have been implicated in this harmful immune activity and activation of the complement cascade has also been linked to development of sepsis (for a detailed review see Anas *et al.*, 2010). Interestingly, the body's reaction to this hyperinflammation (assuming that the patient survives) is to shut down the immune response (immunoparalysis). Cytokine production decreases and there is marked apoptosis in the immune cells (Anas *et al.*, 2010; Stearns-Kurosawa *et al.*, 2011). Although this may mitigate the effects of the inflammation, it does leave the patient vulnerable to disease caused by other organisms, particularly nosocomial infections (Martin, 2012).

Aging and some underlying medical conditions which affect the immune system such as human immunodeficiency virus (HIV) infection, cancer and diabetes (Stearns-Kurosawa *et al.*, 2011) are known to predispose the patient to more serious outcomes in sepsis. There is also some evidence for genetic variation (Martin, 2012) which accords with the observations of the specific interactions at molecular level between the PRRs and the PAMPs.

2.2.1.2 Epidemiology

Surveillance of bacteraemia in England, Wales and Northern Ireland is conducted through a combination of mandatory notifying of isolation rates (e.g., methicillin-resistant *Staphylococcus aureus*, MRSA) and voluntary reporting by laboratories to the relevant public health agencies. Data collected in this way shows that there are around 94,000 reported cases of bacteraemia each year in England, Wales and Northern Ireland (PHE, Health Protection Report, 2013). The majority of cases are caused by *Escherichia coli* (approximately 24,000 reports), with *Staphylococcus aureus* (around 9,500 per year) and *Streptococcus* spp. (also around 9,500 per year) the next most common isolates. Other reportedly frequent causative organisms include Enterococci (around 5,500) and *Pseudomonas* spp. (just under 3,000). There are about 2,000 cases of fungaemia reported annually and approximately 9% of cases are polymicrobial episodes (PHE, Health Protection Report, 2013; SMI B37, 2014). This is very similar to the pattern reported in other Western European countries (e.g., Luzzaro *et al.*, 2011), while in the United States, epidemiological data suggests that more cases are now caused by Gram positive cocci (Martin, 2012). In Africa, a significant proportion of symptomatic blood stream infections are attributable to *Plasmodium* activity. One study has suggested that over a third of non-malarial cases of community acquired episodes are caused by Gram negative bacilli – although unlike in Europe the main species is *Salmonella enterica* (with about 40% *S. enterica* serotype Typhi) (Reddy *et al.*, 2010). In contrast, *S. typhi* is reported as the most common isolate from patients with community acquired blood stream infection in South and South East Asia (Deen *et al.*, 2012).

Point to consider 2.2: Why might it be useful to know about these epidemiological variations in the causes of sepsis?

2.2.1.3 Diagnostic Considerations

Patient details and history can also be helpful in identifying the cause of septicaemia in individuals. As a general rule there are differences in the most probable causes of sepsis depending on the age and immunocompetence of the patient and whether the original infection was acquired in the community or is healthcare associated. Table 2.1 lists the organisms most often isolated from blood cultures taken from previously apparently

Table 2.1 Organisms most frequently associated with sepsis in previously healthy adults.

Community-Acquired Infections	Healthcare-Associated Infections
Escherichia coli	*Escherichia coli*
Other members of *Enterobacteriaceae* family	Other members of *Enterobacteriaceae* family
	Pseudomonas aeruginosa
Streptococcus pneumoniae	*Streptococcus pneumoniae*
Staphylococcus aureus	*Staphylococcus aureus*
	Coagulase-negative staphylococci
β-haemolytic streptococci	
	Enterococci
Neisseria meningitidis	
	Anaerobic bacteria
	Candida spp. and other yeasts
	Polymicrobial infections

(Data taken from SMI B37, 2014).

healthy adults. It shows that a greater variety of organisms can cause nosocomial sepsis, including anaerobic bacteria and yeasts. *Candida* species are frequently associated with fungaemia. As Figure 2.2a shows, *Candida* spp. have a distinctive appearance on Sabouraud dextrose agar but definitive species identification would require further analysis. Chromogenic agars are available which can distinguish between species on the basis of a colour change during growth. This is illustrated by the appearance of the organisms in Figure 2.2b. The blue colonies are identified by their colour as *C. albicans*, while the pink/white growth is another species of *Candida*. Most community acquired

Figure 2.2a *Candida albicans* grown on Sabouraud dextrose agar. (*See insert for colour representation of the figure.*)

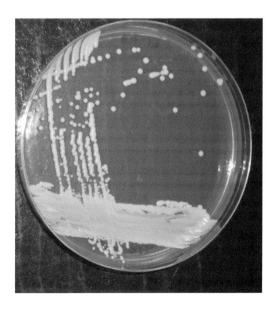

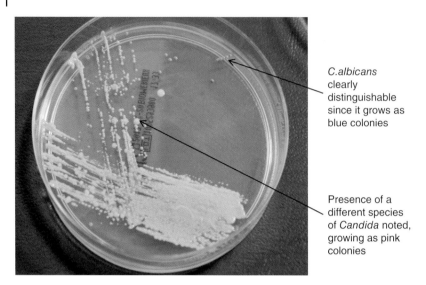

C.albicans
clearly
distinguishable
since it grows as
blue colonies

Presence of a
different species
of *Candida* noted,
growing as pink
colonies

Figure 2.2b Mixed growth of *Candida* spp. on chromogenic agar. (*See insert for colour representation of the figure.*)

paediatric cases involve *Streptococcus pneumoniae, Staph. aureus, Escherichia coli* and *Neisseria meningitidis* (SMI B37, 2014). Nosocomial sepsis infections in children have a similar range of species to those in adults, although anaerobic bacteria and polymicrobial infections are less likely (SMI B37, 2014).

As more patients are successfully treated for a wider range of serious conditions, including HIV infection (which results in them being temporarily or permanently immunocompromised) the list of unusual organisms isolated from patients with sepsis is growing (SMI B37, 2014) and includes *Corynebacterium* spp. (Bernard, 2012). It is therefore important that senior laboratory scientists and clinicians develop (and update) clear protocols for possible isolates according to the patient's profile (e.g., where in the community or hospital they have been and the nature of their immune deficiency) to ensure that specialised media and/or culture conditions are implemented if required. Similarly, 'normal flora' which might be dismissed as contaminants of blood cultures could in fact be opportunistic pathogens. This is illustrated by the recent recognition that coagulase negative staphylococci can cause sepsis in immunologically vulnerable patients (Becker *et al.*, 2014 and see Chapter 7). It is therefore important to discuss any unusual or unlikely findings with senior colleagues and clinicians and not to dismiss an isolate as 'not clinically significant' too quickly.

Sometimes, the patient can be diagnosed with sepsis on clinical grounds, but no organism is isolated from their samples in the laboratory. The most common cause of this 'culture-negative sepsis' is collection of the blood after antibiotics have been administered. It could also be due to a very low level of bacteraemia or fungaemia, meaning that growth of the organism does not reach detectable concentrations *in vitro*. Another reason which must be considered is that the causative organism is unusual or fastidious and thus requires particular, non-standard culture conditions – such as *Mycobacterium* spp. (e.g., Esnakula *et al.*, 2013).

Point to consider 2.3: What factors might account for the differences in the lists of organisms in the two columns in Table 2.1?

2.2.1.4   Treatment

Once the clinical diagnosis of sepsis has been made, broad-spectrum antibiotics are administered to the patient. Where feasible, these are switched for more targeted antimicrobial agents once more information is available about the organism isolated from the blood culture (Inglis, 2013). Reducing the activity of the causative bacterium or fungus in the blood stream should start a decline in the hyperinflammation. Treatment for sepsis also includes supportive interventions, as relevant to the patient's symptoms and condition (Martin, 2012; Inglis, 2013). A number of immunotherapy approaches have been attempted, as the various molecular factors involved in the pathological processes have been identified (Bochud and Calandra, 2003; Anas *et al.*, 2010). Therapies designed to reduce the activity of the cytokines TNF-α and IL-1 were trialled in the 1980s (Martin, 2012). These were largely unsuccessful in reducing morbidity or mortality from severe sepsis and septic shock, highlighting the complexity of the interactions which cause the pathology (Bochud and Calandra, 2003; Anas *et al.*, 2010). Ultimately a combination of immune therapies may prove more effective.

2.2.2   Infective Endocarditis

Cardiac valves which have been structurally weakened in some way can be colonised by microorganisms in the blood stream. This can lead to circulatory problems (ending in heart failure), infection at the site of this colonisation and the release of emboli into the blood stream caused by dislodging of material from the infected valve (SMI B37, 2014; Watkin and Elliot, 2009). The diagnosis of infective endocarditis (IE) is defined by the Duke Criteria (outlined in Text Box 2.1 below) which were developed in the mid-1990s (Murdoch *et al.*, 2009; Watkin and Elliot, 2009).

Confirmed IE is further categorised according to the particular site and nature of the infection. It is classified as left-sided native valve, left-sided prosthetic valve, right-sided or device-related (e.g., on a pace maker – as opposed to actual valve tissue) and then community acquired, healthcare-associated or a consequence of intravenous drug use (Habib *et al.*, 2009; Watkin and Elliot, 2009).

2.2.2.1   Pathology

An anatomical abnormality (e.g., congenital heart defect or prosthetic valve), mitral valve prolapse or degeneration due to age can all interrupt the normal the flow of blood in the area, causing damage to the endothelial tissue of the valve. This produces an inflammatory response which includes accumulation of platelets and fibrin, as part of the body's attempt to heal the wound (Habib *et al.*, 2009; Watkin and Elliot, 2009). As microorganisms pass through the blood stream they can attach to these lesions and accumulate to form a 'vegetation' (Habib *et al.*, 2009). Inflammation can sometimes occur without physical damage to the valve, resulting in the same effect microbiologically; *Staph. aureus* is most frequently identified as the cause of the IE in this scenario (Habib *et al.*, 2009). The infected site can develop into an abscess, causing further damage. The vegetation area can grow and material containing a mixture of tissue, immune cells and microorganisms can fall off into the blood stream as emboli, to be carried to other parts of the body – including the brain. The constant stimulation of the immune system produces other effects, such glomerulonephritis due to the deposition of immune complexes in the kidney (Struthers and Westran, 2005). This collection of microorganisms on the valve can occur rapidly – for example, as a consequence of sepsis – or build-up over a period of years. This

---

Text Box 2.1 Summary of Duke Criteria for Diagnosis of Infective Endocarditis (IE). For details see Murdoch *et al.* (2009)

Pathological criteria
Microbiological and/or histological evidence of an infected vegetation on a cardiac valve, embolus originating from a vegetation or an abscess from that site

Clinical criteria
Major – Consistent positive blood culture result; isolation of an organism associated with IE
Endocardiological investigations show an abnormality characteristic of IE

Minor
Predisposing heart condition or IVDU
Pyrexia >38 °C
Vascular changes associated with IE
Immunological changes associated with IE
Blood culture positive but not consistent or unlikely organism
Endocardiology findings indicate possible IE but inconclusive

Definitions

Definite IE
Pathological criteria met **and**
Two major clinical criteria met **or**
One major clinical criterion and three minor met **or**
Five minor clinical criteria met

Possible IE
Pathological criteria not met/ not investigated **and**
One major clinical criterion and one minor met **or**
Three minor clinical criteria met

Information taken from Murdoch *et al.* (2009)

---

is why previous descriptions of IE as 'acute' or 'subacute' were used; note that the Duke criteria (Text Box 2.1) have now superseded these terms.

2.2.2.2 Epidemiology

Infective endocarditis is a relatively rare condition in the UK, with incidence quoted at < 10 per 100,000 population (Watkin and Elliot, 2009). The age distribution and pattern of associated underlying cardiac defects is changing, with rheumatic fever in young people becoming a less likely cause, while age-related degeneration in the elderly is more common (Habib *et al.*, 2009). The majority of cases are caused by infection with Gram positive cocci, with *Staph. aureus* being most commonly implicated in many parts of the world (Habib *et al.*, 2009; Murdoch *et al.*, 2009). A number of other organisms are associated with IE, including the HACEK group (SMI B37, 2014) of bacteria (*Haemophilus aphrophilus, Aggregatibacter actinomycetemcomitans, Cardiobacterium hominis, Eikenella corrodens and Kingella kingae*) – which are rarely associated with any other diseases in humans – and *Bartonella* spp. which is reported with increasing frequency in this condition, particularly amongst HIV/AIDS patients (SMI B37, 2014). These potential pathogens should be remembered when assessing situations where the patient has clinical indications of IE but negative blood cultures, which is reported to occur in up to 30% of cases (Habib *et al.*, 2009; Fournier *et al.*, 2010). This is because they are slow growing, fastidious organisms. Although relatively rare, these organisms do have characteristic colonial morphologies. As an example, Figure 2.3 is a picture of a blood agar plate on which *Eikenella corrodens* has been grown and the distinctive flat colonies with cleared ('pitted') centres are visible. *Candida* spp. infection is more likely in prosthetic than native value infections (SMI B37, 2014). *Pseudomonas* spp. and fungi are found to cause IE in intravenous drug users; it is known that spores from soil organisms regularly contaminate batches of heroin for injection (Pitt and Gunn, 2012; Health Protection Agency, 2012) and it also is thought that debris in the preparation can settle on the valve and precipitate the damage response (Struthers and Westran, 2005). Another unusual cause to consider is chronic Q fever, where the infection with *Coxiella burnetii* is subclinical and undiagnosed. It was noted that after the outbreak of Q fever associated with goats in the Netherlands in the late 2000s, there was a rise in the number of cases of IE attributed to chronic *C. burnetii* infection (Wegdam-Blans *et al.*, 2012). It has also been suggested that *Tropheryma whipplei* is a more frequent cause of IE than usually suspected (Fournier *et al.*, 2010, Geißdörfer *et al.*, 2012). It should also be noted that the aetiology of IE can be non-infectious, including autoimmune disease (Fournier *et al.* 2010).

Point to consider 2.4: Why is infective endocarditis now a relatively rare condition in the UK?

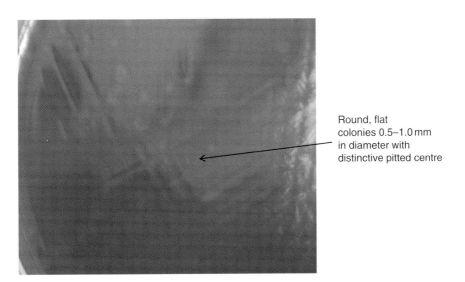

Round, flat colonies 0.5–1.0 mm in diameter with distinctive pitted centre

Figure 2.3 *Eikenella corrodens* on blood agar. (*See insert for colour representation of the figure.*)

2.2.2.3 Treatment

The recommended first line of treatment is combination of a β-lactam antibiotic (since many cases are caused by Gram positive cocci) and gentamycin (Watkin and Elliot, 2009). This can be altered if necessary once the isolate has been fully identified. In very severe cases of accumulation of vegetation, surgical debridement of the valve tissue is sometimes undertaken (Watkin and Elliot, 2009).

2.2.3 Laboratory Diagnosis of Blood Stream Infections

It is important that the responsible organism in any type of blood stream infection is detected accurately as well as rapidly, with the antimicrobial susceptibilities determined as quickly as possible. The first step towards this is the collection of the blood samples, which must be done in accordance with clear guidelines. The standard operating procedures (SOPs) in place should aim to minimise contamination with skin organisms and avoid pseudobacteraemia (SMI B37, 2014), ensure that the correct volume of blood is added to the appropriate blood culture bottle and that the samples are taken before antimicrobial therapy has started, where possible. It is recommended that two sets of specimens are collected (from different sites – to exclude contamination) and that the collection is timed to coincide with the peak of the pyrexia if fever is intermittent (SMI B37, 2014). In adults it is usual to collect samples for both aerobic and anaerobic culture, to optimise the detection of pathogens; for paediatric patients, generally only an aerobic bottle is collected (partly due to the nature of potential isolates and partly due to the difficulty of obtaining several mL of blood from a small child). The samples should be incubated as soon as possible (arrangements across multisite pathology departments and for blood collected outside of routine opening hours will vary).

Automated blood culture systems (which rely on detection of changes in CO_2 and O_2 concentrations in the culture medium due to bacterial metabolism) are very sensitive and can detect microbial growth within hours (SMI B37, 2014). After Gram staining of aliquots from positive blood bottles, subculture onto a suitable set of agar plates can then take place promptly. Subculture sets should include blood agar, a medium for fastidious anaerobes and then other media as suitable; for example, in the case of a Gram negative, MacConkey or CLED plate should be included, if a fungus is found by microscopy, then Sabouraud dextrose agar is required. Faster species identification can be arrived at through the use of chromogenic agar (Figure 2.2b and see also Chapter 7). This means that in a case of serious sepsis, where there is a high microbial load of a fast-growing organism, it is theoretically possible to have a full identification of the organism within 24 hours. In practise, in most cases it is likely to be significantly longer than that and confirming full antimicrobial sensitivities will take a further 18 hours (overnight incubation). Clinicians can treat with broad-spectrum antibiotics at first (after initial blood culture collection) and switch to a drug recommended for the particular species once the identification is reported (Inglis, 2013). Once the full sensitivity pattern is known, further alterations to treatment can be made if necessary. However, given that sepsis can kill within hours (see above) it is clearly better to administer the drug likely to be the most effective within hours rather than days. One solution is to put up antimicrobial sensitivity plates at the same time as the subculture (with antibiotics chosen on the basis of the Gram appearance).

There have been studies comparing kit-based PCR followed by sequencing methods on whole blood samples, to conventional incubation and culture. These have highlighted the greater sensitivity and reduced time to full identification of the organism using the molecular protocol (e.g., Leitner *et al.*, 2013). There are currently some logistical issues which have yet to be resolved, including balancing the greater cost of molecular methods against the faster turnaround times and how to determine antimicrobial sensitivities from a sequence result. In time, these might be addressed since the cost of 'new' technologies tends to fall as they become more widely adopted and with more laboratories using molecular diagnostics it should be possible to build up a databank of sequences of strains of known sensitivities. From a clinical perspective, there remains the problem of how to determine whether the detected sequence represents a viable organism which could indeed be causing the illness or not, particularly in the case of low copy numbers of genetic material for opportunistic pathogens (Leitner *et al.*, 2013).

Diagnostic laboratories are making increasing use of matrix-assisted laser desorption/ionisation-time of flight (MALDI-TOF) in bacterial identification (e.g., El-Bouri *et al.*, 2012) as this can provide a rapid and species-specific answer in a single step from solid media plate culture. This has also proved effective in blood stream infection protocols (Christner *et al.*, 2010; El-Bouri *et al.*, 2012) and it is has been shown that in some cases, sufficient organism can be present on the plate for MALDI-TOF to provide a reliable identification with a few hours of inoculation onto the agar plate (Haigh *et al.*, 2013). It has been demonstrated that MALDI-TOF can detect organisms directly from positive blood culture bottles without the need for culture on agar (Christner *et al.*, 2010; Haigh *et al.*, 2013) and this has the potential to reduce turnaround times for the full species identification result considerably. While this method has a number of advantages over PCR (e.g., detection of live, viable organisms, fewer steps in the technique, lower overall cost), its efficacy is limited to some extent by the array of species in the MALDI-TOF database. Antimicrobial sensitivities would usually still need to be set up from the positive blood culture, meaning an extra day's wait for that result. It is also reported to be more reliable in identification of Gram positive than Gram negative organisms (Christner *et al.*, 2010; El-Bouri *et al.*, 2012).

2.3 Blood-Borne Infections

Blood-borne infections include microorganisms which replicate inside red or white blood cells, others which move into the blood following reproduction elsewhere in the body and some vector-borne diseases. In all cases, a significant concentration of the organism in the blood is required to facilitate transmission to new vectors or human hosts. Depending on the organism and the host's response to the infection, the patient may show obvious signs of an acute and serious disease, or they may not know that they have been infected until years later. There are examples of blood-borne infections involving all groups of microorganism. In some cases laboratory investigations will be undertaken in response to general symptoms but specific history (e.g., vector-borne disease), while in others there will be particular symptoms of a blood-borne infection but relating to another part of the body (e.g., hepatitis viruses). There are also a number of scenarios where the infection is detected through screening (e.g., after risk of exposure or prior to certain medical procedures).

Table 2.2 UK screening protocols for a range of procedures which might transmit blood-borne organisms.

	Blood donation	Organ/tissue donation	Pre-organ/ tissue transplant	Pre-dialysis	Ante natal
HIV	X	Y	X	Y	X
HBV	X	Y	X	X	X
HCV	X	Y	X	X	Y
HEV	X				
Plasmodium spp.	X				
Trypanosoma cruzi	X				
HTLV-I	X	X			
WNV	X				
Treponema pallidum[1]	X				X
CMV		X	X		

X = applies to all patients.
Y = only selected individuals tested, as guided by patient history (e.g., a live organ donor may have documented evidence of successful anti-HBV vaccination).
[1] *T. pallidum* antibodies were found in 85 blood donations in 2011(NHSBT/HPA, 2012).

The approach taken here will be to consider one blood-borne virus of major public health interest (human immunodeficiency virus) in detail and to discuss specific aspects of pathology and laboratory investigations for a selection of other blood-borne organisms. Where appropriate, the reader is directed to other chapters in this book or other references for more information.

Point to consider 2.5: Look at Table 2.2, **which gives the blood-borne organisms usually screened for in a range of situations. Notice that some are common to all the lists (X), but think about why others are only tested for in particular circumstances (Y).**

2.3.1   Human Immunodeficiency Virus

The story of the discovery of human immunodeficiency virus (HIV) is an intriguing one, involving Nobel Prize–winning science, competition amongst the scientists and the fear of a fatal disease (Gallo and Montagnier, 2003). In the late 1970s, previously apparently fit young people (mostly male and homosexual) were presenting at clinics and hospitals in the United States (particularly New York and San Francisco) with odd combinations of symptoms including lymphadenopathy, night sweats, weight loss and unusual opportunistic infections. These included severe and disseminated forms of herpes infection and candidiasis, as well as Kaposi's sarcoma – which had hitherto been a rare skin disease seen in elderly men with genetic or residential connections with the Mediterranean – and an usual atypical pneumonia caused by an organism then called *Pneumocystis carinii* (e.g., Gottlieb *et al.*, 1981). These infections suggested that patients were severely immunocompromised and they often died soon after presentation. A few years later, a similar trend was reported in major European cities (e.g., Lundgren *et al.* 1994). It was recognised that there was likely to be a common source for these symptoms and the disease was named acquired immune deficiency syndrome (AIDS) (Jaffe *et al.*, 1983). Epidemiological connections suggested that the cause might be infectious, but investigations were hampered by the nature of the organism, which we now know to be the human immunodeficiency virus. Laboratory work was not easy, as it only could only be grown in white blood cells in suspension, rather than the conventional monolayer cultures of fibroblasts and monkey kidney cells which were widely used in diagnostic and research virology laboratories at the time. Eventually, a positive sense single-stranded RNA retrovirus was isolated (Pillay *et al.*, 2009) and through genetic analysis it was found to be related to the already recognised human T cell lymphotropic viruses (HTLV) -I and -II (see below). Montagnier and Barré-Sinousi who were at the Institut Pasteur in Paris, isolated and characterised the virus and reported their findings in 1983 (Barré-Sinousi *et al.*, 1983); they named the virus lymphadenopathy virus. At more or less the same time, Gallo and his team at the National Institutes of Health in the United States published very similar results (Gallo *et al.*, 1983); due to its similarity to HTLV-I and -II, they named it HTLV-III. They went on to report evidence of the virus in patients with AIDS (Gallo *et al.*, 1984). Robust discussions ensued about who had actually found the virus first and who therefore had the right to name it. In 1986, the International Committee on the Taxonomy of Viruses (ICTV) decided to resolve this dispute by adopting a completely new name – human immunodeficiency virus (HIV). The discovery of the virus allowed investigations into pathogenesis, which enhanced the development of diagnostic test methods and of anti-viral treatments, all of which happened relatively rapidly (Pillay *et al.*, 2009). In 2008, Montagnier and Barré-Sinousi were awarded the Nobel Prize for their work (Cohen and Enserink, 2008). It transpired that while Gallo had contributed significantly to demonstrating the connection between HIV and AIDS, it was the French team who had actually discovered the virus. Two species of the virus are now recognised (HIV-1 and HIV-2) and they are classified in the Family: *Retroviridae*, subfamily: *Orthoretrovirinae*, Genus: *Lentivirus*.

2.3.1.1 Epidemiology

After the first laboratory confirmed cases of HIV/AIDS were recorded in the early 1980s, the reported numbers increased dramatically across the world. Countries introduced the laboratory test at different rates. Therefore, amongst the initial tranches of people coming forward for testing were many who had been infected some years previously. The infection seemed to spread quickly and easily so that by 2001, 30 million people had been diagnosed as HIV positive and in that year it was estimated that 1.9 people died from AIDS. (www.unaids.org/en/dataanalysis/datatools/aidsinfo). The global transmission was undoubtedly helped by the fact that the virus is transmitted via blood and body fluids and through sexual intercourse. With greater awareness of the mode of transmission of the virus, screening of people at risk and also the checking of blood and blood products prior to donation, the global incidence decreased. The advent of anti-viral drugs extended the time between diagnosis as HIV positive and the development of AIDS, with the result that by the mid-2010s there were over 36 million people living with HIV/AIDS (www.unaids.org) and around 1.6 million deaths each year. The majority of people infected with the virus live in sub-Saharan Africa (Girard *et al.*, 2011). The calculated prevalence in adults was 0.8% in both 2001 and 2013; however, the prevalence in young people (15- to 24-year-olds) had decreased from 0.4% to 0.3% in males and from 0.7% to 0.5% in females (www.unaids.org/en/dataanalysis/datatools/aidsinfo). For more information, see the UNAIDS website (www.unaids.org). In the UK, available data shows that the number of confirmed new cases of HIV infection acquired within the country was recorded as 7928 in 2005. Incidence has been in steady decline since then (https://www.gov.uk/government/collections/hiv-surveillance-data-and-management). The overall prevalence of HIV is calculated to be 1.6 per 100,000 population. This figure takes into account a significant minority of infected people who have not yet been diagnosed with the infection. In the years since 1998, the highest number of AIDS diagnoses were recorded in 2003 (1034) with 390 in 2012; deaths from AIDS in the UK are at a rate of about 500 per year on average (https://www.gov.uk/government/collections/hiv-surveillance-data-and-management).

While transmission of HIV-1 occurs throughout the world – albeit with geographic variation in prevailing subtypes – HIV-2 appears to be restricted to West Africa (Pillay *et al.*, 2009). It is important to note that HIV-2 is also found in West African communities in other regions, including Europe. The highest prevalence of HIV-2 in Europe is reported from Portugal, which has strong cultural and political links to some West African countries (Valadas *et al.*, 2009).

2.3.1.2 Pathogenesis

The pathological consequences of infection with HIV are related to its effects on the host immune system. The virus infects CD4+ cells, which include monocytes, macrophages, dendritic cells and some types of brain cells, in addition to CD4+ T lymphocytes (Pillay *et al.*, 2009). The viral surface glycoprotein gp120 recognises the CD4 glycoprotein and entry of the virus into the cell is effected through 'co-receptors' (associated with binding of chemokines), such as CCR5 or CXCR4. (Pillay *et al.*, 2009). Interestingly, some people carry a 32-base pair deletion in both copies of the CCR-5 gene (which codes for the CCR5 receptor) and this appears to afford some resistance to HIV-1 infection (Girard *et al.*, 2011). This Δ 32 CCR5 mutation is not found in African and Asian populations, but it is present in about 10% of Europeans (Girard *et al.*, 2011). It has been suggested that the gene deletion was selected for as it conferred protection against the plague or smallpox (Hopkin, 2011).

The virus particle carries two copies of the positive sense ssRNA genome, which codes for the *gag, pol* and *env* genes (see Figure 2.4), along with other genes which produce regulatory proteins (Pillay *et al.*, 2009). The *gag, pol* and *env* gene products are all polyproteins which are cleaved to make smaller structures with specific roles (Girard *et al.*, 2011). The gag proteins include the p24 capsid antigen, while env produces viral envelope glycoproteins gp41 and gp120 (Figure 2.4). Reverse transcriptase enzyme is derived from the pol protein and this makes a 'provirus' cDNA copy of its genome, from which the complementary strand is made before it is integrated into the host cell DNA via a mechanism involving long terminal repeats (LTR) of DNA at the 5' and 3' ends (Pillay *et al.*, 2009). Replication and infection of additional cells occurs rapidly during the acute phase and blood viral loads are usually at least 10^6 copies per mL (Girard *et al.*, 2011), though they fall subsequently during the chronic, latent phase of the infection. Since CD4+ T cells are 'helper cells', involved in the cellular immune response and also in stimulating B cells to produce antibodies, infection inevitably leads to immunosuppression in the host and lays them open to opportunistic infections. The body appears to respond to the presence of HIV antigens by activating more CD4+ T cells. Thus, the consequence of the body's reaction intended to try and eliminate HIV is to increase the number of cells available for the virus to infect (Silvestri, 2013).

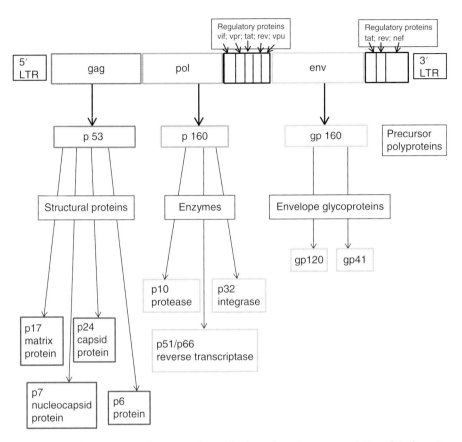

Figure 2.4 HIV-1 genome and gene products. (*See insert for colour representation of the figure.*)

Once a person has been infected with HIV they appear to be unable to clear it. Since the function of the reverse transcriptase enzyme is to copy RNA, it does not have a proofreading capacity, which means like other RNA viruses, HIV is prone to mutation. Coupled with the fast replication rate, this property of the virus gives rise to many mutants, some of which can become viable variant strains. Indeed, in a long-standing infection a number of different strains are often found in a single host (Girard *et al.*, 2011). The uniqueness of a person's HIV virus populations has been exploited in criminal investigations (Gunn and Pitt, 2010). The reader is also directed to volume 254 (2013) of the journal *Immunological Reviews*, which provides a series of papers considering the 'complexity of HIV immunology'.

2.3.1.3 Clinical Aspects

The rapid viral replication and overwhelming viraemia which occurs when a person is first infected with HIV, is accompanied by a symptomatic illness in up to 90% of cases (Pillay *et al.*, 2009). Unfortunately, the symptoms are rather non-specific and include some or all of the following: pyrexia, tiredness, lethargy, aching and lymphadenopathy (particularly in the neck glands) and a rash. Some people do experience opportunistic infections (such as pneumocystis pneumonia) at this stage, but in general, this primary infection is short-lived (up to 14 days) and self-limiting (Pillay *et al.*, 2009). As with other chronic infections such as hepatitis B and syphilis, if no investigations or treatment are sought at this point, then the virus will persist, while the person is largely asymptomatic and possibly unaware of their condition. The virus is still replicating and destroying CD4+ cells, but at a slower rate than in the acute phase and the host is able to keep the levels of these cells sufficiently high to allow a relatively normal immune response to continue (Pillay *et al.*, 2009). Without treatment the CD4+ count does gradually fall and after a period of several years (generally between 5 and 15, but it may be more or less than that), it declines to the point where the person is severely immunocompromised. This is usually quantified as a CD4+ count < 200 cells μL^{-1} of blood (Pillay *et al.*, 2009) and is manifest through a range of unusual, or unusually severe, infections including *Pneumocystis jirovecii* pneumonia; primary infection or re-activation of *Mycobacterium tuberculosis;* meningitis or encephalitis associated with *Cryptococcus* spp., *Toxoplasma gondii* or cytomegalovirus; disseminated candidiasis and Herpes simplex infection; cryptosporidial or microsporidial diarrhoea; malignancies such as Kaposi's sarcoma, lymphomas and ano-genital carcinomas are also observed at a higher rate than the general population. HIV viral load is typically around 10^3–10^5 copies mL^{-1} (Pillay *et al.*, 2009) and the person will be clinically diagnosed as having acquired immune deficiency syndrome (AIDS). Without treatment, this susceptibility to infection (which the body is progressively less able to fight) leads to deterioration of the patient's condition and AIDS is fatal within a few years. Anti-viral treatments were developed remarkably quickly after the discovery of HIV and a range of drugs with different biochemical activities are now available (Davey *et al.*, 2015). This allows for multiple drug treatment regimes (called highly active anti-retroviral therapy, HAART), which help to overcome the tendency of the virus to develop resistance through mutation, improving the quality as well as the longevity of the lives of people who are HIV positive.

Due to the effect that the HIV virus has on the host's immune system, some rather unusual and more severe presentations of other infections can be found in patients. Co-infection with other blood-borne microorganisms is not infrequent. For example, HBV and HCV co-infection can occur, primarily due to the common transmission route (sexually and through contaminated needles and blood products), while the majority of HIV cases are reported from sub-Saharan Africa, where most malaria transmission also occurs. There is evidence that in people with HIV who are co-infected with HBV or HCV, the retrovirus affects the body's ability to mount a suitable immune response to the hepatitis viruses. Measurable HCV RNA or HBV DNA viral loads are higher in HIV-positive patients, the infection is less likely to be cleared and in the case of HBV, detectable serum HBe antigen is typical (see Chapter 4); progression to end-stage liver disease is more probable (e.g., Operskalski and Kovacs, 2011; Thio *et al.*, 2002). This means that awareness of the co-infection could be important when undertaking laboratory assessments, including monitoring. Viral loads for the hepatitis viruses might be higher than expected after anti-viral treatment and resolution of the HCV or HBV infection is less likely (Operskalski and Kovacs, 2011; Coffin *et al.*, 2013).

Interactions between HIV and some blood-borne protozoa are notable. Malaria is one of the most important infectious diseases worldwide. There are five species of *Plasmodium* which are human pathogens and the parasite is transmitted in blood via the bite of *Anopheles* spp. mosquitoes (Figure 2.1). Over 200 million people have at least one episode of malaria each year and around 500,000 people die from the disease (http://www.who.int/malaria/en/).

Most of the deaths occur in people who are infected with *P. falciparum* and many of them are children in sub-Saharan Africa. While malaria is not commonly seen in Europe, there is some transmission of *P. vivax* in Southern Europe (http://ecdc.europa.eu/en/healthtopics/malaria/Pages/epidemiological-updates.aspx; WHO/ECDC, 2013).

Figure 2.5 shows images from thick and thin blood films taken from a patient with *P. vivax* malaria. In the thick film (Figure 2.5a), the red blood cells have been lysed, since the slide is not fixed prior to staining. The trophozoites shown here are from quite an early stage in the parasite development, but have a larger cytoplasm than would be expected for other *Plasmodium* spp. The thin film slide (Figure 2.5b) has been fixed before staining and therefore the red blood cells remain intact. This allows the enlargement and unusual shapes characteristic of red blood cells infected with *P. vivax* to be noticeable.

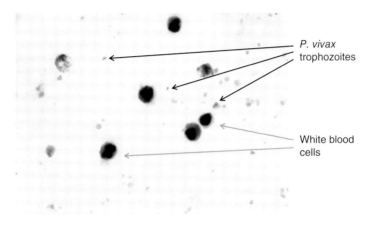

P. vivax trophozoites

White blood cells

Figure 2.5a Thick blood film stained with Giemsa stain showing early trophozoites of *Plasmodium vivax*. (*See insert for colour representation of the figure.*)

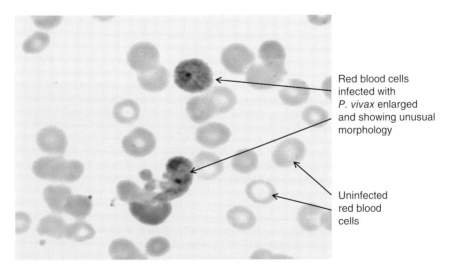

Red blood cells infected with *P. vivax* enlarged and showing unusual morphology

Uninfected red blood cells

Figure 2.5b Thin blood film stained with Giemsa stain showing late trophozoites of *Plasmodium vivax*. (*See insert for colour representation of the figure.*)

Blood donors who have a travel history indicating possible contact with malaria are screened (UKBTS; www.transfusionguidelines.org.uk/Index.aspx?Publication=HTM& Section=9&pageid=1102). In the context of clinical laboratory microbiology, malaria diagnosis usually falls under the remit of the haematology department in the UK, so it will not be considered in detail in this book. For more details, the reader is directed to a diagnostic haematology textbook, such as Bain *et al.* (2012) or a diagnostic parasitology text book, such as Shore Garcia (2007). In the context of HIV it is interesting that the two organisms appear to facilitate each other's activities within the host. Acute *Plasmodium* spp. infection in a person who is HIV positive results in a rise in plasma viral load, while co-infection during pregnancy appears to increase the chances of both severe pregnancy-associated malaria and vertical transmission of the virus, although the mechanisms are uncertain (Gunn and Pitt, 2012; see Chapter 5). Since the majority of the HIV/AIDS patients in the world live in sub-Saharan Africa, where malaria is holoendemic, the relationship between the two organisms is an issue of concern (Kublin *et al.*, 2005).

More is understood about the relationship between HIV and *Leishmania* spp. during co-infection, in particular the species which cause visceral leishmaniasis (VL), *L. infantum* and *L. donovani*. Both organisms infect CD4+ T cells and establish latency. This means that when a T cell with an existing *Leishmania* spp. parasite is infected by an HIV virion, the protozoan can be activated to grow and reproduce more vigorously and *vice versa*. Their respective effects on the host's immune system can lead to more rapid progression in both diseases (Monge-Maillo *et al.*, 2014). Although *Leishmania* spp. are transmitted by specific sandfly vectors and incidence is therefore usually geographically restricted, VL has been spreading in many areas of the world, including southern Europe (Desjeux and Alvar, 2003; Monge-Maillo *et al.*, 2014). In this region, unwitting co-transmission through blood transfusions and shared needles appears to have occurred. It is estimated that at least a third of adults with VL in the Mediterranean area also have HIV/AIDS (Monge-Maillo *et al.* 2014). In a diagnostic microbiology situation it may occasionally be useful to know that co-infection with one of these parasites could account for unexpected HIV viral load results.

2.3.1.4   Laboratory Diagnosis

Screening tests involve detection of anti-HIV antibodies (anti-p24, anti-gp 41 and anti-gp 120) in blood (or serum) by immunoassay (SMI V11, 2016). Although the so-called 'third-generation' anti-HIV assays are capable of detection IgM, it is now recommended that a fourth-generation method, which can also detect p24 antigen, is used in any diagnostic laboratory setting. It is also important to use an assay which can detect both HIV-1 and HIV-2 (SMI V11, 2016). Any positive results from the initial assay must be confirmed from the primary tube, preferably using at least one different type of test (e.g., Western Blot or molecular assay). It should be noted that in late stages of infection (AIDS), the immune system is compromised to such an extent that even IgG antibody production may be at too low a level to be detected by diagnostic tests – and this includes the anti-HIV antibody assay! In this case a test which includes antigen detection would be helpful, but if AIDS is clinically indicated, then a viral load assay (see below) would probably be more useful (SMI V11, 2016).

When the result is confirmed, then further samples must be obtained for the HIV viral load test (by reverse transcriptase PCR) and the CD4+ count. Regular monitoring of patients to assess these parameters is a key laboratory function in this situation.

Once the CD4+ count is < 350 cells μL^{-1} and /or the viral load is > 10^4 copies mL^{-1}, anti-retroviral chemotherapy should be started. Successful treatment would be expected to increase the CD4+ count by 100–150 cells μL^{-1} in the first year and thereafter for a few years before reaching a consistent level (which tends to vary according to the individual). Similarly, the aim is for the viral load to fall to 'unde-tectable' levels, which is around <50 copies mL^{-1} (Pillay *et al.*, 2009). Genotyping and checking isolates for the possible development of anti-viral resistance through sequencing are also important parts of the long term follow up of patients with HIV/AIDS.

Laboratory screening tests for HIV are carried out for a range of reasons in people who are at relatively low risk of infection. A number of immunochromatographic assays (point-of-care tests) have become commercially available and they are being implemented and evaluated in certain settings, such as GUM clinics and sexual health centres in the com-munity (e.g., Surah *et al.*, 2009). They are likely to be useful for screening and as well as providing a rapid result. Some kits utilise whole blood from a finger prick or oral fluid, so are less invasive than laboratory tests which require venous blood (Delaney *et al.*, 2011). However, they are reported as having lower sensitivity and specificity than third- (Delaney *et al.*, 2011) and fourth-generation laboratory tests (Rosenberg *et al.*, 2012), so can there-fore detect established HIV infections but not the early stages of the acute illness.

Point to consider 2.6: Find out about some case studies where co-infections with HIV and other microorganisms gave rise to unusual laboratory results and or an unexpected clinical picture. Either ask your laboratory-based colleagues and/or search the literature for examples.

2.3.2   Human T-Cell Lymphotrophic Viruses

Within the sub-family *Orthoretrovirinae*, there is another genus, *Deltaretrovirus*, which contains two viruses which have been found in humans, the human T-cell lym-photrophic viruses HTLV-1 and HTLV-2 (sometimes written as HTLV-I and HTLV-II). These viruses appear to be transmitted vertically and through sexual contact, as well as contaminated blood and blood products (Taylor, 2009), but serological surveys tend to show high prevalence in specific populations. For instance HTLV-1 is com-monly found in Japan and the Caribbean and thus in the UK, infection is rarely seen in people who do not have genetic or geographical connections to those regions. In contrast, HTLV-2 seems to have a more disparate epidemiology and in addition it has found its way into the intravenous drug user (IVDU) populations both in Europe and North America (Taylor, 2009). HTLV-1 is associated with a number of malignancies including tropical spastic paraparesis (TSP) – also called HTLV-1-associated myelop-athy (HAM) – and adult T-cell leukaemia-lymphoma (ATLL). These conditions are relatively rare; for example, the risk of someone carrying HLTV-1 developing HAM at some point during their life is usually reported as significantly below 10% (and in Japan it is estimated as 0.25%). Like HIV, this virus infects CD4+ cells but in this case, evidence suggests that it transforms them to be immortal (Taylor, 2009). The patho-logical mechanisms which lead to HAM or ATLL are not well understood. In contrast, evidence of relationships between HTLV-2 and any disease syndromes at all has so far proved inconclusive.

If there are grounds to suspect HTLV-1 associated malignancy, then laboratory investigations would usually comprise serology for antibodies, followed by PCR for viral load (Taylor, 2009). Routine screening is not carried out in the UK, mainly due to the low prevalence and because asymptomatic carriers are not very likely to become ill (and may not want to know!). The exception to this is testing donated blood prior to transfusion (UKBTS; www.transfusionguidelines.org.uk/Index.aspx?Publication=HTM& Section=9&pageid=1102). In 2011, HTLV-I antibodies were detected in blood from 14 donors in the UK (NHSBT/HPA, 2011). Since the immune system of a recipient of a blood product or donated organ may already be compromised, the risk of developing an HTLV-1–related disease is higher than in the general population. A case reported from Germany (Glowaska *et al.*, 2013) highlighted this: two out of three recipients of organs transplanted from an HTLV-1 seropositive donor developed a T-cell lymphoma associated with this virus within 3 years of the procedure. As might be expected from the examples mentioned above, co-infection with HIV seems to have a detrimental effect on the patient's prognosis. There is evidence of more a more rapid course to from HIV positivity to AIDS than might otherwise be expected, while HTLV-1-associated HAM is more likely (Taylor, 2009).

2.3.3 *Trypanosoma cruzi*

Trypanosoma cruzi, the causative agent of Chagas disease is a protozoan parasite in the Phylum Kinetoplastida, Genus *Trypanosoma*. It is transmitted in blood through the bite of the Triatomid bug ('kissing bug') and is generally restricted to areas where this insect vector lives, namely North and South America. Chagas disease is a chronic life-threatening condition, so blood donors who have visited this region and are at risk of possible infection are screened (UKBTS; www.transfusionguidelines.org.uk/Index.aspx?Publication= HTM&Section=9&pageid=1102. Since the disease is caused by a blood-borne parasite and laboratory investigations are usually the remit of the haematology department, it will not be discussed further here. The reader is directed to http://www.cdc.gov/dpdx/ trypanosomiasisamerican/.

However, *T. cruzi* is of interest because it can be transmitted through blood transfusion. In endemic countries in South America, the prevalence of the parasite is so high that it is considered more cost effective to pre-treat donations with an anti-trypanosomal agent, rather than screening donations for antibodies against *Trypanosoma* spp. The compound of choice, which is found to be effective in killing the organisms, but with minimal toxicity to the patient, is crystal (Gentian) violet. While it does mean that the blood is safe, it does have the rather disconcerting effect of turning the patient purple for a while (Gunn and Pitt, 2012)!

2.3.4 Creutzfeldt-Jakob Disease

Creutzfeldt-Jakob Disease (CJD) is a neurological condition. The 'transmissible protein' or 'prion' which is associated with CJD is known to have unwittingly been passed between human hosts through medical procedures such as surgery with contaminated instruments and corneal transplants (Collinge, 2009). The incubation time for the disease can be many years and a reliable pre-mortem test suitable for screening is not available. It is therefore UK Blood and Tissue Transfusion Service policy to exclude anyone who might be a risk of carrying CJD (or the more recent variant vCJD). This now includes anyone who has received blood or blood components since January 1980 (UKBTS. www.trans fusionguidelines.org.uk/Index.aspx?Publication=HTM&Section=9&pageid=1102).

2.3.5 West Nile Virus

West Nile fever is caused by the West Nile virus (WNV) which is a positive sense ssRNA virus in the Family *Flaviviridae*, Genus *Flavivirus* – which also includes dengue virus and yellow fever virus. The usual reservoir hosts for the West Nile virus (WNV) are birds of the *Corvidae* family (e.g., crows, jackdaws, magpies) and it is transmitted between them via the bite of *Culex* spp. mosquito. Corvids are medium-sized birds found throughout most of the world. They are intelligent and versatile and live in urban and rural areas (Figure 2.6). *Culex* mosquitos are 4–10 mm in length, with clear wings. Their maxillary palps are noticeably shorter than their proboscis which is a feature distinguishing *Culex* spp. from *Anopheles* spp. Another way to tell the difference between these two species is that *Culex* mosquitoes tend to rest parallel to the surface rather than at an angle. Evidence from transmission during the twenty-first century in the United States suggests that other mosquito species can occasionally be viable vectors for WNV. Humans and also horses are occasional accidental hosts and before the turn of the twenty-first century, sporadic cases of human disease were reported in the Middle East, Africa and Europe (Hubálek and Halouzka, 1999). The virus had no noticeable effect on the host birds in these regions and the majority of human infections were asymptomatic. This is still the case, although in about 20% of cases an overt illness involving fever, lethargy, lymphadenopathy, rigours, diarrhoea, vomiting and sometimes a rash occurs. About 1% of patients develop serious neurological symptoms including meningitis and encephalitis (Lloyd, 2009). Transmission in Europe is seasonal – occurring between June and September (http://ecdc.europa.eu/en/healthtopics/west_nile_fever/pages/index.aspx).

Figure 2.6 Crow (*Corvus cornix*) in a rural habitat. Source: Courtesy of Dr A Gunn. (*See insert for colour representation of the figure.*)

In the 1990s, a mutation appears to have occurred to make the virus more virulent and more neurotropic (Lloyd, 2009) and it also spread to North America. The story of West Nile fever in the United States since 1999 provides an ideal example of an emerging infectious zoonotic disease. The tale involves people developing unexplained neurological diseases and crows falling out of the trees! It is reviewed in Colpitts *et al.* (2012). It is fascinating and also worrying to notice how quickly the infection spread across the United States. There were around 60 cases in New York State in 1999, which turned into over 9800, with cases being reported from every state in the country, by 2003. The maps and data provided by the Centres for Disease Control and Prevention show this very effectively and the interested reader is directed to www.cdc.gov/westnile/statsMaps/. Blood donors with a history of travel to areas where West Nile fever transmission is known to be occurring are questioned about possible exposure and advised accordingly. From a diagnostic microbiological point of laboratory, when a patient presents with meningitis or encephalitis which was preceded by a 'flu-like' illness, a check on their holiday destination may indicate that a test for WNV RNA may be useful.

Point to consider 2.7: What factors contributed to the rapid spread of WNV across the United States?

2.3.6 Blood-Borne Herpes Viruses

Epstein-Barr virus (human herpes virus-4) and cytomegalovirus (human herpes virus-5) both invade white blood cells and establish latency. They can also exploit the immunosuppression caused by HIV infection to produce more severe disease in their hosts. They are in the family *Herpesviridae* and as such are relatively large, dsDNA viruses. Epstein-Barr virus (EBV) is in the subfamily *Gammaherpesvirinae*, genus *Lymphocryptovirus*. It infects B lymphocytes, inducing polyclonal activation (meaning that the patient could be producing spurious, detectable levels of antibodies to other organisms or other antigens during acute infection) and in some circumstances transformation to malignancy (Haque and Crawford, 2009). In HIV-positive patients, there is a much higher incidence than usual of non-Hodgkin's lymphoma associated with EBV, as well as a range of rare and unusual malignancies such as primary central nervous system lymphoma (Haque and Crawford, 2009). These conditions arise due to the immunosuppression caused by the HIV which allows EBV-transformed B cells to proliferate; the underlying retroviral infection also makes these lymphomas much harder to treat.

Cytomegalovirus (CMV) is in the subfamily *Betaherpesvirinae*, genus *Cytomegalovirus*. As a herpes virus it shares the characteristic ability of latency and re-activation. In spite of extensive research, the site of the latency is a yet uncertain (unlike for most of the other *Herpesviridae*), although evidence points to cells of the myeloid lineage in the bone marrow being a possibility (Sinclair and Sissons, 2006). The virus is ubiquitous and is transmitted through blood and body fluids (including saliva and breast milk). Serological studies have shown that in developing countries, most children have acquired the virus before their teenage years, while in richer countries, about 40% of adults have detectable serum CMV IgG antibodies (Griffiths, 2009); once infected, a person sheds the virus intermittently, producing detectable viraemia and viruria. Transmission occurs congenitally and perinatally, through childhood contacts (when infection is largely asymptomatic or unremarkable), oral contact, such as kissing, later in life (sometimes producing a glandular fever-like illness), sexual contact, and iatrogenically through blood transfusion and organ and tissue transplantation (Griffiths, 2009). CMV can cause serious and sometimes life-threatening disease in immunocompromised patients (see Chapter 7). This can be as a result of primary infection when the virus is transmitted via blood or tissues to seronegative patients as well as re-activation due to the effects of immunosuppressive chemotherapy or HIV infection.

Although CMV infection/re-activation in immunocompromised patients may be asymptomatic, it is common for some effects to be noted even if it is only pyrexia. Leucopaenia can transpire and subsequently a number of possible conditions including pneumonitis and gastroenteritis and secondary bacterial or fungal infections (Griffiths, 2009). In common with other microorganisms mentioned in this chapter, the course of CMV disease is usually different when the person is co-infected with HIV. Certain manifestations such as retinitis, encephalopathy, dementia, polyradiculopathy (weakness of lower limbs and bladder) and infection of the adrenal glands (usually found at post-mortem) are much more likely to be seen in patients with AIDS.

Screening for CMV status prior to a procedure involves serological detection of IgG antibody. In general, diagnosis of acute infection can be done by serum IgM testing, but antigen detection using PCR would be considered wherever possible, particularly when the patient has a lowered immune response (Binnicker and Espy, 2013). For patients who are immunocompromised and in hospital (e.g., post-transplant, AIDS) then regular screening of blood and urine for the presence of CMV by PCR is important and is undertaken routinely (Razonable and Hayden, 2013). The anti-viral agent gancyclovir is effective in the treatment of CMV disease (Griffiths, 2009); in some situations it is used prophylactically (e.g., after a kidney transplant involving a seropositive donor and a seronegative recipient) while in others it would be given as treatment for symptoms related to a rise in detectable CMV levels in blood and urine; where appropriate, a respiratory sample such as a broncho-alveolar lavage (BAL) would also be tested. In either of these approaches to patient management, the laboratory's role is vital in detecting virus activity as soon as possible so that treatment can be started or reviewed in order to reduce the chances of serious CMV disease occurring. Standardisation of methods to inform management is a current issue in this area (Hirsch *et al.*, 2013).

2.4   Exercises

2.1   **A** Without consulting any sources of information, write out an algorithm for the diagnosis of sepsis.

B Then read the PHE SMI and or your laboratory's SOP and compare these with your algorithm.

C Did your version agree with the recommended protocol? If not, reflect on the points of disagreement.

D Can you think of any ways in which the recommended protocol could be improved (i.e., in the light of the recently published research about organisms or diagnostic techniques)?

2.2 **A** Construct an algorithm for the use of a POCT test for HIV testing in GUM clinics

B What are the benefits and risks of allowing the use of home testing kits for HIV?

2.3 **A** Find a copy of your laboratory's testing algorithm for screening for blood-borne infections prior to a selected procedure (e.g., transplant, dialysis).

B Review it carefully and decide whether it is up to date with current knowledge both of potential transmission blood-borne pathogens and laboratory techniques.

2.5 Case Studies

2.1 Mrs AB is a 66-year-old woman with multiple myeloma. She is a hospital inpatient as she is currently on cytotoxic chemotherapy treatment which is being delivered intravenously. Yesterday her temperature was slightly elevated and today she has a marked pyrexia of 39.5 °C, a heart rate of 95 beats minute^{-1} and a decreased white cell count of 3200 cells μl^{-1}.

Blood culture bottles are collected in the appropriate way and at suitable times and after 20 hours incubation in an automated blood culture system, the bottles are flagged as positive.

Gram stain shows that one organism is growing and it is a Gram positive coccus. Further tests indicate that it is coagulase negative.

a) Would you report this result to the haematology/oncology clinicians at this stage?

b) The identity of the isolate is confirmed as *Staphylococcus epidermidis*. It is assumed to be the cause of Mrs AB's sepsis and anti-microbial susceptibility plates are then set up. Treatment is therefore started empirically. What should be taken into account when deciding on appropriate antibiotics?

2.2 Mr FG is a 42-year-old man with a history of past and current intravenous drug use. He is known to be HIV and HCV positive. All investigations to date indicate that he has never been infected with HBV. He attends the hospital for regular monitoring of his condition, which includes taking blood for virological investigations.

Three months ago, his results were:

HIV RNA viral load: 10^{5} copies mL^{-1} blood; CD 4+ count: 300 cells μL^{-1}
HCV RNA viral load: 8, 300 IU/mL
Hepatitis B surface antigen: Not detected

a) What did these results suggest?

Now, three months later his HIV and HCV results are as follows:

HIV RNA viral load: 10^{4} copies mL^{-1} blood; CD 4+ count: 370 cells μL^{-1}
HCV RNA viral load: 7,400 IU/ml

b) How do you explain these results?

References

Anas AA *et al.* (2010). Recent insights into the pathogenesis of bacterial sepsis. *The Netherlands Journal of Medicine*, **68**: 147–152.

Bain BJ *et al.* (2012). *Dacie and Lewis Practical Haematology, 11th edn*. London: Churchill Livingstone Elsevier.

Barré-Sinousi F *et al.* (1983). Isolation of a T-lymphotropic retrovirus from a patient at risk for acquired immune deficiency syndrome. *(AIDS) Science*, **220**: 868–871.

Bearman GML and Wenzel RP (2005). Bacteremias: A leading cause of death. *Archives of Medical Research*, **36**: 646–659. doi: 10.1016/j.arcmed.2005.02.005

Becker K, Heilmann C and Peters G (2014). Coagulase-negative staphylococci. *Clinical Microbiology Reviews*, **27(4)**: 870–926. doi: 10.1128/CMR.00109-13

Bernard K (2012). The genus Corynebacterium and other medically relevant coryneform-like bacteria. *Journal of Clinical Microbiology*, **50**: 3152–3158. doi: 10.1128/JCM.00796-12

Binnicker MJ and Espy ME (2013). Comparison of six real-time PCR assays for the qualitative detection of cytomegalovirus in clinical specimens. *Journal of Clinical Microbiology*, **51**: 3749–3752. doi: 10.1128/JCM.02005-13

Bochud P-Y and Calandra T (2003). Pathogenesis of sepsis: New concepts and implications for future treatment. *British Medical Journal* **326**: 262–266 doi: 10.1136/bmj.326.7383.262

Christner M *et al.* (2010). Rapid identification of bacteria from positive blood culture bottles by use of matrix-assisted laser desorption-ionization time of flight mass spectrometry fingerprinting. *Journal of Clinical Microbiology*, **48**: 1584–1591. doi: 10.1128/JCM.01831-09

Coffin CS *et al.* (2013). Virology and clinical sequelae of long-term antiviral therapy in a North American cohort of hepatitis B virus (HBV)/human immunodeficiency virus type 1 (HIV-1) co-infected patients. *Journal of Clinical Virology*, **57**: 103–108. doi: 10.1016/j.jcv.2013.02.004

Cohen J and Enserink M (2008). HIV, HPV researchers honoured, but one scientist is left out. *Science*, **322**: 174–175. doi: 10.1126/science.322.5899.174

Collinge J (2009). Human prion diseases. In: *Principles and Practice of Clinical Virology*, 6th edn. Zuckerman AJ, Banatvala JE, Schoub BD, Griffiths PD and Mortimer P, eds. Chichester: Wiley-Blackwell.

Colpitts TM *et al.* (2012). West Nile virus: Biology, transmission, and human infection. *Clinical Microbiology Reviews*, **25**: 635–648. doi: 10.1128/CMR.00045-12

Davey, P, Wilcox MH, Irving W and Thwaites G (2015). *Antimicrobial Chemotherapy*, 7th edn. Oxford: OUP.

Deen J, von Seidlein L, Anderson F, Elle N, White NJ and Lubell Y (2012). Community-acquired bacterial blood stream infections in developing countries in south and southeast Asia: a systematic review. *Lancet Infectious Diseases*, **12**: 480–487. doi: 10.1016/S1473-3099(12)70028-2

Delaney K *et al.*, (2011). Evaluation of the performance xharacteristics of 6 rapid HIV antibody tests. *Clinical Infectious Diseases*, **52**: 257–263. doi: 10.1093/cid/ciq068

Delves PJ, Martin SJ, Burton DR and Roitt IM (2011). *Roitt's Essential Immunology, 12th edn*. Chichester: Wiley-Blackwell.

Desjeux, P and Alvar, J (2003). Leishmania/HIV coinfections: Epidemiology in Europe. *Annals of Tropical Medicine and Parasitology*, **97(suppl 1)**: S3–S5. doi: 10.1179/000349803225002499

El-Bouri *et al.* (2012). Comparison of bacterial identification by MALDI-TOF mass spectrometry and conventional diagnostic microbiology methods: agreement, speed and cost implications. *British Journal of Biomedical Science*, **69**: 47–55.

Esnakula AK, Mummidi, SK, O'Neal PA and Naab TJ (2013). Sepsis caused by Mycobacterium terrae complex in a patient with sickle cell disease. *BMJ Case Reports*, **May 2.** doi: 10.1136/bcr-2013-009159

Fournier P-E *et al.* (2010). Comprehensive diagnostic strategy for blood culture-negative endocarditis: A prospective study of 819 new cases. *Clinical Infectious Diseases*, **51**: 131–140 doi: 10.1086/653675

Gallo RC *et al.* (1983). Isolation of human T cell leukemia virus in acquired immune deficiency syndrome (AIDS). *Science*, **220**: 865–867.

Gallo R *et al.* (1984). Frequent detection and isolation of cytopathic retroviruses (HTLV-III) from patients with AIDS and at risk from AIDS. *Science*, **224**: 500–503.

Gallo RC and Montagnier L (2003). The discovery of HIV as the cause of AIDS. *New England Journal of Medicine*, **349**: 2283–2285. doi: 10.1056/NEJMp038194

Geißdörfer W *et al.* (2012). High frequency of *Tropheryma whipplei* in culture-negative endocarditis. *Journal of Clinical Microbiology* **50**: 216–222. doi: 10.1128/JCM.05531-11

Girard MP *et al.* (2011). Human immunodeficiency virus (HIV). Immunopathogenesis and vaccine development: A review. *Vaccine* **29**: 6191–6218. doi: 10.1016/j. vaccine.2011.06.085

Glowaska I *et al.* (2013). Delayed seroconversion and rapid onset of lymphoproliferative disease after transmission of HTLV-1 from a multi organ donor. *Clinical Infectious Diseases*, **57**. doi: 10.1093/cid/cit545

Gottlieb MS *et al.* (1981). *Pneumocystis carinii* pneumonia and mucosal candidiasis in previously health homosexual men: Evidence of a new acquired cellular immunodeficiency. *New England Journal of Medicine*, **305**: 1425–1431.

Griffiths PD (2009). Cytomegalovirus. In: *Principles and Practice of Clinical Virology*, 6th edn. Zuckerman AJ, Banatvala JE, Schoub BD, Griffiths PD and Mortimer P, eds. Chichester: Wiley-Blackwell.

Gunn A and Pitt S (2010). Application of virology to forensic science. *Biomedical Scientist*, **54**: 341–347.

Gunn A and Pitt SJ (2012). Parasitology: An integrated approach. Chichester: Wiley-Blackwell.

Habib G *et al.* (2009). Guidelines on the prevention, diagnosis and treatment of infective endocarditis (new version 2009). *European Heart Journal*, **30**: 2369–2413. doi: 10.1093/ eurheartj/ehp285

Haigh JD *et al.* (2013). Rapid identification of bacteria from bioMérieux BacT/ALERT blood culture bottles by MALDI-TOF MS. *British Journal of Biomedical Science*, **70**: 149–155.

Haque T and Crawford DH (2009). Epstein-Barr virus. In: *Principles and Practice of Clinical Virology*, 6th edn. Zuckerman AJ, Banatvala JE, Schoub BD, Griffiths PD and Mortimer P, eds. Chichester: Wiley-Blackwell.

Health Protection Agency (2012). Shooting Up: Infections Among People Who Inject Drugs in the UK 2011. An Update: November 2012. London: Health Protection Agency.

Hirsch HH *et al.* (2013). An international multicenter performance analysis of cytomegalovirus load tests. *Clinical Infectious Diseases*, **56**: 367–373. doi: 10.1093/cid/cis900

Hopkin M (2011). Did black death boost HIV immunity in Europe? *Nature*, **March 11.** doi: 10.1038/news050307-15

Hubalék Z and Halouzka J (1999). West Nile fever—a reemerging mosquito-borne viral disease in Europe. *Emerging Infectious Diseases*, **5**: 643–650.

Inglis TJJ (2013). A clinical laboratory approach to severe sepsis: The changing role of laboratory medicine in clinical decision support during management of septicaemia. *Sri Lankan Journal of Infectious Diseases*, **3**: 2–8. doi: 10.4038/sljid.v3i1.5150

Jaffe, HW, Bregman DJ and Selik RM. (1983). Acquired immune deficiency syndrome in the United States: The first 1,000 cases. *Journal of Infectious Diseases*, **148**: 339–345.

Kublin JG *et al.* (2005). Effects of *Plasmodium falciparum* on concentration of HIV-1 RNA in the blood of adults in rural Malawi: A prospective cohort study. *Lancet*, **365**: 233–240. doi: 10.1016/S0140-6736(05)17743-5

Leitner E *et al.* (2013). Comparison of two molecular assays with conventional blood culture for diagnosis of sepsis. *Journal of Microbiological Methods*, **92**: 253–255. doi: 10.1016/j.mimet.2012.12.012

Lloyd G (2009). *Alphaviruses.* In: *Principles and Practice of Clinical Virology, 6th edn.* Zuckerman AJ, Banatvala JE, Schoub BD, Griffiths PD and Mortimer P, eds. Chichester: Wiley-Blackwell.

Lockhart PB (2012). Antibiotic prophylaxis for dental procedures: Are we drilling in the wrong direction? *Circulation* **126**: 11–12. doi: 10.1161/CIRCULATIONAHA.112.115204

Lundgren JD *et al.* (1994). Survival differences in European patients with AIDS, 1979 to 1989. *British Medical Journal*, **308**: 1068–1073.

Luzzaro F *et al.* (2011). Prevalence and epidemiology of microbial pathogens causing blood stream infections: Results of the OASIS multicenter study. *Diagnostic Microbiology and Infectious Disease*, **69**: 363–369. doi: 10.1016/j.diagmicrobio.2010.10.016

Manguin S *et al.* (2010) Review on global co-transmission of human *Plasmodium* species and *Wuchereria bancrofti* by *Anopheles* mosquitoes. *Infection, Genetics and Evolution*, **10**: 159–177. doi: 10.1016/j.meegid.2009.11.014

Martin GS (2012). Sepsis, severe sepsis and septic shock: changes in incidence, pathogens and outcomes. *Expert Review of Anti-infective Therapy*, **10**: 701–706. doi: 10.1586/eri.12.50

Monge-Maillo B *et al.* (2014). Visceral leishmaniasis and HIV coinfection in the Mediterranean region. *PLoS Negl Trop Dis*, **8(8):** e3021. doi: 10.1371/journal.pntd.0003021

Murdoch DR *et al.* (2009). Clinical presentation, etiology, and outcome of infective endocarditis in the 21st century. *Archives of Internal Medicine*, **169**: 463–473. doi: 10.1001/archinternmed.2008.603

NHSBT/HPA (2012). *Safe Supplies: New Horizons. Annual Review from the NHS Blood and Transplant/Health Protection Agency Colindale Epidemiology Unit.* London: Health Protection Agency.

Operskalski EA and Kovacs A (2011). HIV/HCV co-infection: Pathogenesis, clinical complications, treatment and new therapeutic technologies. *Current HIV/AIDS Report*, **8**: 12–22. doi: 10.1007/s11904-010-0071-3

Pillay D, Geretti AM and Weiss RA (2009). Human immunodeficieny viruses. In: *Principles and Practice of Clinical Virology, 6th edn.* Zuckerman AJ, Banatvala JE, Schoub BD, Griffiths PD and Mortimer P, eds. Chichester: Wiley-Blackwell.

Pitt SJ and Gunn A (2012). Contribution of bacteriology to forensic investigations. *Biomedical Scientist*, **56**: 458–464.

PHE, Health Protection Report (2013). Health Protection Report. Weekly Report, V7, Issue 51. http://webarchive.nationalarchives.gov.uk/20140714084352/http://www.hpa.org.uk/hpr/archives/2013/hpr50-5113.pdf

Razonable RR and Hayden RT (2013). Clinical utility of viral load in management of cytomegalovirus infection after solid organ transplantation. *Clinical Microbiological Reviews*, **26**: 703–727. doi: 10.1128/CMR.00015-13

Reddy EA, Shaw, AV and Crump JA (2010). Community-acquired blood stream infections in Africa: A systematic review and meta-analysis. *Lancet Infectious Diseases*, **10**: 417–432. doi: 10.1016/S1473-3099(10)70072-4

Rosenberg NE *et al.* (2012). Detection of acute HIV infection: A field evaluation of the Determine® HIV-1/2 Ag/Ab combo test. *Journal of Infectious Diseases*, **205**: 528–534. doi: 10.1093/infdis/jir789

Shore Garcia L (2007). *Diagnostic Medical Parasitology, 5th edn.* Washington DC: ASM Press.

SMI B37: Public Health England. (2014). Investigation of blood cultures (for organisms other than *Mycobacterium* species). UK Standards for Microbiology Investigations. B37 Issue 8. https://www.gov.uk/uk-standards-for-microbiology-investigations-smi-quality-and-consistency-in-clinical-laboratories

SMI V11: Public Health England. (2016). HIV screening and confirmation. UK Standards for Microbiology Investigations. V11 Issue 4. https://www.gov.uk/uk-standards-for-microbiology-investigations-smi-quality-and-consistency-in-clinical-laboratories/

Silvestri, G (2013). Embracing the complexity of HIV immunology. *Immunological Reviews* **254**: 5–9. doi: 10.1111/imr.12081

Sinclair J and Sissons P (2006). Latency and reactivation of human cytomegalovirus. *Journal of General Virology*, **87**: 1763–1779. doi: 10.1099/vir.0.81891-0

Stearns-Kurosawa, DJ, Osichowski, MF and Remick, DG (2011). The pathogenesis of sepsis. *Annual Review of Pathology*, **6**: 19–48. doi: 10.1146/annurev-pathol-011110-130327

Struthers JK and Westran RP (2005). *Clinical Bacteriology*, (2nd impression). London: Manson Publishing.

Surah S *et al.* (2009). Utilization of HIV point-of-care testing clinics in general practice and genitourinary medicine services in south-east London. *International Journal of STD and AIDS*, **20**: 168–169. doi: 10.1258/ijsa.2008.008220

Taylor GP (2009). The human T-lymphotropic viruses. In: *Principles and Practice of Clinical Virology*, 6th edn. Zuckerman AJ, Banatvala JE, Schoub BD, Griffiths PD and Mortimer P, eds. Chichester: Wiley-Blackwell.

Thio CL *et al.* (2002). HIV-1, hepatitis B virus, and risk of liver-related mortality in the Multicenter Cohort Study (MACS). *Lancet*, **360**: 1921–1926. doi: 10.1016/S0140-6736(02)11913-1

UKBTS (2014). Transfusion Handbook. www.transfusionguidelines.org.uk/Index.aspx?Publication=HTM&Section=9&pageid=1102

Valadas E, França L, Sousa S and Antunes F (2009). 20 years of HIV-2 infection in Portugal: Trends and changes in epidemiology. *Clinical Infectious Diseases*, **48**: 116–117. doi: 10.1086/597504

Watkin RW and Elliot TSJ (2009). The prophylaxis and treatment of bacterial infective endocarditis in adults. *Medicine*, **37**: 582–585. doi: 10.1016/j.mpmed.2009.08.002

Wegdam-Blans MCA *et al.* (2012). Chronic Q fever: Review of the literature and a proposal of new diagnostic criteria. *Journal of Infection*, **64**: 247–259. doi: 10.1016/j.jinf.2011.12.014

WHO/ECDC (2013). *Joint WHO-ECDC mission related to local malaria transmission in Greece, 2012*. Stockholm: European Centre for Disease Prevention and Control.

3

Respiratory Tract Infections

3.1   Introduction

Acquisition of microorganisms through the mouth and nose is a daily hazard of interaction with other people. Environmental air and droplets can contain high concentrations and a wide variety of microorganisms. The respiratory tract has a range of anatomical and biological features to protect it against infection from microorganisms and other debris which are inhaled. These include the mucociliary escalator – which works by moving particles from the lower respiratory tract back up to the throat to be swallowed – and macrophages in the alveoli (Mims *et al.*, 2008). Some organisms acquired by droplet infection are able to set up infections, through overcoming these non-specific defences, as well as the more specific immune response. One method is attachment to respiratory epithelial cells using specific receptors (e.g., the haemagglutinin and neuraminidase of influenza virus), while another is to be adapted to survive and grow inside macrophages, which is the mechanism used by *Mycobacterium tuberculosis*. Some organisms produce biochemical factors which adversely affect ciliary activity (such as *Pseudomonas aeruginosa*), while others cause respiratory infections opportunistically when the patient is already anatomically or immunologically comprised, including *Streptococcus pneumoniae* (Murray *et al.*, 2013; Mims *et al.*, 2008).

Most people experience a few, relatively mild, respiratory tract infections each year. These usually have an acute onset, are commonly caused by viruses and rarely require medical intervention in healthy adults. Nonetheless, some respiratory viral infections can advance to serious disease quite rapidly – such as respiratory syncytial virus (RSV) infection in children. Also the presence of a pathogenic virus can disrupt the mucociliary escalator and depress the immune response, leaving the patient vulnerable to secondary (usually bacterial) infections, which can progress to bronchitis or pneumonia. The interplay between organisms in the respiratory tract is considered to be an important factor in determining whether the presence of a particular organism results in symptomatic infection in the first place and also the severity of the consequent disease (e.g., Bosch *et al.*, 2013; Falsey *et al.*, 2013; van den Bergh *et al.*, 2012). A study of 842 adults with serious respiratory tract infection requiring hospital treatment, found evidence of viral and bacterial co-infection in approximately 40% of them (Falsey *et al.*, 2013). Understanding the interactions between organisms could therefore be important in guiding treatment and management of patients. While some mechanisms have been elucidated, this remains an area of uncertainty and of ongoing research. It has been shown that *Streptococcus pneumoniae* can generate H_2O_2

Clinical Microbiology for Diagnostic Laboratory Scientists, First Edition. Sarah J. Pitt.
© 2018 John Wiley & Sons Ltd. Published 2018 by John Wiley & Sons Ltd.

during growth and metabolism in the respiratory tract. It uses this ability to compete with other bacteria by producing the H_2O_2 in such high concentrations that even other bacteria which possess catalase (e.g., *Staphylococcus aureus* and *Haemophilus influenzae*) are susceptible to its adverse effects (Bosch *et al.*, 2013). Thus, commensal carriage of *Staph. aureus* may not protect against colonisation by a virulent strain of *Str. pneumoniae*. Both the innate and adaptive immune responses of the host are thought to promote competition between pathogenic bacterial species, limiting their effects. Interestingly, sometimes the organisms appear to co-operate to overcome this. Studies have shown that *Moraxella* (previously *Branhamella*) *catarrhalis* can produce proteins which inhibit complement factor C3; this production can be induced in the presence of *H. influenzae*, thus protecting the latter from complement-mediated antibacterial activity (Bosch *et al.*, 2013). Similarly, investigation of respiratory samples in patients (particularly children) with suspected viral infections reveal that co-infection with a number of potential pathogens is fairly common. How important interaction between such viruses is for the outcome of symptomatic disease is not well understood (Bosch *et al.*, 2013). Interestingly a significant proportion (up to 30%) of healthy, asymptomatic children may be carrying combinations of pathogens such as respiratory syncytial virus and rhinovirus (Bosch *et al.*, 2013) and/or *H. influenzae* (van den Bergh *et al.*, 2012) although the significance of this is not clear. The effects of viruses on host cells can pre-dispose to secondary bacterial infection, including through cell lysis within the respiratory epithelium, suppression of defensins, upregulation of adhesins and altering immune responses (Bosch *et al.*, 2013). Influenza virus activity is thought to create conditions suitable for secondary infection with bacteria such as *H. influenzae* and *Str. pneumoniae* in a variety of ways. These include damage to epithelial cells, alteration to factors affecting lung function (surfactant, cilia) and breaking sialic acid residues on respiratory epithelial cells – to expose glycoproteins which can act as receptors for bacteria – via the influenza neuramindase. The virus and the bacterium both seem to stimulate changes in components of the immune response which enhance pathogenesis (McCullers, 2006). Rhinoviruses have been shown to upregulate the expression of intracellular adhesion molecule-1 (ICAM-1) on the surface of respiratory epithelial cells (Bosch *et al.*, 2013). Both the virus and *H. influenzae* use this receptor to enter host cells, so each microorganism benefits from the viral metabolism.

Point to consider 3.1: The interplay between viruses and bacteria within the respiratory tract appears to be important in causing symptomatic infection. Which host factors might contribute to a poorer outcome for the patient?

Respiratory tract infections (RTIs) account for significant mortality and morbidity in all parts of the world. This can be attributed to a combination of the ease of transmission (there is usually a period of several days or more during which a person with an RTI is ambulant but shedding highly infectious droplets), the rapid rate of progression of some infections (e.g., pandemic influenza), the severity of some infections (e.g., pneumonia), lack of suitable antimicrobial treatments (e.g., most respiratory viruses) and the development of resistance to available antimicrobials (e.g., extensively drug-resistant tuberculosis, 'XDR-TB'). Diarrhoea and acute respiratory infection (ARI) resulting in pneumonia and are the most common causes of morbidity and mortality in children under the age of 5 across the world (Walker *et al.*, 2013). Most of the deaths should be preventable, which is why WHO and UNICEF introduced the Global Action Plan for Pneumonia and Diarrhoea (GAPPD), which set a target for reduction of fatalities from these two types of infection by 2025 (http://www.who.int/mediacentre/news/releases/2013/pneumonia_diarrhoea_plan_20130412/en/).

Respiratory infections are usually categorised as upper respiratory tract infections (URTIs) or lower respiratory tract infections (LRTIs). For healthy adults, the most common experience of URTIs, such as colds and sore throats, is that is they are a regular occurrence and mostly an inconvenience.

Point to consider 3.2: There are hundreds of different cold viruses and it is thought that we are continually exposed to them, while we usually only experience disease occasionally. What factors might pre-dispose someone to developing the clinical symptoms of a cold?

3.1.1 Upper Respiratory Tract Infections

The upper respiratory tract essentially comprises the ear, nose and throat. Therefore as well as the generally mild, short-lived rhinitis infections, URTIs include the more serious conditions of pharyngitis, sinusitis, mastoiditis and otitis media, which can also arise as complications from a seemingly uneventful cold. Table 3.1 lists the main pathogens implicated in URTIs, along with the clinical conditions each is usually associated with. It shows the variety of possible clinical outcomes of URTIs and some patterns. For example viral infections usually result in rhinitis, while bacteria are more commonly implicated in ear infections.

Table 3.1 Main microorganisms associated with human upper respiratory tract infections.*

Organism	Clinical condition associated with infection with this organism
Streptococcus pneumoniae	Sinusitis, mastoiditis and otitis media
Haemophilus influenzae	Sinusitis, mastoiditis and otitis media
Moraxella catarrhalis	Sinusitis, mastoiditis and otitis media
Staphylococcus aureus	Sinusitis, mastoiditis and otitis media
Group A *Streptococcus* spp.	Sinusitis, mastoiditis and otitis media, pharyngitis
Fusobacterium spp.	Sinusitis, mastoiditis and otitis media
Group C *Streptococcus* spp.	Pharyngitis
Corynebacterium diphtheriae	Pharyngitis
Chlamydia pneumoniae	Pharyngitis
Mycoplasma pneumoniae	Pharyngitis
Rhinovirus	Rhinitis, laryngitis, pharyngitis
Coronavirus	Rhinitis, laryngitis, pharyngitis
Influenza virus	Pharyngitis
Parainfluenza virus	Rhinitis, pharyngitis
Adenovirus	Pharyngitis
Metapneumovirus	Rhinitis, pharyngitis
Respiratory syncytial virus	Rhinitis, pharyngitis

*Information on this table are taken from a range of sources including Murray *et al.* (2013) and Greenwood *et al.* (2012).
Source: Courtesy of Greenwood (2012). Reproduced with permission of Elsevier.

3.1.2 Lower Respiratory Tract Infections

The components of the lower respiratory tract are the trachea, bronchi, bronchioles and lungs. Infections of these areas can be acute, but may take weeks or months to become clinically apparent. They often become chronic, deep-seated and life-threatening. LRTIs can arise as complications of URTIs, but some pathogens are only associated with infections of the deeper tissues. The main presentations of LRTI are pneumonia, pneumonitis, bronchiolitis, bronchitis and lung abscess (Woodhead *et al.*, 2011). The most important pathogens associated with these clinical presentations are shown in Table 3.2. Note that there is some overlap between organisms listed in Tables 3.1 and 3.2. For these examples, a combination of host and pathogen factors influence whether an LRTI develops.

Table 3.2 Main microorganisms associated with human lower respiratory tract infections.*

Organism	Clinical condition associated with infection with this organism
Streptococcus pneumoniae	Community-acquired pneumonia
Haemophilus influenzae	Community-acquired pneumonia
Moraxella catarrhalis	Community-acquired pneumonia
Staphylococcus aureus	Community-acquired pneumonia, hospital-acquired pneumonia, lung abscess
Klebsiella pneumoniae	Community-acquired pneumonia hospital-acquired pneumonia, lung abscess
Chlamydia pneumoniae	Atypical community-acquired pneumonia
Chlamydia psittaci	Atypical community-acquired pneumonia
Mycoplasma pneumoniae	Atypical community-acquired pneumonia
Coxiella burnetii	Atypical community-acquired pneumonia, Q fever
Legionella pneumophila	Atypical community-acquired pneumonia, Legionnaires' disease
Citrobacter spp.	Hospital-acquired pneumonia
Enterobacter spp.	Hospital-acquired pneumonia
Pseudomonas aeruginosa	Hospital-acquired pneumonia
Methicillin-resistant *Staphylococus aureus*	Hospital-acquired pneumonia
Mycobacterium spp.	Tuberculosis
RSV	Bronchiolitis, pneumonia
Parainfluenza virus	Bronchiolitis, pneumonia
Paragonimus spp.	Tuberculosis-type disease
Pneumocystis jirovecii	Atypical pneumonia, interstitial pneumonitis
Aspergilllus spp.	Bronchopulmonary disease, interstitial pneumonia

*Information on this table are taken from a range of sources including Murray *et al.* (2013) and Greenwood *et al.* (2012).
Source: Greenwood (2012). Reproduced with permission of Elsevier.

It is estimated that acute LRTIs (pneumonia and bronchiolitis) are responsible for over 1 million deaths in young children each year (Nair *et al.*, 2013). Patients with severe disease usually require hospital treatment and the majority of deaths occur in children who do not go to hospital or whose care is inadequate (Nair *et al.*, 2013). A range of organisms are associated with paediatric LRTIs, but there are four which currently stand out as the most prevalent. Bacterial childhood pneumonia is most commonly caused by *Str. pneumoniae*, with a significant contribution by *H. influenzae* type b where the Hib vaccine is not part of the routine immunisation programme (http://www. who.int/mediacentre/factsheets/fs331/en/). The most important viral cause is RSV. Immunocompromised children, particularly those with human immunodeficiency virus (HIV) infection, are at risk of pneumonia caused by the fungus *Pneumocystis jirovecii*. The symptoms of pneumonia in children include cough, difficulty in breathing, wheezing and fever (though the temperature may not be markedly raised). It may be hard to persuade the patient to drink or eat and this can be dangerous – especially in very young babies who are unable to feed.

In adults, pneumonia is categorised as community acquired or hospital (sometimes 'healthcare') acquired infection. People who have not been in a healthcare setting (either as a visitor, inpatient or outpatient) in the previous weeks are considered to have community-acquired pneumonia (CAP). Although for severe infections, hospital treatment is warranted (Musher and Thorner, 2014) it is usual to manage CAP patients in the primary care system wherever possible. The most important risk factor for contracting CAP is being older than 65 years (Torres *et al.*, 2013). People with underlying conditions compromising the cardiovascular and /or immune system (e.g., HIV infection, chronic obstructive pulmonary disease, diabetes mellitus) are more likely to develop pneumonia, as are those who smoke or drink heavily (Torres *et al.*, 2013). The most common cause of CAP is *Str. pneumoniae* (Musher and Thorner, 2014; Torres *et al.* 2014) and this is one of the reasons why empirical treatment (without microbiology laboratory diagnosis) is advocated in the first instance. In Europe, it is usual to use penicillin or amoxicillin, with other drugs and combination therapies reserved for seriously ill, hospitalised patients (Torres *et al.*, 2014). In North America a macrolide or doxycycline are suggested (Musher and Thorner, 2014) and these would offer more broad-spectrum activity. Other pathogens often associated with CAP include *H. influenzae, Staph. aureus* and influenza virus (Musher and Thorner, 2014). Bacterial CAP could also be caused by *M. catarrhalis, Pseudomonas aeruginosa* and some *Enterobacteriaceae* (Musher and Thorner, 2014; Torres *et al.*, 2014). Atypical pneumonia can also be community acquired; this is associated with a range of organisms, including *Mycoplasma pneumoniae, Legionella pneumophila, Chlamydophila pneumoniae, Chlamydophila psittcai* and *Coxiella burnetii* (Musher and Thorner, 2014; Torres *et al.* 2014). When the patient's condition is serious enough to need hospital management, microbiology laboratory investigations would be conducted, to determine the exact pathogen and antimicrobial susceptibilities. The exposure history is important in this situation and less- frequent causative agents, such as Middle East respiratory syndrome coronavirus (MERS-CoV), should be borne in mind. Hospital or healthcare-acquired pneumonia (HAP or HCAP), usually leads to more serious outcomes in terms of morbidity and mortality by comparison with CAP (Maruyama *et al.*, 2013). This is partly due to the patients being ill enough to require healthcare intervention, which sometimes includes mechanical ventilation (thus creating the conditions for the exposure) and partly a function of the nature of hospital-acquired pathogens, which may be more virulent and are more likely to be multidrug-resistant (Maruyama *et al.*, 2013). Nevertheless, *Str. pneumoniae* is again the most commonly identified pathogen in HAP (Maruyama *et al.*, 2013; Polverino *et al.*, 2013). Respiratory viruses and *Ps. aeruginosa* are found in patients' samples more frequently in HAP than CAP. Another notable difference is the rate of isolation of *Klebsiella pneumoniae*, *Acinetobacter* spp. and *Staph. aureus* (particularly methicillin-resistant *Staphylococcus aureus*), which are not often seen in CAP (Maruyama *et al.*, 2013; Polverino *et al.*, 2013).

3.1.3   Respiratory Pathogens

Some respiratory tract infections are caused by microorganisms which are specifically respiratory pathogens, including influenza virus (e.g., Girard *et al.*, 2010), *Mycoplasma pneumoniae* (e.g., Waller *et al.*, 2014; Waites and Talkington, 2004), *Legionella pneumophila* (e.g., Carratala and Garcia-Vidal, 2010; Reuter *et al.*, 2013) *Chlamydophila pneumoniae* (e.g., Blasi *et al.*, 2009) and *C. psittaci* (e.g., West, 2011). Others are associated with a range of clinical presentations, which can include lung infections; this type of organism is typically opportunistic and examples include *Ps. aeruginosa* (e.g., Fujitani, *et al.* 2011; Fothergill *et al.*, 2012) and *Staph. aureus* (e.g., Gould *et al.*, 2012; Kallen *et al.*, 2009). This chapter will consider a selection of microbiologically different respiratory pathogens in detail, to illustrate points about pathogenesis, epidemiology and laboratory diagnosis. These are respiratory syncytial virus, *Streptococcus pyogenes* and *Str. pneumoniae*, *Mycobacterium tuberculosis*, *Pneumocystis jirovecii* and *Paragonimus westermani*. Other respiratory pathogens are considered elsewhere in the book (influenza virus in Chapter 5 and *Aspergillus* spp. in Chapter 7).

3.2   Respiratory Syncytial Virus

Order: *Mononegavirales*
Family: *Paramyxoviridae*
Subfamily: *Pneumovirinae*
Genus: *Pneumovirus*
Species: *Human Respiratory Syncytial Virus*

3.2.1   Introduction

Respiratory syncytial virus (RSV) infection is very common, but it usually only causes clinically significant lower respiratory tract (LRT) disease in children under the age of three years (Hall, 2009) or individuals with underlying pre-dispositions to serious LRTI (Collins and Melero, 2011). Transmission of the virus is seasonal; in temperate climates most RSV disease is recorded between November and March with a peak occurring in December and January (PHE Respiratory Data, 2016). By contrast in tropical countries, the season is much less marked and outbreaks tend to occur all year round (Hall, 2009). Most children will have experienced RSV by the time they reach three years of age (Hall, 2009; Collins and Melero, 2011). Recurrent infection is a regular occurrence throughout life – and sometimes in the same season! (Meng *et al.*, 2014). In older children and adults, the infection is symptomatic in 85% of cases, but usually manifests as a mild upper RTI. Nevertheless, some people can experience a more debilitating flu-like illness (Hall, 2009).

3.2.2 Pathogenesis and Clinical Symptoms

The virus is a negative sense, single-stranded RNA virus. The genome is known to code for 11 proteins including the structural envelope surface proteins F and G (Hall, 2009; Meng et al., 2014). Glycoprotein G is thought to have a role in both facilitation of attachment of the virus to the host respiratory epithelial cell and in viral replication (Hall, 2009; Collins and Melero, 2011). The F protein appears to be involved in attachment, viral entry into the cell and promoting the fusion of the viral and host cell membranes (Hall, 2009; Collins and Melero, 2011). During virus isolation in monolayer cell culture, this membrane fusion and other processes lead to formation of the characteristic syncytia which give the virus its name (Hall, 2009). Two strains of the virus are recognised – RSV A and RSV B – on the basis of differences in sequences of the G protein; whether there are notable differences between the virulence or epidemiology of the strains has yet to be determined (Hall, 2009; Collins and Melero, 2011).

RSV is considered to be highly infectious (Collins and Melero, 2011). Transmission occurs through large droplets and contamination of fomites (Hall, 2009). The virus has been found to be viable in respiratory secretions for up to 6 hours (Hall et al., 1980). It is therefore very likely to be associated with nosocomial infections. Thus rapid and accurate diagnosis is important to allow suitable infection control procedures to be implemented – particularly in paediatric secondary care settings (SMI G8, 2014). Co-infection with other respiratory viruses such rhinovirus, parainfluenza, adenovirus, human metapneumovirus and influenza is commonly seen (Hall, 2009), especially in young children. This is likely to be due to the common transmission routes and overlapping seasons (Hall, 2009).

The incubation time for RSV ranges from a few days to up to a week (Hall, 2009). Primary infection is usually manifested as an URT with a fever; if the infection progresses to a LRT, the pyrexia usually subsides (Hall, 2009). Bronchiolitis is a major cause of hospitalisation during the peak season (or outbreaks) and tends to be most common amongst children younger than six months old (Collins and Melero, 2011; SMI G8, 2014). Although otherwise healthy children do sometimes develop severe bronchiolitis and require hospital treatment, it is infants who were born prematurely and those with underlying lung or heart conditions who are more likely to be admitted (and to be later re-admitted) (Hall, 2009; SMI G8, 2014). They are also at greater risk of death, although case fatality rates are generally low (Hall 2009; Collins and Melero, 2011). Patients are usually ill for 1–3 weeks, but their condition tends to improve within about seven days; children who need hospital treatment are often discharged after a few days (Hall, 2009).

The role of the host's immune response in pathogenesis of the disease is interesting and somewhat ambiguous. When the virus enters the respiratory epithelium, it triggers the innate immune response involving chemokines and pro-inflammatory cytokines. This attracts polymorphonuclear monocytes and eosinophils, along with Interferon – α (IFN-α) and natural killer (NK) cells to the area (Collins and Melero, 2011). The specific immune response includes making detectable levels of antibody to both the F and G glycoproteins (Hall, 2009). Research from the 1980s suggested that specific IgG antibody (e.g., made in response to a previous infection) interacted with the virus in subsequent infections to enhance the pathology. It became clear in later years that this is not actually the case and that indeed maternal antibody has a protective effect in their infants (Hall, 2009). This understanding led to the development of a prophylactic treatment in the form of a monoclonal antibody treatment, Palivizumab (Hall, 2009; Collins and Melero, 2011; SMI G8, 2014).

While the antibody-mediated response is now not implicated in influencing the course of the disease, it is thought that the individual's cell-mediated response to the RSV infection may have an effect. The NK cells produce IFN-γ, stimulating the differentiation of the CD4+ T cells into populations which produce interferons and the CD8+ cells into cytotoxic T cells (Collins and Melero, 2011). This Th1 response appears to be important for clearance of the virus (Hall, 2009). In the Th2 response, the CD4+ cells produce interleukin 4 (IL-4), IL-5 and IL-13, along with IgE and is associated with eosinophila, causing restriction to the airways and a worse clinical outcome (Hall, 2009). This is similar to the immunological picture found in asthma (Collins and Melero, 2011). Evidence collected from trials of an unsuccessful vaccine, along with subsequent research, suggests that the balance between the Th1 and the Th2 response is crucial in determining the course of the disease – but the exact mechanisms are still unclear (Hall, 2009). There is some evidence that a higher viral load may be linked to more serious symptoms of disease, at least where RSV is the only detectable respiratory pathogen (Houben *et al.*, 2010), which implies that the virus may cause damage directly (Collins and Melero, 2011). Viral proteins NS1 and NS2 may decrease IFN production and block apoptosis of infected cells, enhancing viral replication (van Drunen-Littel-van den Hurk and Watkiss, 2012). The observed pathology could also be attributable to the more vigorous immune response in the presence of higher concentrations of replicating virus.

Difficulty in breathing is a key feature of serious RSV disease. Very young babies tend to develop apnoea (Hall, 2009), particularly if they were born prematurely. Bronchiolitis is associated with accumulation of inflammatory cells in the bronchioles, which leads to necrosis of the respiratory epithelium and sloughing off of tissue (Meng *et al.*, 2014). This causes a physical obstruction of the airways, which is more dangerous in younger babies simply due to the smaller size of their respiratory tract (Collins and Melero, 2011). There is some evidence that environmental factors affecting airways such as exposure to cigarette smoke or traffic pollution may exacerbate RSV symptoms (van Drunen Littel-van den Hurk and Watkiss, 2012). Progression from URT to LRT is characterised by wheezing, coughing and then symptoms of pneumonia (Hall, 2009). Infants tend to have a stronger Th2 response, but research has not conclusively linked this with worsening of RSV disease (Collins and Melero, 2011); indeed some studies indicate that the deterioration appears to be a function of the virus itself (Meng *et al.*, 2014). Otitis media is a common complication of RSV infection (Hall, 2009). Although this can sometimes be associated with a secondary bacterial infection (Pettigrew *et al.*, 2011), the virus itself is often active in the ear and can be isolated from exudate (Hall, 2009; SMI G8, 2014).

The impact of co-infection with other respiratory viruses on the severity of the disease is difficult to elucidate; studies are hampered by the fact that when two or more viruses are found in a patient's respiratory sample it is hard to know whether the infection occurred simultaneously or (more likely) the RSV was acquired earlier or later than the other organism (Hall, 2009). Severe infections, particularly in children, can result in pneumonia. In developing countries this is often due to a secondary bacterial infection, whereas in developed countries the pneumonia is often caused by the virus alone (Hall, 2009). In fact the most common complicating bacterial infection in children with acute RSV disease from developed countries is reported to be a urinary tract infection (!) (Hall, 2009) although this has been disputed (Kaluarachchi *et al.*, 2014).

3.2.3 Epidemiology

A study estimated that 34 million cases of acute LRTI in children younger than five years old were attributable to RSV infection worldwide in 2005 (Nair *et al.*, 2010). About 10% of these are likely to have received hospital treatment and up to 0.6% of the children may have died (Nair *et al.*, 2010). Those most at risk of life-threatening complications from RSV are premature babies and children with underlying immune or cardio-pulmonary conditions (SMI G8, 2014). Older children and adults – especially the elderly – who happen to be in hospital during peak RSV season are also vulnerable to infection and can experience severe respiratory complications (SMI G8, 2014), which can sometimes be fatal (Hall, 2009).

   As mentioned above, RSV can be seasonal and/or associated with outbreaks – depending on geographical area (Bloom-Feshbach *et al.*, 2013). The total number of cases each year can vary considerably. This illustrated by the fact that in England and Wales laboratory confirmed cases of RSV have been over 800 every year since 2010, but there were over 1,000 in 2010, 2014 and 2015 (PHE Respiratory Data, 2016). It is important to remember that this data relates to the small proportion of patients who were ill enough to require medical intervention. Thus, these are either children who were admitted to hospital (with croup, bronchiolitis or pneumonia) or nosocomial infections amongst inpatients with other conditions. Nevertheless, it gives an indication of extent of the transmission in the community.

3.2.4 Laboratory Diagnosis

Laboratory diagnosis is based on detection of the virus in respiratory cells. It is therefore important to collect samples from the lower respiratory tract, which will contain adequate cellular material. These include nasopharyngeal aspirate (NPA), nasal washings and bronchoalveolar lavage (BAL). The most sensitive and specific assay for RSV is real time, reverse transcriptase polymerase chain reaction (PCR) (Hall, 2009), which is often included as part of a multiplex respiratory virus panel (e.g., Hymas *et al.*, 2010). Figure 3.1 is the output from a multiplex PCR panel for respiratory viruses. The grid on the left-hand side shows the location of the individual samples in the test plate; a red circle indicates a sample where a respiratory virus was detected. The amplification curves indicate both the species and concentration of viral genomic material (Figure 3.1). The results require careful interpretation in order to identify the virus(es) present in a particular specimen, bearing in mind that there could be more than one. The clinical significance of any results must also be considered in the light of the patient's history. For example, a dual infection of RSV and a rhinovirus in a child with bronchiolitis may affect the course of the infection, but the management may not be different for a similar

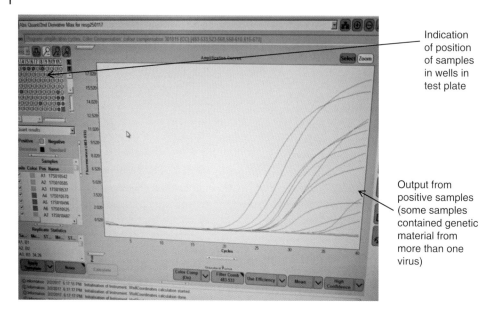

Indication of position of samples in wells in test plate

Output from positive samples (some samples contained genetic material from more than one virus)

Figure 3.1 Screenshot of computer output from respiratory virus multiplex PCR assay.

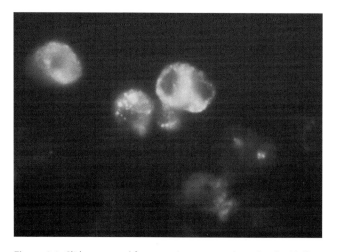

Figure 3.2 Slide prepared from respiratory sample stained with fluorescent-labelled anti-RSV antibody and viewed under UV microscope showing cells infected with the virus. Source: Courtesy of Mr I. Phillips. (*See insert for colour representation of the figure.*)

patient with evidence of RSV alone. Where PCR is not readily available, detection of glycoproteins F and G on the surface of infected cells, using fluorescent labelled mono-clonal antibodies, is a suitable alternative (SMI G8, 2014) which can provide relatively rapid results. As Figure 3.2 illustrates infected cells are marked by the bright green fluo-rescence visible when the slide is viewed in a UV microscope. Although it is not very obvious in Figure 3.1, the staining procedure includes a red counter stain which clearly shows up respiratory cells. This allows the observer to assess whether adequate cellular material is present on the slide, which can be important in confirming the decision that no infected cells are present. However, there is some subjectivity in the reading of the results and sensitivity and specificity of immunofluorescent methods are reported to be slightly inferior to PCR (e.g., Jokela *et al.*, 2010).

In a clinical situation, where there is a risk of nosocomial spread, even a few hours may be too long to make the decision about whether to isolate a patient (Mills *et al.*, 2011). A number of point of care test (POCT) assays for RSV are now available and their use as screening tools for patients on paediatric wards has been evaluated (e.g., Miernyk *et al.*, 2011; Mills *et al.*, 2011). Taking PCR as the gold standard, POCT methods are generally reported to be less specific and sensitive, but this does not preclude their use. At the sensitivity and specificity rates of 83% which Mills *et al.* (2011) reported, most patients with RSV would be identified. Since the practice on paediatric wards is to nurse known RSV positive children in 'cohorts', it would be reasonable to consider including those with a positive result from the POCT in the cohort, while referring samples from those with negative or equivocal results to the laboratory for further investigations (Miernyk *et al.*, 2011; Mills *et al.*, 2011). Multiplex PCR for respiratory pathogens would have greater sensitivity and potentially provide more information. Input from laboratory scientists into decision making about policy would be invaluable in such a case – there is a clear need to consider the evidence and assess the situation in terms of both patient and resource management.

3.2.5 Management and Treatment

In most cases, patients with RSV recover uneventfully. Amongst those with severe disease who require hospitalisation, supportive therapy (including administration of fluids and oxygen) is important for a good patient outcome (Collins and Melero, 2011). Treatment of the symptoms, for example, using bronchodilators to open the airways, is sometimes used, particularly when the patient has an underlying immunosuppressive condition (SMI G8, 2014).

Since viral load has been correlated with disease severity, it might be expected that a specific antiviral drug might be beneficial. Ribavirin gave promising results against RSV *in vitro* and in animal tests, but has not proved consistently clinically effective (Collins and Melero, 2011; SMI G8, 2014). However it is used, in conjunction with other measures, in the form of an aerosol for severely affected patients (Collins and Melero, 2011; SMI G8, 2014). Prophylaxis for infants at risk of severe RSV disease is also available. Palivizumab is a preparation which contains a monoclonal antibody raised against the F glycoprotein – which has a more conserved sequence than the G protein. It is given intramuscularly (SMI G8, 2014). It has been available since the late 1990s and has so far proved to be efficacious; resistant virus strains do not appear to be an issue (Collins and Melero, 2011).

Point to consider 3.3: The fact that people experience recurrent respiratory disease associated with RSV suggests that a long-lasting protective immune response is not produced in response to infection with this virus. Why does this present a challenge for researchers attempting to develop a successful vaccine?

3.3 *Streptococcus pyogenes* and *Streptococcus pneumoniae*

Classification
Kingdom: Bacteria
Phylum: Firmicutes
Class: Bacilli
Order: Lactobacillales
Family: Streptococcaceae
Genus: *Streptococcus*
Species: *Str. agalactiae,Str. mitis, Str. mutans, Str. pneumoniae, Str. pyogenes*

3.3.1 Introduction

At least 70 species of *Streptococcus* have been recognised and the genus includes some important human and animal pathogens (Kilian, 2012). Characterisation of an isolate can be based on the type of haemolysis displayed on blood agar plates and Lancefield grouping. This is not always entirely helpful, since not all species lyse blood cells (so called 'γ haemolysis') or have the requisite cell wall polysaccharide for Lancefield grouping (Murray *et al.*, 2013). Molecular analysis and phylogenetics suggest the species group into five clusters – Pyogenic, Mitis, Anginosus, Bovis and Mutans; *Str. pyogenes* is in the first of these and *Str. pneumoniae* is in the second (Kilian, 2012). *Str. pyogenes* is associated with a range of clinical conditions, particularly upper respiratory tract and soft tissue infections, while *Str. pneumoniae* causes lower respiratory tract disease and also ear infections and meningitis (Murray *et al.*, 2013; Kilian, 2012). This section will focus on the role of these two species in respiratory tract infections. For considerations of the other clinical conditions, the reader is directed to text books and reviews – for example, Murray *et al* (2013); Kilian (2012); Walker *et al.* (2014); Mehr and Wood (2012).

3.3.2 Pathogenesis and Clinical Symptoms

Streptococcus spp. are known to carry a variety of virulence factors. These include bacterial cell surface attachments and destructive enzymes – some of which are used in identification and characterisation of species and strains. The potential mechanisms of action have been described for some of these factors. In general, the exact ways in which they interact with the host cells, tissues and immune system (and each other) to produce the damage which is manifest in the clinical symptoms are not yet well understood. A full discussion of the pathogenesis of all streptococcal disease is beyond the scope of this chapter. For more information, the reader is directed towards consulting review articles for each species – for example, Walker *et al.*, (2014) for *Str. pyogenes*, Mitchell and Mitchell (2010) for *Str. pneumoniae* and Mitchell (2011) for *Str. mitis* (one of the 'viridans' streptococci). For more information about *Str. agalactiae* (GBS) see Chapter 5.

3.3.2.1 Pathogenesis of *Streptococcus pyogenes* Infection

Str. pyogenes (Group A streptococci, 'GAS') causes pharyngitis and scarlet fever, along with a range of debilitating skin conditions such as cellulitis, erysipelas and necrotising fasciitis (Murray *et al.*, 2013). The virulence factors are known to contribute to GAS pathogenicity include those given in Text Box 3.1:

Text Box 3.1 Virulence Factors Found in *Streptococcus pyogenes*

Bacterial cell surface components:

- M protein
- M-like proteins
- F protein
- Lipoteichoic acid
- Pili
- Capsule

Toxins and destructive enzymes:

- C5a peptidase
- Streptolysins O and S
- Streptococcal pyrogenic exotoxins (SPEs) A, B and C
- Hyaluronidase
- Streptokinase
- Deoxyribonucleases

The M protein compromises two linked polypeptide chains which run through the cell wall, fixed in the cytoplasm and projecting out from the surface (Murray *et al.*, 2013). This surface 'N terminal' region is immunogenic and hypervariable and thus forms the basis of serological strain differentiation (Walker *et al.* 2014). Over 150 variants of the M protein have been recognised by serology and genetic sequencing. This explains why some people experience recurrent infections. The M protein, along with other 'M-like' proteins, is able to inhibit the activity of phagocytes, through binding to plasma proteins such as antibodies, complement factors, fibrinogen and albumin – thus making them appear to be 'host', not bacterial. The DNAases are also thought to be used in evasion of phagocytosis (Kilian, 2012).

Several pathogenic factors are involved in adhesion to host epithelial cells – the M protein, lipoteichoic acid, pili and the F protein. Although it is not completely understood, it is thought that the lipoteichoic acid may bring the bacterial and host cells together through weak, hydrophobic bonds, allowing more specific interactions to take place (Walker *et al.*, 2014). The F protein is considered to be the most important factor for entry by streptococci into host cells. It binds to fibronectin in the extracellular matrix, thus bringing the bacteria into close contact with the epithelium (Kilian, 2012), affording access; this appears to be the mechanism by which streptococci persist in the host (Murray *et al.*, 2013). The M protein can also attach to the fibronectin, after being cleaved from the cell wall by SPE-B, which is a cysteine protease. The resulting complexes activate an inflammatory response within the tissues, causing disruption of the fluid balance and cell damage (Kilian, 2012). A number of strains of *Str. pyogenes* are capsulated and this seems to afford extra virulence properties. This could be due to the capsule being made of a hyaluronic acid which is the same as that found in mammalian connective tissue – so again allowing the bacterium to masquerade as 'host' (Murray *et al.*, 2013). The enzyme C5a peptidase is present on the cell surface of all GASs. It inactivates the C5a protein, which is produced by the body at the point of an infection, to attract neutrophils and macrophages (Murray *et al.*, 2013). The bacteria also produce enzymes when metabolically active and there is some strain variation in this, which accounts for different clinical manifestations. The streptolysins lyse red blood cells, polymorphs and platelets, thus initiating tissue damage (Kilian, 2012). Streptolysin S causes the β-haemolysis seen when GAS are cultured on blood agar plates, while streptolysin O is immunogenic (the basis of the anti-streptolysin O 'ASO' titre assay). While all strains of the bacterium possess SPE-B, the SPE-A and SPE-C are only found in isolates from patients with toxic shock-type presentations and are important in causing with the rash associated with scarlet fever (Kilian, 2012). The hyaluronidase and streptokinase are used by the pathogen to promote invasion and spread within the host tissues (Kilian, 2012).

The respiratory tract infections associated with *Str. pyogenes* are pharyngitis and scarlet fever. Pharyngitis is a fairly common, usually mild, upper respiratory tract infection. Symptoms include headache, lethargy, pyrexia and a sore throat. It is often referred to as 'strep throat' (although the majority of cases labelled as this are actually viral!). On examination, the patient will have extreme redness of the pharynx, swelling of the tonsils and enlarged cervical lymph nodes. This usually self-limiting infection can develop into tonsillitis. If an abscess forms on the tonsils, this can then provide the bacteria with a route into the blood stream, resulting in septicaemia (Kilian, 2012). When the patient is infected with a strain of *Str. pyogenes* which is capable of producing SPE -A and -C, the pharyngitis can develop into scarlet fever (or 'scarlatina'). They activate regulatory T cells (T_{regs}), causing a release of pro inflammatory cytokines (IL-1, IL-2, tumour necrosis factor (TNF)-α IFN-γ and the outcome of the process is a characteristic diffuse red rash, which is rough to the touch ('sandpaper rash') and pruritic ('itchy') (Kilian, 2012). It usually appears first on the upper abdomen a few days after the start of the pharyngitis symptoms and then spreads to the rest of the body. The tongue is characteristically affected by a noticeable yellow/white covering, which is succeeded by redness and a raw surface ('strawberry tongue'). Given that the throat is already swollen and sore, this can be very distressing for the patient and makes it difficult to eat and drink. Without treatment, scarlet fever can be fatal. Antibiotics and epidemiology (i.e., change in prevalence of toxigenic strains) markedly reduced the morbidity and mortality due from this infection during the twentieth century, but significant outbreaks still occur (see below).

Acute rheumatic fever is a potential sequela to pharyngitis and/or scarlet fever; it is usually seen several weeks after the initial infection has resolved. It is characterised by inflammation in various sites including the joints, heart and central nervous system (Kilian, 2012). It is thought to be due to an autoimmune reaction. The arthralgia or arthritis is possibly caused by the build of immune complexes; joints are typically affected sequentially rather than simultaneously (Murray *et al.*, 2013). Since the host immune system reacts in a destructive (rather than protective) way to the presence of the bacterium, repeated *Str. pyogenes* infections can result in irreversible damage (Kilian, 2012) particularly to cardiac muscle, resulting in rheumatic heart disease (Murray *et al.*, 2013). Another possible outcome is acute post-streptococcal glomerulonephritis, which is characterised by haematuria, oedema (especially in the face and limbs) and circulatory problems. Again, this condition is thought to be caused by the host's immune reaction and is probably associated with development of immune complexes (Kilian, 2012). The pathogenic mechanism(s) are uncertain, but may involve build-up of immune complexes within the renal glomeruli, damage to renal tissue by bacterial products such as streptokinase, streptococcal components triggering inappropriate complement or antibody responses (Kilian, 2012).

3.3.2.2 Pathogenesis of *Streptococcus pneumoniae* Infection

Str. pneumoniae ('pneumococcus') is the most common causative agent of pneumonia worldwide. It is also frequently involved in childhood otitis media and is one of the main pathogens associated with bacterial meningitis (Kilian, 2012).

The virulence factors identified for pneumococci include those given in Text Box 3.2:

Text Box 3.2 Virulence Factors Found in *Streptococcus pneumoniae*

Bacterial cell surface components:

- Capsule
- Lipoproteins
- Pilus
- Choline-binding proteins

Toxins and destructive enzymes:

- Pneumolysin
- Secretory IgA protease
- Hyaluronidase
- Neuraminidase

The polysaccharide capsule affords resistance to phagocytosis and complement. It is immunogenic and there are over 90 different serotypes, which vary in their pathogenicity (Kilian, 2012). The bacterium can overcome the effects of the mucociliary escalator by secreting IgA protease (thus preventing the IgA-mediated binding of the organism to mucin) therefore facilitating its movement through the respiratory tract (Murray *et al.*, 2013). Similarly, spread is aided by pneumococcal hyaluronidase, which destroys the hyaluronic acid in the extra cellular matrix (Mitchell and Mitchell, 2010). The pilus, lipoproteins choline-binding proteins and neuraminidase have all been implicated in pneumococcal virulence. Naturally occurring strains, or laboratory-designed mutants, which lack these components have been demonstrated to be less pathogenic (Mitchell and Mitchell, 2010), although the exact mechanisms are not well understood. Pneumolysin (PLY) is known to be an important factor for the pathogenesis of pneumococci, but its activity has not been fully elucidated. It is thought to bind to cholesterol in the membrane of host respiratory epithelial cells; PLY molecules attach to each other, creating a pore in the membrane, thus leading to cell lysis (Mitchell and Mitchell, 2010). It has also been shown to activate complement and to connect with toll-like receptor 4 (TLR-4) on the surface of white blood cells, thus activating an inflammatory response within the lungs (Mitchell and Mitchell, 2010). Interestingly, the activity of the various virulence factors appears to vary depending on the site of infection – thus they have different contributions to clinical conditions. This point is demonstrated by PLY, which is not thought to be important in otitis media, but it is key to pathology in respiratory and central nervous system (CNS) infections.

The onset of pneumococcal pneumonia is usually sudden, with a high fever (up to 41 °C), although the patient may have experienced what appeared to be a viral upper respiratory tract infection for a few days beforehand (Murray *et al.*, 2013). The bacteria are aspirated from the upper to the lower respiratory tract due to a disturbance in the usual innate protective processes. Therefore patients who develop pneumococcal pneumonia usually have an underlying pre-disposition, such as co-infection with influenza, chronic bronchial disease or a situation where consciousness and breathing are disturbed (e.g., epilepsy, alcoholism). People with reduced resistance to infection in general are particularly vulnerable, including those with chronic cardiac, renal or pulmonary disease, diabetes mellitus or immunocompromising conditions including being HIV positive (Kilian, 2012). *Str. pneumoniae* is able to cause more damage in such patients and death may be the outcome.

3.3.3 Epidemiology

Pharyngitis is a relatively common infection, which usually resolves uneventfully without the intervention of healthcare professionals – and therefore no laboratory diagnosis is involved. This makes it difficult to know the annual incidence, but it has been estimated that >600 million cases of pharyngitis per year worldwide are attributable to *Str. pyogenes* (Walker *et al.*, 2014). A proportion of these infections will develop into scarlet fever, which is notifiable in some countries (including the UK). The criteria can include clinical instead of laboratory diagnosis, which increases the reported rates. Most cases are in children over 2 years but below 10 years of age. It is seasonal (more cases in colder months), with incidence and prevailing *emm* subtype (see below) tending to vary each year. For instance, in 2011 in Hong Kong the number of reported cases of scarlet fever was 996, compared to rates of between 100 and 200 in the preceding three years. The prevailing type was found to be *emm*12 (Luk *et al.* 2012). In England there was a dramatic increase in scarlet fever during the spring of 2014, with over 13,000 notifications (equivalent to 24.5 per 100,000 population); however, in this case, the main genotypes were *emm*3 and *emm*4 (Turner *et al.*, 2016).

Estimates suggest that there are over 15 million people with streptococcal-associated rheumatic heart disease (RHD) worldwide, with over 250,000 new cases of acute rheumatic fever and over 200,000 deaths each year (Dale *et al.*, 2013). Although RHD is probably under-diagnosed and therefore under-reported, it is thought that over 2 million children between the ages of 5 and 14 years have the condition (Walker *et al.*, 2014), making it an important cause of paediatric cardiac disease. This complication of *Str. pyogenes* infection is relatively rare in Europe, but reported incidence is rising in some parts of the world (Kilian, 2012). Acute post-streptococcal glomerulonephritis is an even less common outcome, with about 470,000 cases each year globally and 5,000 deaths (Walker *et al.*, 2014)

Str. pneumoniae is considered to be the most frequent cause of community-acquired bacterial pneumonia, 'CAP' (Drijkoningen and Rohde, 2014; Kilian, 2012). Reported incidence rates across the world vary from <0.2 to over 10 per 1,000 population, but they are all likely to be underestimates since a significant proportion of people experiencing CAP are treated symptomatically outside of hospital with a standard antibiotic; thus no respiratory specimens would be collected for laboratory analysis (Drijkoningen and Rohde, 2014). Estimated annual incidence is around 3 per 1,000 population in the UK and the United States each year (Drijkoningen and Rohde, 2014; Kilian, 2012), with an overall case fatality rate of approximately 5% in Europe. This may be higher in other parts of the world – particularly in situations where suitable antimicrobials are not readily available or the start of the course of treatment is delayed (Feldman and Anderson, 2014). Death rates amongst patients with pneumococcal pneumonia who are admitted to hospital for treatment are reported to be 10% to 15% (Drijkoningen and Rohde, 2014; Kilian, 2012), with the rate slightly higher in South America and slightly lower in North America (Drijkoningen and Rohde, 2014). In the longer term, patients who recover from the pneumococcal episode may have an up to 50% chance of dying within the subsequent five years as a result of the damage caused (Drijkoningen and Rohde, 2014). In about half of all adult patients diagnosed with invasive pneumococcal disease (IPD), the original infection was pneumonia, but data is usually collated as IPD in total, which means that it can an include bacteraemia or meningitis where prior respiratory symptoms have not been identified. Case fatality rates for IPD are reported at up to 30% (Feldman and Anderson, 2014).

3.3.4 Laboratory Diagnosis

When the patient has recurrent pharyngitis or tonsillitis and in cases of invasive streptococcal infection, isolation and identification of the causative agent can be a significant contribution to patient management and treatment. For respiratory tract infections, the bacteria can be isolated from appropriate specimens including throat and pharyngeal swabs and washings, sputum, BAL and – in the case of septicaemic complications – blood cultures (SMI B9, 2015; SMI S2, 2014). *Streptococcus* spp. are Gram positive, catalase negative cocci. Samples should be plated out onto blood agar and a selective medium such as CLED. Chromogenic 'Staph/Strep' agar which allows selection of and differentiation between the two species is also available (SMI ID4, 2014).

Although the clinical syndromes discussed above for *Str. pyogenes* and *Str. pneumoniae* are broadly quite different, it should be noted that both are associated with pneumonia. Under those circumstances species differentiation and determination of antimicrobial susceptibility would be important. In Gram stains prepared directly from clinical specimens, *Str. pyogenes* will usually appear as short chains, while *Str. pneumoniae* is generally found in the form of diplococci. *Str. pyogenes* is β-haemolytic on blood agar plates and *Str. pneumoniae* exhibits α-haemolysis, which is shown in Figure 3.3. In complete β-haemolysis, the red blood cells are fully lysed, leading to zones of clearing around bacterial colonies (Figure 3.3a). In contrast, partial α-haemolysis is characterised by green areas in the agar where the bacteria have grown (Figure 3.3b). Also note that *Streptococcus* species which grow on blood agar without affecting the red blood cells are sometimes called γ haemolytic. Serendipitously, *Str. pyogenes* is resistant to the antibiotic optochin and sensitive to bacitracin (as shown in Figure 3.3c) and for *Str. pneumoniae*, the susceptibility is the opposite. This can form the basis of a simple test to separate the two species (SMI ID4, 2014) should it be necessary. Where antibiotic treatment is required then susceptibility assays can be set up. Strain determination can be achieved through serotype testing, but matrix-assisted laser desorption/ionisation-time of flight (MALDI-TOF) and PCR protocols have been established (SMI ID4, 2014). These have the advantage of allowing both species and strain determination to be done relatively quickly and, in some cases, the output will indicate antimicrobial susceptibility. MALDI-TOF is reported to perform well against conventional identification methods for *Str. pyogenes* (e.g., Cherkaoui *et al.*,

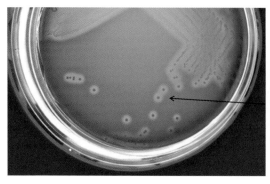

Clear zones of complete red blood cell lysis around bacterial colonies

Figure 3.3a *Streptococcus* spp. grown on blood agar exhibiting beta-haemolysis. (*See insert for colour representation of the figure.*)

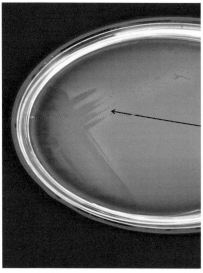

Figure 3.3b *Streptococcus* spp. grown on blood agar exhibiting alpha-haemolysis. (*See insert for colour representation of the figure.*)

Greenish zones around areas of bacterial growth due to incomplete lysis of red blood cells

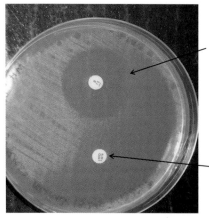

Clear zone of inhibition around Bacitracin disc

Figure 3.3c *Streptococcus pyogenes* grown on Mueller-Hinton agar and tested for susceptibility to bacitracin. (*See insert for colour representation of the figure.*)

No zone of inhibition around Optochin disc

2011). Appropriate primers for *Streptococcus* spp. can be incorporated into multiplex PCR assays (SMI ID4, 2014). In this case, it is important to note that in molecular and biochemical sequencing terms *Str. pneumoniae* is extremely close to other members of the 'mitis' group. This is at least partly due to the fact that *Str. pneumoniae* readily acquires genetic material from closely related species such as *Str. mitis* through transformation and gene recombination (Mitchell and Mitchell, 2010). Thus some technical difficulties can be experienced with precise identification by MALDI-TOF (e.g., Werno *et al.*, 2012) and it means that the usual target for bacterial PCR – the 16S ribosome gene sequence – is not useful, so that different sequences must be found (e.g., Zbinden *et al.*, 2011). Thus for *Streptococcus* spp. the more conventional identification methods still have value. Once the species has been established, strains are distinguished by the 5′ variable region of the gene coding for the M protein (rather than detection of the antibody raised against it). This gene is called *emm* and a small number of *emm* types are most often associated with serious GAS disease, although molecular analysis has shown that the prevalent type in a particular geographic region can vary considerably from year to year (Walker *et al.*, 2014).

3.3.5   Management and Treatment

If antimicrobial treatment is indicated for *Str. pyogenes* infection, penicillin would usually be the drug of choice, since resistance has never been reported (Kilian, 2012; Walker *et al.*, 2014). Patients with rheumatic fever can also safely take penicillin or amoxicillin prophylactically. As these patients are usually children, the recommendation is that they should take the antibiotic for five years (and then review) or until they reach the age of 21 (Kilian, 2012; Walker *et al.*, 2014). For those who are hypersensitive (allergic) to penicillin, macrolides such as erythromycin, clarithromycin or azithromycin are advocated as suitable alternatives, but surveillance data suggests that up to 10% isolates may be resistant (Kilian, 2012; Walker *et al.*, 2014), so susceptibility testing of individual bacterial cultures would be indicated.

In contrast, penicillin resistance is widespread amongst *Str. pnemoniae* strains (Kilian, 2012) and susceptibility to macrolides is also reduced. Vancomycin, a cephalosporin or fluoroquinolone would be indicated as treatments for pneumonia. It is suggested that the sensitivity of each isolate should be determined in the laboratory (Murray *et al.*, 2013), although in CAP, samples may not always be collected.

The severe sequelae which may result from infection with both of these species of *Streptococcus* has provided the impetus to develop vaccines against each of them. A number of the *Str. pyogenes* pathogenic factors have been targeted and particular success has been achieved using the M protein. This is highly immunogenic – so the aim of the vaccine would be to elicit protective anti-M protein antibodies – and the basis of the vaccine is a short peptide sequence from the N-terminal region of the protein. This is not conserved between the different strains, so the preparation requires multiple M protein peptides (Dale *et al.*, 2013) and a good understanding of the epidemiology of the most virulent *Str. pyogenes* serotypes. A 26-valent vaccine, which covers the majority of variants known to be associated with invasive GAS disease, has undergone clinical trials and a 30-valent version has been developed (Dale *et al.*, 2013). These may not be universally effective, since the 30-valent anti-M protein vaccine was reported to have induced antibodies which would protect recipients against 90% of the invasive serotypes found in the United States, but only 78% of those circulating in Europe (Dale *et al.*, 2013).

The capsule is also the target for vaccines against *Str. pneumoniae*. Two formulations are currently available, namely the pneumococcal polysaccharide vaccine (PPV) and the pneumococcal conjugate vaccine (PCV). Due to the strain variation in the capsule proteins, these are also polyvalent – PPV23 and PCV13 (Feldman and Anderson, 2014). The PPV23 is given to adults over 65 years old and patients with underlying factors which pre-dispose them to IPD, particularly pneumonia (e.g., immunocompromised, chronic cardio-pulmonary conditions, renal disease). There are some issues with the efficacy of the vaccine (Kilian, 2012). Studies suggest that it is protective in otherwise healthy adults, but that its effectiveness tends to decrease with age in 65 year olds and older and it may not be particularly useful in preventing serious pneumococcal disease in the immunocompromised, such as HIV-positive patients (Feldman and Anderson, 2014; Murray *et al.*, 2013). The main problem appears to be with study designs – making it difficult to reliably interpret outcomes of individual research programmes and even harder to use the power of meta-analysis to combine findings (Feldman and Anderson, 2014). The PPV23 formulation does not work in babies under two years of age, since it requires a fully functioning humoral immune response. The solution to this was to conjugate the immunogenic proteins to a carrier, to stimulate the T-cell response, which is the design of the PCV13 vaccine. This is included in the childhood vaccination programme in some countries including UK and the United States and is intended to protect against pneumococcal meningitis (Feldman and Anderson, 2014). In terms of a vaccine, it is considered useful and cost effective, but as it only contains capsular proteins for 13 serotypes, there is a potential risk that other strains of *Str. pneumoniae* will move in to the available ecological niche. Analysis of bacterial isolates from patients with pneumococcal pneumonia and IPD suggest that this has indeed happened, with rates of detection of some serotypes included in PCV13 noticeably decreasing (Feldman and Anderson, 2014). Thus, while these two vaccines have reduced morbidity and mortality from pneumococcal disease, ongoing work to refine them is required (Feldman and Anderson, 2014).

Point to consider 3.4: Do you think that concentrating on the most prevalent and virulent strains of *Str. pyogenes* and *Str. pneumoniae* in the vaccine preparations was the best approach? Are there any short-term or long-term risks to this strategy? If so, how could they be minimised?

3.4   *Mycobacterium tuberculosis*

Classification
Kingdom: Bacteria
Phylum: Actinobacteria
Order: Actinomycetales
Suborder: Corynebacterinae
Family: Mycobacteriaceae
Genus: Mycobacterium
Species: *M. tuberculosis, M. avium, M. ulcerans, M. leprae*

3.4.1 Introduction

Robert Koch published his work demonstrating that tuberculosis (TB) is caused by a bacterium in 1882 (Gaynes, 2011). This research contributed to the growing body of evidence for the 'germ theory' of infectious diseases, as well as forming the basis of the rules he developed to identify the infectious agent of a particular disease, now called Koch's postulates (Gaynes, 2011). The causative organism was named *Mycobacterium tuberculosis*. Despite advances in antimicrobial therapy and vaccination in the proceeding century, tuberculosis remains a major cause of mortality and morbidity world wide – and in some geographic areas is a re-emerging infection (e.g., Lee *et al.*, 2015; Borgdorff and Soolingen, 2013).

Mycobacterium spp. are non-motile rods which do not form spores. Their cell walls are about two-thirds lipid and contain a thick peptidoglycan layer, along with mycolic acids and other polysaccharides and proteins (Murray *et al.*, 2013). This complexity makes them essentially refractive to the decolourising effect of the acid/alcohol step in the Gram's stain, which is why they are sometimes referred to as acid-alcohol fast bacilli – 'AAFBs' or 'AFBs' (Grange, 2012). The Ziehl-Neelson stain was developed in the 1880s by colleagues of Koch in Germany (Gaynes, 2011). This method uses heat to encourage the bacterium to take up the pink carbol fuchsin stain (Grange, 2012) which then binds to the mycolic acid. The bacteria can be cultured *in vitro;* they grow aerobically, but require a specialised egg-based medium such as Löwenstein-Jensen and usually take several weeks to appear (Grange, 2012).

Over a hundred species of *Mycobacterium* have been described. Those known to be pathogenic to humans broadly fall into three groupings, namely the '*Mycobacterium tuberculosis* (MTB) complex', the 'slow-growing nontuberculous mycobacteria' and the 'rapidly growing nontuberculous mycobacteria' (Murray *et al.*, 2013). The last two groups contain organisms of interest to the medical microbiologist, including the opportunistic pathogens in the *M. avium* complex ([MAC] – also known as the *Mycobacterium avium- intracellulare* group [MAI]) (SMI B40, 2014) and *M. ulcerans* which causes Buruli ulcer (http://www.who.int/mediacentre/factsheets/fs199/en/). The most pathologically and epidemiologically important organisms of this species are found in the MTB complex. As well as *M. tuberculosis* itself, the MTB complex grouping includes *M. leprae* (Rodrigues and Lockwood, 2011) and *M. bovis* (Corner *et al.*, 2011).

3.4.2 Pathogenesis and Clinical Symptoms

The clinical features of *M. tuberculosis* have been well known since ancient times and were to some extent romanticised as the 'consumption' of writers and poets in the eighteenth and nineteenth centuries (e.g., Daniel, 2006). While the incubation time is usually between a few weeks and a few months, the onset of symptoms occurs relatively slowly. Patients experience lethargy, weight loss, night fevers and a cough – which is usually productive and may include visible blood (Murray *et al.*, 2013). While primary infection is generally confined to a LRT (although other organs could be infected), in some patients the bacteria are later spread to other sites via the blood stream, leading to disseminated or military tuberculosis (Grange, 2012; Murray *et al.*, 2013). The disease is transmitted via aerosols and people with untreated infection are known to shed the bacteria in high concentrations. An intriguing observation is that sometimes known close (e.g., household) contacts of confirmed cases do not develop tuberculosis, which suggests that innate immunity is important in initial protection (Schwander and Dheda, 2011).

In situations where the bacteria are able to enter the lower respiratory tract, they travel to the alveoli; here they are ingested by the alveolar macrophages via phagocytosis (Murray, 2013). They initially appear to be able to evade non-specific immunity, such as killing by reactive nitrogen species and fusion of the macrophages with lysosomes. The infected macrophages express mycobacterial cell surface proteins, which are recognised by TLRs and nucleotide oligomerisation domain receptors (Schwander and Dheda, 2011; Pawlowski *et al.*, 2012), which triggers a local inflammatory reaction. The presence of the *M. tuberculosis* has also been found to stimulate the production of IFN-γ, TNF-α and IL-12 (Murray *et al.*, 2013; Schwander and Dheda, 2011). This leads to the Th1 immune response, further inflammation and the involvement of CD4+, CD8+ and NK T cells (Murray *et al.*, 2013). The cytokines help the body's reaction against the *M. tuberculosis*, since the IFN-γ promotes the fusion of infected phagosomes with liposomes, while TNF-α increases the production of reactive nitrogen species. The inflammatory cellular response traps the infected macrophages into a granuloma. If there are only a small number of bacteria in the lesion, then they are usually destroyed, with little damage to the surrounding tissue. A greater inoculum of organisms at the site leads to a larger granuloma, which cannot be dispersed, so is enclosed in a capsule to prevent further spread; the bacteria still remain viable despite being dormant.

It is known that CD4+ T cells are crucial in the effective elimination or at least containment of *M. tuberculosis* infection (Ottenhoff, 2012; Pawlowski *et al.*, 2012); evidence for this was starkly provided in the early years of the acquired immune deficiency syndrome (AIDS) epidemic by the reactivation of underlying tuberculosis infection in these patients (Ottenhoff, 2012; Pawlowski *et al.*, 2012). Other T-cell groups appear to be important, including Th17 cells which are involved in the pro-inflammatory response and CD8+ T cells, which play a part in the continued latency of an infection (Pawlowski *et al.*, 2012). There are some particular interactions between *M. tuberculosis* and the host immune system which are interesting. One example is that the bacterium seems to use activation of T_{regs} for its own purposes, to delay the Th1 immune response, although the mechanism for this is not yet clear (Ottenhoff, 2012). In patients who are co-infected with TB and HIV the severity of both infections is enhanced. In the case of someone previously infected with TB who acquires HIV and develops AIDS, the reduction in CD4+ cell count leads to the reactivation of any tuberculous lesions. The chances of someone who is not HIV positive reactivating a latent TB infection is estimated at 5% to 10% throughout their life time; this can be due to depletion in the immune response through factors such as malnutrition and old age (Pawlowski *et al.*, 2012; Murray *et al.*, 2013). In a person who is HIV-positive, this risk is between 5% and 15% per year (Pawlowski *et al.*, 2012; Murray *et al.* 2013) and again is associated with the immune suppression cause by CD4+ count falling below <200 cells μL^{-1} (i.e., AIDS). The TB lesions in AIDS patients tend not to be as pus-filled and necrotic as expected (this effect is thought to be a manifestation of dead CD4+ cells) and are rarely confined to the lungs (Pawlowski *et al.*, 2012). Conversely, *M. tuberculosis* has been demonstrated to encourage HIV replication; while there are a number of possible explanations for this, the pathological process is an area of ongoing research (Pawlowski *et al.*, 2012).

3.4.3 Epidemiology

It is estimated that about one third of the global population is infected with *M. tuberculosis* (Murray *et al.*, 2013). The WHO estimates that in 2012, 8.6 million people were newly diagnosed with tuberculosis and over 1 million died. Multi-drug resistant (MDR) TB was found in 450,000 patients and was thought to have caused at least 170,000 deaths (WHO, 2013). The disease is more common in males than females (just under 3 million of the new cases in 2012 were found in adult women) and remains a significant risk to children, with over half a million diagnosed annually (WHO, 2013). Most cases are reported from Southeast Asia, Africa and the Western Pacific region – with 26% from India in 2012. At least 1 million of the new patients and over 300,000 of those who died from the disease were HIV-positive and three quarters of these were in the WHO African region (WHO, 2013).

Tuberculosis is considered to be a re-emerging infection in most areas of the world (Borgdorff and Soolingen, 2013). In the United Kingdom, tuberculosis is a notifiable disease and this illustrates the important role that laboratory confirmation can play in public health. The incidence of TB in 2013 was 12.3 per 100,000 population (7892 cases) (PHE TB Report, 2014). The majority of people with newly diagnosed with TB in the UK were born outside of the country and so could have acquired the bacterium while travelling overseas or could be reactivating a previous infection. In line with reported global figures (WHO, 2013), the highest rates were reported in people born in India (9.8%) and Pakistan (20.4%). Although only 15% of new infections were found in people who had come to live in the UK in the previous two years, it should be noted that 130 cases were identified as part of a policy of screening for active TB (PHE TB Report, 2014); this is conducted as part of a health check given on entry to the country and is clearly an effective measure, as it allows treatment to be started and potentially prevents further spread of the disease. The disease is currently most commonly found in deprived inner city areas, with London accounting for 2,985 of the newly reported cases in 2013. A diagnosis of TB is more likely in men than women (58% of cases in UK in 2013 were in males). The highest incidence (24.4 per 100,000) is reported in adults aged 25 to 34 years, while children accounted for 3% of patients (PHE TB Report, 2014). It is interesting that from the clinical details noted, just over half of patients (52.1%) had LRT infection only and in the remainder, TB bacilli were found in a wide range of sites including elsewhere in the thorax, bone, central nervous system, gastrointestinal tract and genitourinary tract (PHE TB Report, 2014). This high proportion of extra-pulmonary infections highlights the requirement for laboratory scientists to be vigilant in identifying *M. tuberculosis* from unusual sites and in unlikely patients.

3.4.4 Laboratory Diagnosis

Where pulmonary TB is suspected, a chest X-ray can be invaluable and where there is a possibility of the disease in other sites, a reaction in the tuberculin skin test can indicate a likely infection (Murray *et al.*, 2013). Laboratory diagnosis is essential to meet the criterion for a confirmed case and also to investigate drug susceptibility in each isolate (see below).

Appropriate samples depend on the site of infection and they might include respiratory specimens (such as sputum or BAL) cerebrospinal fluid (CSF), biopsy tissue, urine, blood and bone marrow aspiration specimens (SMI B40, 2014). In the UK, *Mycobacterium tuberculosis* is classed as a Hazard Group 3 – often referred to as category ("cat") 3 organism (Health and Safety Executive, 2013) and work on samples which might contain this bacterium should be done in containment-level 3 facilities (SMI B40, 2014). In situations where this is not feasible, extreme care must be taken to ensure the safety of the laboratory staff.

The mainstay of laboratory investigation is microscopic examination of a suitably prepared, fixed (heat or alcohol) and stained smear from the patient's sample, following by culture in specialised media (SMI B40, 2014). While Ziehl-Neelson (Z-N) staining is still widely used, auramine-phenol stain is considered to be more sensitive and therefore more useful as the initial test (SMI B40, 2014). This is followed by bacteriological culture of an aliquot (which has been de-contaminated in NaOH for safety reasons) on a slope of an egg-based culture medium such as Löwenstein-Jensen where facilities allow for this. Although a rapid diagnosis can be made by microscopy, the limit of detection is 5,000 to 10,000 bacteria mL^{-1} sample (Cheesbrough, 2006). This translates into one or two *M. tuberculosis* bacilli visible per slide after examining several fields.

Culture, which is more sensitive (SMI B40, 2014) also affords the advantage of producing an isolate which can be tested for antimicrobial sensitivities and possibly sent to a reference laboratory for sequencing, which can be useful for epidemiological purposes (e.g., Seoudi *et al.*, 2012) Unfortunately, it often takes two to three months for MTB to grow in conventional culture (SMI B40, 2014) and up to a further two weeks to complete antimicrobial sensitivity tests (Seoudi *et al.*, 2012). Thus, by the time the results are available, the patient will be several weeks into a six-month course of treatment which might actually be ineffective. Considerable research has therefore taken place in recent years to improve the laboratory diagnosis of tuberculosis, initially by enhancing and speeding up the culture process and latterly through molecular methods (SMI B40, 2014). Automated culture systems, which detect products of mycobacterial metabolism such as CO_2 are available. Since they are not reliant on detection of visible colonies, they are reported to provide results in half the time of conventional culture (SMI B40, 2014) and some methods can also determine susceptibilities to anti-mycobacterial drugs.

Nevertheless, the turnaround time for results is still weeks rather than days and TB diagnosis is clearly a situation which lends itself to the application of molecular diagnostics. The line probe assay (LPA) which uses PCR and a colourimetric method to detect the amplification product appears to be useful in this context (Seoudi *et al.*, 2012; Wilson, 2013). It was originally developed to confirm culture results (Seoudi *et al.*, 2012; Wilson, 2013); when used for this purpose, sensitivity is reported to be over 90% and specificity over 80%, depending on the study design (Seoudi *et al.*, 2012; Wilson, 2013). As the assay obtains the microbial genetic sequence, the isolate can be identified accurately to species level; other information, such as strain variation and drug resistance, can also be determined. Identification of rifampicin susceptibility is reported at sensitivities of over 90% (Seoudi *et al.*, 2012). A method for extracting mycobacterial DNA directly from sputum samples for use in the test has successfully been developed. Where the LPA method is not automated, the turnaround time is still a few days, but this qualifies it as a 'rapid test' in this context and it has been recommended by the WHO (Seoudi *et al.*, 2012; Wilson, 2013). A real-time PCR method is also under evaluation (e.g., Boehme *et al.*, 2011; Wilson, 2013). This can provide results within a day, but the sensitivity as a diagnostic (rather than confirmatory) tool is reported as fairly disappointing and in some studies was found to be as low as 55% (Wilson, 2013). Another method for characterising bacterial isolates is MALDI-TOF, which has the potential to accurately identify to species level as well as detect multidrug-resistant (MDR) and extreme multidrug-resistant (XDR) strains once the database is populated (Saleeb *et al.*, 2011; SMI B40, 2014).

Attempts to develop a POCT exploiting the fact that patients with active pulmonary TB infection produce detectable 'volatile organic compounds' in their breath have also been reported (e.g., Phillips *et al.*, 2012). The equipment used by Phillips *et al.* (2012) compromised a breath collection apparatus, a gas chromatography analyser and customised software; they suggested that a result could be obtained in only 6 minutes. Sensitivity and specificity were around 70% and while the NPV was 98%, the PPV was cited as only 13%. This method probably has limited application at the moment, but is an interesting development. Another diagnostic problem is the recognition of patients with latent TB infection, particularly those who are at risk of re-activating, for example, due to HIV co-infection. The tuberculin skin test is widely used, but people who have previously received the BCG will also give a positive reaction (Diel *et al.*, 2011). This has been addressed by assays which detect IFN-γ production by white cells in peripheral blood, on stimulation with specific MTB antigens. A meta-analysis of data from evaluation studies of two commercially available kits indicated that while they have limitations – particularly with respect to HIV positive patients – their negative predictive value is around 98% or higher (Diel *et al.*, 2011) which is an improvement on the tuberculin test. Thus these assays may have a useful application in TB diagnosis and control because they can at least determine who does not have a latent infection.

Point to consider 3.5: In TB diagnosis, a balance between speed and high sensitivity and specificity is needed. If there still a place for microscopy in that balance?

3.4.5 Control, Treatment and Management

In many countries in the world, the incidence of and mortality from TB infection declined markedly during the twentieth century (Lienhardt *et al.*, 2012). This is attributable to a combination of an enhanced understanding of the microbiology and improved standards of living for most people; it is arguable that the latter was more important. Microbiological advances brought the means for laboratory confirmation of the clinical diagnosis, recognition of the role of infection control measures (specifically isolation hospitals) anti-mycobacterial chemotherapy and a vaccine (Lienhardt *et al.*, 2012). Socioeconomic changes led to a better quality of housing, along with a lower density of people per dwelling, improved nutrition and more protection from the harmful substances which could damage lungs (e.g., coal dust).

In the 1920s, the attenuated Bacille Calmette-Guérin (BCG) was created and was found to be effective as a vaccine. It was widely introduced into national vaccination programmes after the Second World War (Lienhardt *et al.*, 2012) and remains part of many childhood immunisation schedules (Mangtani, *et al.*, 2014). Used in conjunction with screening (e.g., pre-employment chest X-rays) and isolation and treatment of confirmed cases, the BCG has made a significant contribution to control of TB (Lienhardt *et al.*, 2012). Nevertheless, doubts have been raised about its efficacy, including the noted (but puzzling) observation that while the BCG is protective in around 80% in immunocompetent adults in the UK, this percentage reduces the nearer to the equator the vaccinated population is living (Smith, 2014). The possible explanation for this is the complexity of the associated immunological response, with both bacterial and host genetics influencing the activity of the vaccine (Smith, 2014) and it is suggested that the solution is a better designed vaccine. Interestingly, research has also pointed to co-infection with helminths and non-tuberculous mycobacteria as factors which can reduce the effectiveness (McShane, 2014); exposure to these is more likely in regions near the equator such as sub-Saharan Africa, the Indian sub-continent and South America. A possible clarification has been offered by a systematic review conducted by Mangtani *et al.* (2014) of clinical trials of the BCG published between 1930s and 1980s from across the world. The authors concluded that previous exposure to *Mycobacterium tuberculosis* or other non-tuberculous mycobacteria had the most significantly adverse effect on the usefulness of the vaccine (Mangtani *et al.*, 2014). The practice of screening potential vaccinees using the tuberculin test (Murray *et al.*, 2013) and not giving the BCG to those who react is therefore justified and the recommendation is to give the vaccine as soon as possible after birth (Mangtani *et al.*, 2014). The BCG is no longer given routinely to children in the UK on the grounds of the very low transmission rate unless they live in an area of high risk such as inner London (see above; http://www.nhs.uk/conditions/vaccinations/pages/bcg-tuberculosis-tb-vaccine.aspx). As a consequence there are occasional outbreaks in schools, colleges or familial contacts of a patient (e.g., Estée Török *et al.*, 2013), in which case there will be targeted vaccination.

The ethos of the TB isolation hospital ('sanatorium') was that fresh air was the best cure (Daniel, 2006). This meant that patients with a debilitating respiratory disease lived in buildings where all the windows were open and were taken to sit outside even on very cold days! The introduction of the antimycobacterial agents streptomycin and para-aminosalicylic acid in the late 1940s meant that drug treatment superseded braving the elements. The advent of isoniazid in the 1950s allowed for a multidrug treatment regime, which was found to be more effective and was also intended to reduce the chances of resistant strains developing (Lienhardt *et al.*, 2012). Although the patient had to take the drugs for up to two years, they did not need to be isolated and the TB hospitals were eventually closed. Rifampicin and pyrazinamide were added to the arsenal of weapons against TB and also reduced the usual course of treatment to six to eight months.

However, even six months is a long time to continue with a course of treatment especially when it is for an infection and the drugs can have unpleasant side effects (http://www.tbalert.org/about-tb/global-tb-challenges/side-effects/). Compliance amongst people who had been sent home with the drugs became an issue. Not only were people not being fully cleared of their TB infection – thus posing a public health risk – but the suboptimal dosing allowed the bacterium to develop resistant strains. The simple, but very effective, solution to this was DOTS ('directly observed therapy, short course'), which began in the late 1980s/early 1990s (Lienhardt *et al.*, 2012). It is based on the principle of patients being observed taking their daily anti-mycobacterial drugs. Thus, they are administered by a healthcare professional either when the patient attends a clinic or during a home visit. Unfortunately, the lack of comprehensive implementation of DOTS in TB endemic areas and the HIV epidemic meant that initial success with TB control stalled somewhat in the 1990s. In addition, 'multidrug resistant' strains of *Mycobacterium tuberculosis* (MDR-TB) (defined as being resistant to both rifampicin and isoniazid) and later 'extensively multidrug resistant' strains XMDR-TB (additionally resistant to the second-line treatments) have arisen (http://www.who.int/tb/challenges/mdr/en/), and are a particular problem in immunocompromised people, including HIV patients. Incidence has been increasing and therefore the WHO started a 'Stop TB' campaign in 2006, which aimed to stabilise the incidence of TB by 2015 and also to reduce the prevalence and mortality rates to 50% of the levels reported in 1990 by that time (http://www.who.int/tb/strategy/stop_tb_strategy/en/). Laboratory support is vital to this programme as rapid, accurate diagnosis, surveillance for drug resistant strains and epidemiological investigations are all required. By 2014 it was reported that the mortality rate for TB had fallen by 45% between 1990 and 2013, with a decline in prevalence of 41% (Global TB report; http://www.who.int/tb/publications/global_report/en/). This is a little short of the 50% goal, although the Americas and Western Pacific WHO regions had met their targets. For updates, the reader is directed to updated information from the WHO: (http://www.who.int/tb/strategy/en/).

3.5 *Pneumocystis jirovecii*

Classification
Kingdom: Fungi
Phylum: Ascomycota
Class: Pneumocysidomycete
Order: Pneumocystidales
Family: Pneumocystidaceae
Genus: *Pneumocystis*
Species: *P. jirovecii*

3.5.1 Introduction

Pneumocystis jirovecii is an interesting opportunistic pathogen which affects severely immunocompromised people. It was recognised as one of the unusual presenting features of acquired immune deficiency syndrome (AIDS) in the early 1980s. Thus it was one of the infections included in the clinical diagnostic criteria used for AIDS before HIV was identified (Pillay, 2009). Until the late 1980s, *Pneumocystis* spp. were considered to be protozoa and the pathogen found in AIDS patients was called *P. carinii;* the respiratory infection it is associated with was named *P. carinii* pneumonia (PCP). Subsequent genetic analyses indicated that these organisms are more closely related to fungi, specifically the yeast *Schizosaccharomyces pombe*, which is in the Phylum Ascomycota (Murray *et al.*, 2013). They were subsequently re-classified and the human pathogen was re-named *P. jirovecii* after Jirovec who had originally described it in the 1950s (Stringer *et al.*, 2002; Carmona and Limper, 2011). While the new name has been widely adopted (albeit with some confusion about whether it should be *P. jiroveci* or *P. jirovecii*!), the term PCP for the clinical syndrome has lingered.

3.5.2 Pathogenesis and Clinical Symptoms

The life cycle of *P. jirovecii* involves both asexual and sexual reproduction: (www.cdc.gov/dpdx/pneumocystis/). Detailed investigations have been hampered by uncertainty about the environmental reservoir for the fungus, coupled with the lack of a suitable system for *in vitro* culture (Carmona and Limper, 2011). Humans are infected by inhalation of the fungus either from the environment or via airborne transmission from an infected patient (Coyle *et al.*, 2012). Several life cycle stages are recognised and they tend to be referred to using protozoal rather than mycological terminology, which reflects its previous classification as a protozoan. In clinical specimens, the asexual trophic stage, as well as the sexual sporocysts (single nucleus) and cysts (multinucleate) have all been isolated (Murray *et al.*, 2013). It is thought that the life cycle involves rupture of the multinucleate cyst (which is 8–10 μm in diameter) to release the smaller (around 2 μm) trophic forms (Carmona and Limper, 2011). In normal healthy adults, inhalation of this fungus does not lead to disease, but in individuals with AIDS (CD4+ count <200 cells μL^{-1} of blood) (Pillay *et al.*, 2009) or profound immunosuppression (e.g., post- transplant) it can cause 'atypical' pneumonia, characterised by interstitial pneumonitis (Murray *et al.*, 2013). Patients with underlying lung disease are also at risk of *Pneumocystis* pneumonia (Maini *et al.*, 2013). The symptoms of PCP include (dry) cough, shortness of breath, fever, night sweats and weight loss. HIV-positive patients tend to present with gradual onset of difficulty in breathing, non-productive cough and low-grade pyrexia (Carmona and Limper, 2011), while patients who do not have HIV often have a sudden onset with more severe effects (Carmona and Limper, 2011).

Since the organism's biology is not well understood, there are many aspects of the pathogenesis which require further research (Maini *et al.*, 2013). It is known that on entering the body, *P. jirovecii* moves through the respiratory system to the alveoli; migration to other parts of the body is rarely observed (Carmona and Limper, 2011). When the cysts burst open to release the trophic forms, the contents of the cyst wall are discharged into the lung tissue. The main components of the fungal cyst wall are polysaccharides called β-glucans, which appear to trigger a strong inflammatory response from the respiratory epithelial cells and the macrophages (Carmona and Limper, 2011). This inflammation leads to restriction of the airways (manifest as dyspnoea). The β-glucans are also thought to affect the T-cell response; clearance of the fungus is inhibited when the person's T-cell count is abnormally low (Carmona and Limper, 2011).

3.5.3 Epidemiology

Although it was described as an organism at the turn of the twentieth century, *P. carinii* was not recognised as a human pathogen until the 1940s (Carmona and Limper, 2011). Incidence of PCP rose markedly during the 1980s since it was a common opportunistic infection in AIDS (Stringer *et al.*, 2002) and it was also recognised in other severely immunocompromised patients (Maini *et al.*, 2013). The introduction of HAART (see Chapter 2) has led to improved outcomes for people living with HIV infection and fewer cases of PCP in this population are noted. However, the scope, range and short-term survival rates of medical procedures which require the patient to be immunosuppressed has increased in the last 20 years. This is another group of patients susceptible to opportunistic infections, including PCP (Coyle *et al.*, 2012). Data collected in England indicates that the annual rate of reported cases of *P. jirovecii* infection more than doubled at the beginning of the twenty-first century from 157 cases in 2,000 to 352 in 2010 (Maini *et al.*, 2013). Documented infections in HIV positive patients decreased by about 7% each year over the course of this 10-year period, while the number of patients with PCP who had malignant haematological conditions rose (Maini *et al.*, 2013). Nosocomial infection, particularly amongst renal transplant patients, is being increasingly noted (e.g., Coyle *et al.*, 2012; Chapman *et al.*, 2013).

3.5.4 Laboratory Diagnosis

The most suitable specimens to collect for the diagnosis of *P. jirovecii* infection are BAL, bronchial brushings or lung biopsy. Fixed smears of respiratory material and histological sections from lung tissue can be treated with silver stain or fluorescent-labelled monoclonal antibodies. As indicated in Figure 3.4, the histological stain can highlight the fungal cells. As mentioned above, there are several recognisable life-cycle stages, but the one particularly shown up by the silver stain is the cyst. These are found within the alveoli, along with a characteristic foamy exudate containing eosinophils and macrophages as well as proliferating fungi (Figure 3.4). While these staining methods are reported to have high specificity (e.g., Botterel *et al.*, 2011; McTaggart *et al.*, 2012), they are not sensitive enough to detect relatively low concentrations of the organism and so may produce false-negative results (Carmona and Limper, 2011; Botterel *et al.*, 2012). Due to the seriousness of PCP and the poor immune condition of the majority of the patients, this leads to practice where the clinical diagnosis takes precedence (Carmona and Limper, 2011) which is not ideal as it could lead to incorrect management and treatment. Although the immunofluorescent antibody assay is currently taken as the 'gold standard' and has the advantage of giving results fairly

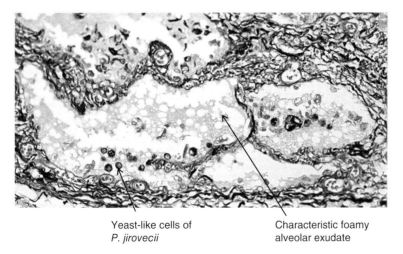

Yeast-like cells of　　　　　　Characteristic foamy
P. jirovecii　　　　　　　　　alveolar exudate

Figure 3.4 Section of lung biopsy stained with Grocott's methenamine silver stain, showing pneumocystis. Source: Courtesy of Ms L. Dixon; photos and interpretive annotations by Dr A. Gal. (*See insert for colour representation of the figure.*)

quickly, it is regularly outperformed by real-time PCR targeting the mitochondrial RNA large subunit gene in the fungal DNA (e.g., Botterel *et al.*, 2011; McTaggart *et al.*, 2012). Thus wherever possible, PCR should be included in the diagnostic algorithm – either as the initial diagnostic test or as a confirmation for negative results where clinical suspicion of PCP is high.

3.5.5   Management and Treatment

If untreated, PCP is likely to be fatal (Carmona and Limper, 2011). Hospitalisation in order to give supportive treatment including oxygen therapy (and sometimes mechanical ventilation) is usually required. Somewhat counterintuitively, the standard antibiotic treatment co-trimoxazole (trimethoprim and sulfamethoxazole) is effective in treating *P. jirovecii* infection (Carmona and Limper, 2011). It is also used prophylactically in certain situations (Maini *et al.*, 2013). Some people are allergic to co-trimoxazole and AIDS patients are particularly prone to poor tolerance. In these situations, alternatives are available including trimethoprim plus dapsone, atovaquone or pentamidine and the treatment chosen will depend on the patient's underlying condition and the severity of the respiratory disease (McTaggart *et al.*, 2012).

3.6   *Paragonimus westermani*

Phylum: Platyhelminths
Class: Trematoda
Family: Paragonomidae
Species *P. westermani, P. africanus, P. mexicanus, P. kellicotti*

3.6.1   Introduction

There is an eclectic mix of protozoa and helminths which can be found causing pathology in the lung (Kunst *et al.*, 2010); although most of them are not commonly seen in Western Europe, they should be borne in mind during diagnosis. For most parasites, the lungs are a transitory or ectopic site, which can mean that they might be overlooked as possible causative agents in respiratory infection. The trophozoites of virulent strains of *Entamoeba histolytica* usually migrate and cause ulcers in the liver ('amoebic liver abscess'), but they can find their way to the lungs, resulting in pneumonia or abscess formation there (Kunst *et al.*, 2010). As an illustration, Musa *et al.* (2013) reported the case of a woman in the UK who had a lung abscess, apparently caused by infection with *Fasciola* spp., presumed to have been acquired in Turkey a year previously.

One species of helminth for which the pulmonary tissue is the normal site is *Paragonimus* spp. (Murray *et al.*, 2013). Most human infections are attributable to *P. westermani* (Kunst *et al.*, 2010), though sporadic cases of disease caused by *P. mexicanus* and *P. kellicotti* have been reported from Central America and the United States respectively (Gunn and Pitt, 2012). The life cycle of this fluke (www.cdc.gov/dpdx/paragonimiasis/index.html) is relatively complex. It involves humans and other fish-eating mammals as the definitive host, particular species of freshwater snails – such as *Melania* spp. and

Brotia spp. – as the first intermediate host and freshwater crabs and crayfish as the second intermediate host (Gunn and Pitt, 2012). An infected definitive host excretes eggs orally as well as via faeces.

3.6.2 Pathogenesis and Clinical Symptoms

Humans acquire the infection through eating raw or undercooked crustaceans infected with the parasite metacercariae (Kunst *et al.*, 2010). Excystation takes place in the duodenum and from there the immature worms penetrate the gut wall and migrate into the coelomic cavity, pairing up if a mate is available. They then move through the diaphragm into the lungs (Greenwood, 2012), where they take two to three months to develop to maturity. Adult worms (flukes) can grow up to be 10–15 mm long and up to 8 mm wide (depending on the species) and be several millimetres thick. They have recognisable features such as oral and ventral suckers, ovary and testes (Figure 3.5a). The flukes can live for many years in the host if undetected and they induce an immune response, which creates a granuloma which comprises a group of up to three helminths plus dead tissue. In a lung X-ray, the lesions are similar to those observed in tuberculosis and malignancy (Kunst *et al.*, 2010). The adult females start to release fertilised eggs before the granuloma completely encloses it and some of those eggs can also be the basis of granulomatous reaction (Murray *et al.*, 2013). The majority of eggs reach the bronchi and are thus coughed up in sputum; they are either spat out or swallowed. During acute infection with *P. westermani*, the patient may experience diarrhoea, abdominal pain and an itchy rash, as well as a raised temperature, cough and breathing problems. It is notable that the symptoms of chronic paragonimiasis are very similar to tuberculosis – weight loss, fever and prolonged productive cough, often with haemoptysis.

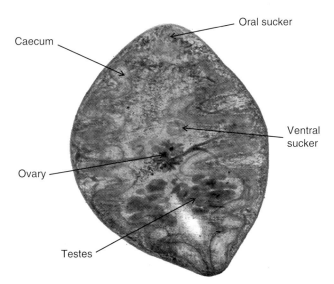

Figure 3.5a Mounted preparation of adult fluke of *Paragonimus* spp. Source: Courtesy of Dr A Gunn. (*See insert for colour representation of the figure.*)

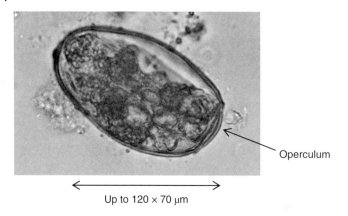

Up to 120 × 70 µm

Figure 3.5b Egg of *Paragonimus* spp. Source: Courtesy of Dr A Gunn. (*See insert for colour representation of the figure.*)

3.6.3 Epidemiology

Data on incidence of paragonimiasis does not appear to be collected and collated systematically. Efforts are hampered because only a minority (estimated at 26%) of infections are symptomatic (Fürst *et al.*, 2012). Also rates of laboratory identification of the parasite may be sub-optimal, due to inappropriate test method selection and sometimes because laboratories concentrate solely on detection of *Mycobacterium* spp. based in the reported clinical symptoms (Fürst *et al.*, 2012). Prevalence in endemic areas (Southeast Asia, South America) is reported to be between <0.1 to around 5% of the population (Fürst *et al.*, 2012).

3.6.4 Laboratory Diagnosis

The main laboratory investigations would be to detect the eggs in sputum or BAL by direct microscopy. *P. westermani* eggs are orangey brown, 80–100 µm by 45–70 µm and ovoid, with a distinct operculum at one end (Figure 3.5b). In the absence of suitable respiratory samples, serological tests for specific antibodies are available (Kunst *et al.*, 2010). Eosinophilia is seen in both peripheral blood and the lung fluids. It has been reported from a study in Laos, that *Paragonimus* spp. can be detected and identified in smears prepared from sputum samples and stained with Ziehl-Neelson stain (Slesak *et al.*, 2011). This was a serendipitous and rather handy finding, because in many clinical situations, tuberculosis is one of the differential diagnoses (Kunst *et al.*, 2010; Slesak *et al.*, 2011).

3.6.5 Management and Treatment

Patients should be managed on the basis of their respiratory symptoms, but once diagnosis has been established, anti-parasitic treatment is quite straightforward and comprises a short course of either triclabendazole or praziquantel (Fürst *et al.*, 2012). Given that the infection is food borne, subsequent investigation to identify the source would be a useful public health measure where it is possible. Also, if there appears to be a cluster of cases of paragonimiasis (e.g., they have all consumed crab meat from the same place), then prophylactic/ presumptive treatment of others likely to have been affected might be considered.

Point to consider 3.6: If a sputum or BAL sample was received in the laboratory with the clinical description of: 'Persistent cough, haemoptysis, possible TB' what (if any) other information would be needed before deciding to do tests for microorganisms other than *Mycobacterium* spp.?

3.7   Exercises

3.1   A   Research the main causes of respiratory infection in immunocompromised patients.

 B   Use this information to design a suitable multiplex PCR assay. Then investigate commercially available kits.

 C   If there is a discrepancy between your suggestion and the assays currently in use, think about the possible reasons for this.

3.2   A   Investigate the safety requirements for handling samples which might contain organisms designated as Hazard Group 3 by the Advisory Committee on Dangerous Pathogens, such as *Mycobacterium tuberculosis*.

 B   Use the information you have obtained to write a standard operating procedure (SOP) to guide laboratory staff working with such samples.

 C   Then compare your SOP with an existing one, either in the laboratory you work in or from a publically available source.

3.3   A   Explore POCTs for respiratory pathogens and find out what is currently offered commercially.

 B   Choose one organism and use the literature to write an evaluation of the available assays.

3.8   Case Studies

3.1   Mr AJ is a 53-year-old man who had a bone marrow transplant to treat chronic myeloid leukaemia 10 weeks ago. His transplant was successful and he has been recovering well at home. However, three days ago, he suddenly started to feel short of breath, with a dry cough and a pyrexia of 39 °C. He has not been eating and reports 'sweating during the night'. He is admitted to hospital and since his cough is not productive, a bronchoalveolar lavage is collected and sent to the laboratory for microbiological investigations.

 a)   Which pathogens do you think would be the mostly causes of Mr AJ's symptoms?

 After 24 hours the following results are reported:

Bacterial culture: no significant growth
Cytomegalovirus DNA not detected by PCR
Respiratory virus multiplex PCR: no virus detected
Pneumocystis jirovecii detected by silver stain

b) Discuss these findings and suitable treatment and management for Mr AJ

3.2 BK is a 3-year-old girl who has been admitted to the acute paediatric ward with symptoms of respiratory infection, which include wheezing. She has a temperature of 39.5 °C, a cough, rhinitis and although she is finding is finding it slightly difficult to breathe, she is not yet in distress.
   A nasopharyngeal aspirate is collected and sent to the laboratory for investigation of viral respiratory pathogens.

a) Which viruses would you suspect might be causing BK's symptoms:
b) Which laboratory test would you suggest in this case and why?
c) Under which circumstances would you suggest treatment for BK?

References

van den Bergh MR *et al.* (2012). Associations between pathogens in the upper respiratory tract of young children: Interplay between viruses and bacteria. *PLoS One*, **7**(10): e47711. doi: 10.1371/journal.pone.0047711

Blasi F, Tarsia P and Aliberti S. (2009). *Chlamydophila pneumoniae. Clinical Microbiology and Infection*, **15**: 29–35. doi: 10.1111/j.1469-0691.2008.02130.x

Bloom-Feshbach K *et al.* (2013). Latitudinal variations in seasonal activity of influenza and respiratory syncytial virus (RSV): A global comparative review. *PLOS One*, **8**: e54445 doi: 10.1371/journal.pone.0054445

Boehme CC *et al.* (2011). Feasibility, diagnostic accuracy and effectiveness of decentralised use of the Xpert MTB/RIF test for diagnosis of tuberculosis and multidrug resistance: A multicentre implementation study. *Lancet*, **377**: 1495–1505. doi: 10.1016/S0140-6736(11)60438-8

Borgdorff MW and Soolingen DV (2013). The re-emergence of tuberculosis: What have we learnt from molecular epidemiology? *Clinical Microbiology and Infection*, **19**: 889–901. doi: 10.1111/1469-0691.12253

Bosch AA *et al.* (2013). Viral and bacterial interactions in the upper respiratory tract. *PLoS Pathogens*, **9**: e1003057. doi: 10.1371/journal.ppat.1003057

Botterel F *et al.* (2012). Clinical significance of quantifying *Pneumocystis jirovecii* DNA by using real-time PCR in bronchoalveolar lavage fluid from immunocompromised patients. *Journal of Clinical Microbiology*, **50**: 227–231. doi: 10.1128/JCM.06036-11

Carratala J and Garcia-Vidal C. (2010). An update on Legionella. *Current Opinion in Infectious Diseases*, **23**: 152–157. doi: 10.1097/QCO.0b013e328336835b

Carmona EM and Limper AH. (2011). Update on the diagnosis and treatment of *Pneumocystis* pneumonia. *Therapeutic Advances in Respiratory Disease*, **5**: 41–59. doi: 10.1177/1753465810380102

Chapman JR *et al.* (2013). Post-transplant *Pneumocystis jirovecii* pneumonia: A re-emerged public health problem? *Kidney International*, **84**: 240–243. doi: 10.1038/ki.2013.212

Cheesbrough M (2006). *District Laboratory Practice in Tropical Countries, Part 2, 2nd edn.* Cambridge: Cambridge University Press.

Cherkaoui A *et al.* (2011). Evaluation of matrix-assisted laser desorption/ionization time-of-flight mass spectrometry for the rapid identification of beta-hemolytic streptococci. *Journal of Clinical Microbiology*, JCM-00240. doi: 10.1128/JCM.00240-11

Collins PL and Melero JA (2011). Progress in understanding and controlling respiratory syncytial virus: Still crazy after all these years. *Virus Research* **162**: 80–89. doi: 10.1016/j.virusres.2011.09.020

Corner LAL, Murphy D and Gormley E. (2011). *Mycobacterium bovis* infection in the Eurasian Badger (*Meles meles*): The disease, pathogenesis, epidemiology and control. *Journal of Comparative Pathology*, **144**: 1–24. doi: 10.1016/j.jcpa.2010.10.003.

Coyle PV *et al.* (2012). Rising incidence of *Pneumocystis jirovecii* pneumonia suggest iatrogenic exposure of immune-compromised patients may be becoming a significant problem. *Journal of Medical Microbiology*, **61**: 1009–1015. doi: 10.1099/jmm.0.043984-0

Daniel TM (2006). The history of tuberculosis. *Respiratory Medicine*, **100**: 1862–1870. doi: 10.1016/j.rmed.2006.08.006

Diel R *et al.* (2011). Interferon-γ release assays for the diagnosis of latent *Mycobacterium tuberculosis* infection: a systematic review and meta-analysis. *European Respiratory Journal*, **37**: 88–99. doi: 10.1183/09031936.00115110.

Dale JB *et al.* (2013). Group A streptococcal vaccines: Paving a path for accelerated development. *Vaccine*, **31**: B216–B222. doi: 10.1016/j.vaccine.2012.09.045

Drijkoningen JJC and Rohde GGU (2014). Pneumococcal infection in adults: Burden of disease. *Clinical Microbiology and Infection*, **20**: 45–51.doi: 10.1111/1469-0691.12461.

van Drunen-Littel-van den Hurk S and Watkiss ER (2012). Pathogenesis of respiratory syncytial virus. *Current Opinion in Virology*, **2**: 300–305. doi: 10.1016/j.coviro.2012.01.008

Estée Török M *et al.* (2013). Rapid whole-genome sequencing for investigation of a suspected tuberculosis outbreak. *Journal of Clinical Microbiology*, **51**: 611–614. doi: 10.1128/JCM.02279-12

Falsey AR *et al.* (2013). Bacterial complications of respiratory tract viral illness: A comprehensive evaluation. *Journal of Infectious Diseases*, **208(3)**: 432–441. doi: 10.1093/infdis/jit190.

Feldman C and Anderson R (2014). Review: current and new generation pneumococcal vaccines. *Journal of Infection*, **69**: 309–325. doi: 10.1016/j.jinf.2014.06.006

Fothergill JL, Walshaw MJ and Winstanley C (2012). Transmissible strains of *Pseudomonas aeruginosa* in cystic fibrosis lung infections. *European Respiratory Journal*, **40**: 227–238. doi: 10.1183/09031936.00204411

Fujitani, S *et al.* (2011). Pneumonia due to *Pseudomonas aeruginosa:* Part I: Epidemiology, clinical diagnosis, and source. *CHEST Journal*, **139**: 909–919. doi:10.1378/chest.10-0166

Fürst T, Keiser J and Utzinger J (2012). Global burden of human food-borne trematodiasis: A systematic review and meta-analysis. *The Lancet Infectious Diseases*, **12**: 210–221. doi: 10.1016/S1473-3099(11)70294-8

Gaynes RP (2011). *Germ Theory: Medical Pioneers in Infectious Diseases.* Washington DC: ASM Press.

Girard MP *et al.* (2010). The 2009 A (H1N1) influenza virus pandemic: A review. *Vaccine*, **28**: 4895–4902. doi: 10.1016/j.vaccine.2010.05.031

Gould IM *et al.* (2012). New insights into meticillin-resistant Staphylococcus aureus (MRSA) pathogenesis, treatment and resistance. *International Journal of Antimicrobial Agents*, **39**: 96–104. doi: 10.1016/j.ijantimicag.2011.09.028

Grange JM (2012). *Mycobacterium*. In: *Medical Microbiology, 18th edn*. Greenwood D, Barer M, Slack R and Irving W, eds. London: Churchill Livingstone – Elsevier.

Greenwood D (2012). *Protozoa*. In: *Medical Microbiology, 18th edn*. Greenwood D, Barer M, Slack R and Irving W, eds. London: Churchill Livingstone – Elsevier.

Greenwood D, Barer M, Slack R and Irving W, eds. (2012) *Medical Microbiology: A Guide to Microbial Infections, 18th edn*. London: Churchill Livingston – Elsevier

Gunn A and Pitt SJ (2012). *Parasitology: An Integrated Approach*. Chichester: Wiley-Blackwell.

Hall CB (2009). Respiratory syncytial virus. In: *Principles and Practice of Clinical Virology, 6th edn*. Zuckerman AJ, Banatvala JE, Schoub BD, Griffiths PD and Mortimer P, eds. Chichester: Wiley-Blackwell.

Hall CB *et al.* (1980). Possible transmission by fomites of respiratory syncytial virus. *Journal of Infectious Diseases*, **141**: 98–102. doi: 10.1093/infdis/141.1.98.

Health and Safety Executive (2013). *The Advisory Committee on Dangerous Pathogens Approved List of Biological Agents, 3rd edn*. London: HSE. http://www.hse.gov.uk/pubns/misc208.pdf

Houben ML *et al.* (2010). Disease severity and viral load are correlated in infants with primary respiratory syncytial virus infection in the community. *Journal of Medical Virology*, **82**: 1266–1271. doi: 10.1002/jmv.21771

Hymas WC *et al.* (2010). Development of a multiplex real-time RT-PCR assay for detection of influenza A, influenza B, RSV and typing of the 2009-H1N1 influenza virus. *Journal of Virological Methods*, **167**: 113–118. doi: 10.1016/j.jviromet.2010.03.020

Jokela P *et al.* (2010). Detection of human metapneumovirus and respiratory syncytial virus by duplex real-time RT-PCR assay in comparison with direct fluorescent assay. *Clinical Microbiology and Infection*, **16**: 1568–1573. doi: 10.1111/j.1469-0691.2010.03191.x

Kallen AJ *et al.* (2009). Staphylococcus aureus community-acquired pneumonia during the 2006 to 2007 influenza season. *Annals of Emergency Medicine*, **53**: 358–365. doi: 10.1016/j.annemergmed.2008.04.027

Kaluarachchi D *et al.* (2014). Comparison of urinary tract infection rates among 2- to 12-month-old febrile infants with RSV infections using 1999 and 2011. AAP diagnostic criteria. *Clinical Pediatrics*, **53**: 7420746. doi: 10.1177/0009922814529015

Kilian M (2012). *Streptococcus and enterococcus*. In: *Medical Microbiology: A Guide to Microbial Infections, 18th edn*. Greenwood, D, Barer, M, Slack, R and Irving W, eds. London: Churchill Livingston – Elsevier.

Kunst H *et al.* (2010). Parasitic infections of the lung: a guide for respiratory physicians. *Thorax*, **66(6)**: 528–536. doi: 10.1136/thx.2009.132217

Lee RS *et al.* (2015). Reemergence and amplification of tuberculosis in the Canadian Arctic. *Journal of Infectious Diseases*, **211(12)**: 1905–1914. doi: 10.1093/infdis/jiv011

Lienhardt, C *et al.* (2012). Global tuberculosis control: lesion learnt and future prospects. *Nature Reviews in Microbiology* **10**: 407–416. doi: 10.1038/nrmicro2797

Luk EY *et al.* (2012). Scarlet fever epidemic, Hong Kong, 2011. *Emerging Infectious Diseases*, **18**: 1658–1661. doi: 10.3201/eid1810.111900

McCullers JA (2006). Insights into the interaction between influenza virus and pneumococcus. *Clinical Microbiology Reviews*, **19**: 571–582. doi: 10.1128/CMR.00058-05

McShane H (2014). Understanding BCG is the key to improving it. *Clinical Infectious Diseases*, **58**: 481–482. doi: 10.1093/cid/cit793

McTaggart LR *et al.* (2012). Validation of the MycAssay *Pneumocystis* kit for detection of *Pneumocystis jirovecii* in bronchoalveolar lavage specimens by comparison to a laboratory standard of direct immunofluorescence microscopy, real-time PCR or conventional PCR. *Journal of Clinical Microbiology*, **50**: 1856–1859. doi: 10.1128/JCM.05880-11

Maini R *et al.* (2013). Increasing pneumocystis pneumonia, England, UK, 2000-2012. *Emerging Infectious Diseases*, **19**: 386–392. doi: 10.3201/eid1903.121151

Mangtani P *et al.* (2014). Protection by BCG vaccine against tuberculosis: A systematic review of randomized controlled trials. *Clinical Infectious Diseases*, **58**: 470–480 doi: 10.1093/cid/cit790

Maruyama T *et al.* (2013). A new strategy for healthcare-associated pneumonia: A 2-year prospective multicenter cohort study using risk factors for multidrug-resistant pathogens to select initial empiric therapy. *Clinical Infectious Diseases*, **57**: 1373–1383. doi: 10.1093/cid/cit571

Mehr S and Wood N (2012). *Streptococcus pneumonia*: A review of carriage, infection, serotype replacement and vaccination. *Paediatric Respiratory Reviews*, **13**: 258–264. doi: 10.1016/j.prrv.2011.12.001

Meng J *et al.* (2014). An overview of respiratory syncytial virus. *PLOS Pathogens*, **10**: e1004016. doi: 10.1371/journal.ppat.1004016

Miernyk K *et al.* (2011). Performance of a rapid antigen test (Binax Now ® RSV) for diagnosis of respiratory syncytial virus compared with real-time polymerase chain reaction in a pediatric population. *Journal of Clinical Virology*, **50**: 240–243. doi: 10.1016/j.jcv.2010.11.011

Mills JM *et al.* (2011). Rapid testing for respiratory syncytial virus in a paediatric emergency department: benefits for infection control and bed management. *Journal of Hospital Infection*, **77**: 248–251. doi: 10.1016/j.jhn.2010.11.019

Mims CA *et al.* (2008). *Mims' Pathogenesis of Infectious Disease*: London: Elsevier Academic Press.

Mitchell J (2011). *Streptococcus mitis*: Walking the line between commensalism and pathogenesis. *Molecular Oral Microbiology*, **26**: 89–98. doi: 10.1111/j.2041-1014.2010.00601.x

Mitchell AM and Mitchell TJ (2010). *Streptococcus pneumoniae*: Virulence factors and variation. *Clinical Microbiology and Infection*, **16**: 411–418. doi: 10.1111/j.1469-0691.2010.03183.x

Murray PR, Rosenthal KS and Pfaller MA (2013). *Medical Microbiology, 7th edn.* Philadelphia: Elsevier Saunders.

Musa D *et al.* (2013). Unusual case of a lung abscess. *BMJ Case Reports*. **Apr 16**. doi: 1136/bcr-2012-008306

Musher DM and Thorner AR (2014). Community-acquired pneumonia. *New England Journal of Medicine*, **2014(371)**: 1619–1628. doi: 10.1056/NEJMra1312885

Nair H *et al.* (2010). Global burden of acute lower respiratory infections due to respiratory syncytial virus in young children: A systematic review and meta-analysis. *The Lancet*, **375(9725):** 1545–1555. doi: 10.1016/S0140-6736(10) 60206-1.

Nair H *et al.* (2013). Global and regional burden of hospital admissions for severe acute lower respiratory infections in young children in 2010: A systematic analysis. *The Lancet,* **381(9875):** 1380–1390. doi: 10.1016/S0140-6736(12)61901-1

Ottenhoff THM (2012). New pathways of protective and pathological host defense to mycobacteria. *Trends in Microbiology* **20**: 419–428. doi: 10.1016/j.tim.2012.06.002

Pawlowski A *et al.* (2012). Tuberculosis and HIV co-infection. *PLoS Pathogens,* **8**: e1002464. doi: 10.1371/journal.ppat/1002464

Pettigrew MM *et al.* (2011). Viral-bacterial interactions and risk of acute otitis media complicating upper respiratory tract infection. *Journal of Medical Microbiology,* **49**: 3750–3755. doi: 10.1128/JCM.01186-11

Polverino E *et al.* (2013). Microbial aetiology of healthcare associated pneumonia in Spain: A prospective, multicentre, case–control study. *Thorax,* **68**: 1007–1014. doi: 10.1136/thoraxjnl-2013-203828

PHE Respiratory Data Figure (2016). https://www.gov.uk/government/uploads/system/uploads/attachment_data/file/523119/six_pathogens_-_Jan2004-Apr2016-colour.pdf

PHE TB Report (2014). Tuberculosis in the UK: 2014 report. London: Public Health England. https://www.gov.uk/government/uploads/system/uploads/attachment_data/file/360335/TB_Annual_report__4_0_300914.pdf

Phillips M *et al.* (2012). Point-of-care breath test for biomarkers of active pulmonary tuberculosis. *Tuberculosis,* **92**: 314–230. doi: 10.1016/j.tube.2012.04.002

Pillay D, Geretti AM and Weiss RA (2009) Human immunodeficiency viruses. **In:** *Principles and Practice of Clinical Virology, 6th edn.* Zuckerman AJ, Banatvala JE, Schoub BD, Griffiths PD and Mortimer P, eds. Chichester: Wiley-Blackwell

Reuter S *et al.* (2013). A pilot study of rapid whole genome sequencing for the investigation of a *Legionella* outbreak. *BMJ Open,* **3**: e002175. doi: 10.1136/bmjopen-2012-002175

Rodrigues LC and Lockwood DN (2011). Leprosy now: Epidemiology, progress, challenges, and research gaps. *The Lancet infectious diseases,* **11**: 464–470. doi: 10.1016/S1473-3099(11)70006-8

Saleeb PG *et al.* (2011). Identification of mycobacteria in solid-culture media by matrix-assisted laser desorption ionization–time of flight mass spectrometry. *Journal of Clinical Microbiology,* **49**: 1790–1794. doi: 10.1128/JCM.02135-10.

Schwander S and Dheda K (2011). Human lung immunity against *Mycobacterium tuberculosis:* Insights into pathogenesis and protection. *American Journal of Respiratory Critical Care Medicine,* **183**: 696–707 doi: 10.1164/rccm.201006-0963PP

Seoudi N *et al.* (2012). Rapid molecular detection of tuberculosis and rifampicin drug resistance: retrospective analysis of a national UK molecular service over the last decade. *Thorax,* **67**: 361–367. doi: 10.1136/thoraxjnl-2011-200610.

Slesak G *et al.* (2011). Ziehl-Neelson staining technique can diagnose paragonimiasis. *PLOS Neglected Tropical Diseases,* **5**: e1048. doi: 10.1371/journal.pntd.0001048

SMI B9: Public Health England. (2015). Investigation of Throat Related Specimens. UK Standards for Microbiology Investigations. B9 Issue 9. https://www.gov.uk/uk-standards-for-microbiology-investigations-smi-quality-and-consistency-in-clinical-laboratories

SMI B40: Public Health England. (2014). Investigation of Specimens for Mycobacterium species. UK Standards for Microbiology Investigations. B40 Issue 6.1. http://www.hpa.org.uk/SMI/pdf

SMI G8: Public Health England. (2014). Respiratory Viruses. UK Standards for Microbiology Investigations. G8 Issue 1.3. http://www.hpa.org.uk/SMI/pdf

SMI ID4: Public Health England. (2014). Identification of *Streptococcus* species, *Enterococcus* species and Morphologically Similar Organisms. UK Standards for Microbiology Investigations.ID4Issue3.https://www.gov.uk/uk-standards-for-microbiology-investigations-smi-quality-and-consistency-in-clinical-laboratories

SMI S2: Public Health England. (2014). Pneumonia. UK Standards for Microbiology Investigations. S2 Issue 1.2. http://www.hpa.org.uk/SMI/pdf

Smith S (2014). A new vaccine against tuberculosis. *Microbiology Today*, **41**: 120–123. http://www.microbiologysociety.org/publications/microbiology-today/past-issues.cfm/publication/mycobacteria

Stringer JR *et al*. (2002). A new name for *Pneumocystis* from humans and new perspectives on the host-pathogen relationship. *Emerging Infectious Diseases*, **8**: 891–896. doi: 10.3201/eid0809.020096

Torres A *et al*. (2013). Risk factors for community-acquired pneumonia in adults in Europe: A literature review. *Thorax*, **68**: 1057–1065. doi: 10.1136/thoraxjnl-2013-204282

Torres A *et al*. (2014). The aetiology and antibiotic management of community-acquired pneumonia in adults in Europe: A literature review. *European Journal of Clinical Microbiology & Infectious Diseases*, **33**: 1065–1079. doi: 10.1007/s10096-014-2067-1

Turner CE *et al*. (2016). Scarlet fever upsurge in England and molecular-genetic analysis in North-West London, 2014. *Emerging Infectious Diseases*, **22**: 1075. doi:10.3201/eid2206.151726

Waites KB and Talkington DF (2004). *Mycoplasma pneumoniae* and its role as a human pathogen. *Clinical Microbiology Reviews*, **17**: 697–728. doi: 10.1128/CMR.17.4.697-728.2004

Walker CLF *et al*. (2013) Global burden of childhood pneumonia and diarrhoea. *The Lancet*, **381(9875):** 1405–1416. doi: 10.1016/S0140-6736(13)60222-6.

Walker, M J *et al*. (2014). Disease manifestations and pathogenic mechanisms of group A Streptococcus. *Clinical Microbiology Reviews*, **27**: 264–301. doi: 10.1128/CMR.00101-13

Waller JL *et al*. (2014). Detection and Characterization of *Mycoplasma pneumoniae* during an outbreak of respiratory illness at a university. *Journal of Clinical Microbiology*, **52**: 849–853. doi: 10.1128/JCM.02810-13

Werno AM *et al*. (2012). Differentiation of *Streptococcus pneumoniae* from nonpneumococcal streptococci of the *Streptococcus mitis* group by matrix-assisted laser desorption ionization–time of flight mass spectrometry. *Journal of Clinical Microbiology*, **50**: 2863–2867. doi: 10.1128/JCM.00508-12

West A (2011). A brief review of *Chlamydophila psittaci* in birds and humans. *Journal of Exotic Pet Medicine*, **20**: 18–20. doi: 10.1053/j.jepm.2010.11.006

Wilson ML (2013). Rapid diagnosis of Mycobacterium tuberculosis infection and drug susceptibility testing. *Archives of Pathology and Laboratory Medicine*, **137**: 812–819. doi: 10.5858/arpa.2011-0578-RA.

Woodhead M *et al.* (2011). Guidelines for the management of adult lower respiratory tract infections-Full version. *Clinical Microbiology and Infection*, **17(s6)**: E1–E59. doi: 10.1111/j.1469-0691.2011.03672.x

World Health Organisation (2013). Global Tuberculosis report 2013. http://apps.who.int/iris/bitstream/10665/91355/1/9789241564656_eng.pdf

Zbinden A, Köhler N and Bloemberg GV (2011). recA-based PCR assay for accurate differentiation of *Streptococcus pneumoniae* from other viridans streptococci. *Journal of clinical microbiology*, **49**: 523–527. doi: 10.1128/JCM.01450-10

4

Gastrointestinal Infections

4.1 Introduction

When considering gastrointestinal (GI) infections, it is most useful to apply a wide scope, encompassing the GI tract itself (oesophagus, stomach, small intestine, large intestine, rectum) and the associated organs (gallbladder, pancreas, liver). Pathogens can reach these organs in a variety of ways, which are sometimes interconnected. Hepatitis A and hepatitis E viruses are both acquired via the faecal-oral route and almost certainly replicate in gut epithelial cells (where progeny are excreted in faeces), before entering blood stream and travelling to the liver (Harrison *et al.*, 2009). Amoebic dysentery is caused by *Entamoeba histolytica* which is similarly transmitted via cysts in contaminated food and water; in some cases, the protozoan subsequently moves from the intestine to the liver setting up amoebic liver disease (Murray *et al.*, 2013). On the other hand, while pathogenic strains of *Escherichia coli* mainly cause gastroenteritis – either through direct damage to the gut epithelium or through the effects of toxins – complications can damage other organs, such as the kidneys (e.g., Enterohaemorrhagic *E. coli*); in addition, strains of *E. coli* which are part of the normal flora can become opportunistic pathogens, causing disease in other parts of the body, such as urinary tract infections (Murray *et al.*, 2013).

4.1.1 Normal Gut Flora

To understand the pathogenesis of GI tract infection, it is advantageous to think about the gut environment and the 'normal' microbiome. An individual's normal flora is distinctive (and possibly unique). Acquisition of microorganisms begins at birth and is determined by exposure through close (e.g., familial) contact with others, as well as environment and food (Clemente, 2012). In healthy adults, the composition is fairly stable, though it can be disrupted by drastic alterations to diet, infection, immunological imbalance (leading to opportunistic infection) or use of antibiotics (Murray *et al.*, 2013). Although most knowledge about the microflora of the intestine concerns bacteria, recent evidence suggests that peoples' normal flora also includes eukaryotes (fungi, protozoa) and viruses (the 'virome') and the technology is now available to investigate this in more detail (Clemente *et al.*, 2012). The innate immune system appears to have mechanisms to recognise normal ('commensal') flora, allowing them to colonise the gut tissue. These include toll-like receptors (TLRs), which recognise microbe-associated molecular patterns (MAMPs) in the constituents of bacterial cell walls; interaction between bacteria and TLRs is also thought to suppress the inflammatory response to these microorganisms, so that they are not destroyed (Clemente *et al.*, 2012). Research also suggests that commensal bacteria can influence the specific immune response to other invading bacteria – thus ensuring the protection of their ecological niche, while also helping the host to clear any potential pathogens (Clemente *et al.*, 2012; Leslie and Young, 2015).

Pathogens can overcome the equilibrium created between the normal gut flora and the host immune system, through a combination of virulence factors in the organism and alterations to the host immunity (Leslie and Young, 2015), as well as disruption to the environment resulting from the use of antimicrobials. An extreme example of the consequences of disturbance of this balance is afforded by *Clostridium difficile* infection (see below). Changes in patient management, such as stopping treatment with broad-spectrum antibiotics – thus allowing renewed growth of the normal flora – can resolve the symptoms in many cases. Nevertheless, patients with severe, chronic *C. difficile* may require interventions to replace their normal gut microbiome, which can include a faecal transplant (Ivarsson *et al.*, 2015). It is also worth bearing in mind that within this 'community' of gut organisms, the pathogenic species could influence the outcomes for each other. A documented example of this is the interaction between *Entamoeba histolytica*, the cause of amoebic dysentery and *Shigella dysenteriae*, the cause of bacilliary dysentery (Clemente *et al.*, 2012). During the trophozoite stage, the protozoan uses a cell surface lectin called galactose/N-acetyl D- galactosamine– inhibitable adherence lectin (Gal/GalNAc) to attach to gut epithelial cells. It also ingests bacteria and research suggests that if *S. dysenteriae* is inside the amoeba, it stimulates the production of additional Gal/GalNAc, thus enhancing attachment. The bacteria also promotes the production of cysteine peptidase enzymes within the trophozoite, which are released into the immediate environment and enhance the inflammatory response, which further facilitates the invasion of the host cells by the protozoan; the exact mechanism is not known, but it may be that the enzymes lead to increased concentration of particular pro-inflammatory cytokines (Gunn and Pitt, 2012).

4.1.2 Diarrhoea

Diarrhoea is a commonly experienced condition, characterised by increased frequency of bowel movements (SMI B30), which is caused by disruption to the gut physiology. The consistency of stools can be anything from unusually loose, to very runny or completely unformed. Diarrhoea can be acute or chronic and may be mild and self-limiting, or severe and life threatening. While it is often attributable to an infectious agent, there are other – potentially serious – causes, which must be considered in the clinical diagnosis. These include chemical poisoning, appendicitis, allergy, irritable bowel syndrome and even bowel cancer (SMI B30). In infectious gastroenteritis, along with the bowel symptoms, the patient may experience nausea and/or vomiting, with abdominal pain (ranging from slight through to unbearable) and they may also be pyrexial (SMI B30). Dehydration is a significant problem and this is often the main contributor to death when it occurs. Diarrhoea is characterised as either inflammatory or secretory. As Text Box 4.1 indicates, the nature of the physiological disturbance affects the quality

---

Text Box 4.1 Features of Inflammatory and Secretory Diarrhoea.*

Inflammatory diarrhoea

Reduced absorption across gut epithelial cells
Characterised by frequency of bowel movement
Large volumes of faeces not always produced
Pus and blood cells may be present in faeces
Examples of causes: invasive organisms (e.g., *Entamoeba histolytica*);
   cytotoxin-production by organism (e.g., *Clostridium difficile* toxin B)

Secretory diarrhoea

Electrolyte imbalance causes fluid leaks out of cells
Copious amounts of unformed, very loose or even watery stools produced
Pus and blood cells not present in faeces
Examples of causes: bacterial enterotoxins (e.g., *Vibrio cholerae* toxin; *Bacillus cereus*
   heat labile (food poisoning) enterotoxin)

*Information taken from (Humphries and Linscott, 2015; Murray *et al.*, 2013).

---

and consistency of the faecal sample; this in turn can give an indication of which pathogen(s) could be responsible, which can be useful to the laboratory scientist. In some cases the faeces produced by people infected with certain organisms have very characteristic features. An example is stool samples from people infected with *Giardia duodenalis*, which are usually soft, grey-coloured with the consistency and appearance of modelling clay and smell of hydrogen sulphide ('bad eggs').

  Diarrhoeal disease is a significant burden to healthcare throughout the world (Humphries and Linscott, 2015) and therefore investigation of faecal samples provides plenty of work for diagnostic laboratories. It has been estimated that up to 17 million people in the UK may experience infectious diarrhoea each year (Tam *et al.*, 2012). The majority of infections are acquired in the community; acute disease is most commonly caused by *Campylobacter* or Norovirus (Tam *et al.*, 2012). About 10% of deaths in children under the age of five globally are thought to be caused by diarrhoeal infection (Kotloff *et al.*, 2013), with the majority of fatalities being in sub-Saharan Africa and South Asia. A large multicentre survey involving over 22,500 children (with and without diarrhoeal symptoms) across seven countries (Kotloff *et al.*, 2013), found that the most common pathogens identified were *Cryptosporidium* spp., *Shigella* spp. Enterotoxigenic *Escherichia coli* (*E. coli*) and Enteropathogenic *E. coli*. They also reported that rotavirus infection was important. However, the study was conducted between 2007 and 2011 and since then, the introduction of the rotavirus vaccine has reduced the global incidence of infection with this virus considerably (e.g., Atchison *et al.*, 2016; WHO, 2016). Infectious causes of diarrhoea and vomiting can be bacterial, viral or parasitological, but the most common causes in particular patient groups tend to vary according to age, immune status and exposure. Thus, the procedures for laboratory investigations of the cause must reflect these variables (SMI S7).

Point to consider 4.1: Investigate microbiological causes of diarrhoeal disease and laboratory investigation (e.g., using SMI S7 or Humphries and Linscott, 2015). Construct a table to include symptoms (diarrhoea, blood and pus in faeces, vomiting, pyrexia, etc.) and list likely pathogens for each clinical description along with the optimal laboratory diagnostic methods for each.

4.1.3   Gastrointestinal Pathogens

Although GI tract infection is most obviously associated with bowel disease, it is important to remember that there are significant infections in other organs, including the stomach and liver. This chapter will consider selected examples of GI tract pathogens, with the aim of examining a range of microbiological, diagnostic, clinical and epidemiological points. The organisms covered are *Campylobacter* spp., *Clostridium difficile*, *Cryptosporidium* spp., norovirus, *Helicobacter pylori* and hepatitis viruses.

4.2   *Campylobacter*

Kingdom: Bacteria
Phylum: Proteobacteria
Class: Epsilonproteobacteria
Order: Campylobacterales
Family: Campylobacteraceae
Genus: *Campylobacter*
Species: *C. jejuni; C. coli*

4.2.1   Introduction

Although *Campylobacter* spp. were isolated from animals in the early twentieth century, it was not until suitable media and protocols for optimal culture conditions were developed in the 1970s, that it was recognised as human pathogen. By the late 1980s it was clear that this organism is a major cause of diarrhoeal disease in humans (Ketley and van Vliet, 2012). The bacterium is a Gram negative, curved or spiral shaped motile bacillus (Ketley and van Vliet, 2012) and over 30 species and sub-species of *Campylobacter* have now been recognised (Murray *et al.*, 2013; SMI ID23). A number of them are known to be human pathogens, with *C. jejuni* accounting for the majority of cases and *C. coli* increasingly implicated (Ketley and van Vliet, 2012). Infection with *Campylobacter* spp. is a zoonosis, with poultry being the main source (Murray *et al.*, 2013). The organism is transmitted via faeces of infected animals, so humans can acquire campylobacteriosis through contaminated food and water and direct contact with animals (Allos, 2001).

4.2.2   Pathogenesis and Clinical Symptoms

The incubation period for *Campylobacter* infection is a few days (up to about a week), during which time some people experience 'flu-like symptoms'. The main clinical features are diarrhoea, abdominal pain and pyrexia, along with nausea – although vomiting is not usual (Ketley and van Vliet, 2012). Frequency of bowel movement is noticeable, with loose watery stools and blood is often found in the faeces (Murray *et al.*, 2013). While in children and immunocompromised patients the disease can be very debilitating, in healthy adults, it is generally self-limiting, resolving within a week. Nevertheless, the abdominal pain can be severe and may be mistaken for colitis or acute appendicitis – sometimes leading to unnecessary surgery (Allos, 2001)! A number of complications are reported, including reactive arthritis, bacteraemia, extra-intestinal infection and Guillain-Barré syndrome (Murray *et al.*, 2013).

Point to consider 4.2: There is evidence to suggest that the risk of developing irritable bowel syndrome (IBS) is increased post *Campylobacter* gastroenteritis. What factors might influence this?

It is thought that quite a low infectious dose (500–800 bacteria) is required to set up an infection. The bacteria are susceptible to the effects of stomach acid, but somehow survive passage through the GI tract in sufficient quantities (Ketley and van Vliet, 2012). Pathogenic mechanisms are not well understood, partly due to the difficulty in culturing the organism both *in vitro* and in animal models. It is thought that invasion of gut epithelial cells is mediated by specific adhesins and several types of toxin have been implicated in the disruption to fluid balance which leads to the diarrhoea (Ketley and van Vliet, 2012; Murray *et al.*, 2013). The bacterium initially affects the jejunum and ileum, but organisms move along to the large intestine. Histology shows damage to the mucosa, associated with an inflammatory response and invasion of neutrophils, as well as ulceration (Ketley and van Vliet, 2012; Murray *et al.*, 2013). The clinical symptoms and histological findings indicate bacterial gastroenteritis, but are similar to those seen in salmonellosis, shigellosis and yersiniosis, which means that microbiological investigations are important.

The development of the complications of *Campylobacter* spp. infection appears to be related to both organism and host factors. *C. fetus* is more commonly associated with sequelae in the cardiovascular system (septicaemia, endocarditis) and in other sites – including encephalitis – than other species (Murray *et al.*, 2013). Guillain-Barré syndrome (GBS) is seen in some patients infected with *C. jejuni* and in this case, host factors are also likely to play a role in pathogenesis. Studies have suggested that a strong innate response to the lipo-oligosaccharides on the bacterial cell surface mediated by TLR-4 and involving increased production of CD30+ and CD40+ T-cells is necessary for the development of GBS in this case. It is thought that cross-reactivity with specific anti–*C. jejuni* antibodies can contribute to the damage to peripheral nerves (Huizinga *et al.*, 2015). This pathological mechanism may be caused by a specific response not only to certain *Campylobacter* species but also particular strains (Murray *et al.*, 2013). It should be noted that GBS is a potential complication of a wide range of infectious diseases, including Influenza, Lyme disease and Hepatitis A infection – for a review, see, for example, van den Berg *et al.* (2014).

4.2.3 Epidemiology

Most human cases of campylobacteriosis are sporadic and the majority are linked to eating inadequately cooked chicken or other poultry (Murray *et al.*, 2013). Food and water that has been in contact with infected chicken products (e.g., in a food preparation area) can also be a source of infection (Allos, 2001; Cody *et al.*, 2015). In the UK, there are around 70,000 laboratory confirmed cases each year, but this is thought to be an underestimate – possibly by as much as seven-fold (Cody *et al.*, 2015). Most infections in healthy adults result in an acute gastroenteritis, which is unpleasant and lasts for a few days, but does not require medical intervention. Therefore, the majority of cases are not reported and consequently faecal samples are not collected. Not only does this affect the diagnosis for the individual, it also limits epidemiological studies of the various species. Recorded *Campylobacter* infection shows a seasonal distribution in temperate climates (Allos, 2001; Cody *et al.*, 2015), whereas in tropical countries it is considered to be endemic, with a constant incidence (Allos, 2001; Kaakoush *et al.*, 2015). Where the infection is seasonal, cases are highest in the summer months and this could be attributed to a number of factors. Cody *et al.* (2015) investigated presence and distribution of *C. jejuni* strains amongst wild birds in one area of England and compared these with isolates from patients in three counties. They concluded that bacterial strains which originated in wild birds accounted for between 2% and 3.5% of human cases of campylobacteriosis and that infection rates peaked in August (Cody *et al.*, 2015). This suggests that outdoor activities could bring humans into contact with affected birds more readily at this time of year. It should be noted that July and August is also the time when cooking meat for human consumption on barbeques (i.e., possibly inadequately) is also the most common in the UK!

The reported global incidence of campylobacteriosis has been steadily increasing during the twenty-first century (Kaakoush *et al.*, 2015). In England and Wales the number of confirmed cases was 48,133 in 2002, but by 2012 this had risen to 65,032 (https://www.gov.uk/government/publications/campylobacter-cases-2000-to-2012).

A similar trend has been noted in other European countries. Schielke *et al.* (2014) reported an increased incidence in Germany from 67 per 100,000 population in 2001 to 80 per 100,000 in 2010. There are a number of possible explanations for this rise. Infection rates in chickens and other animals destined for human meat could have risen. There has certainly been a trend towards eating more poultry, as this is considered to be healthier than processed red meat. Reporting rates may have been enhanced by public health policy changes – for example, in Germany *Campylobacter* infection has been a notifiable disease since 2001 (Schielke *et al.*, 2014). Another factor could be improvements in diagnosis such as the introduction of molecular testing, which is more sensitive than conventional culture (see below). It seems likely that a combination of all these factors account for the epidemiological observations.

4.2.4 Laboratory Diagnosis

Campylobacter spp. can be grown from faecal samples, but they are fastidious organisms requiring specialised media, anaerobic or microaerobic conditions and incubation temperatures between 37 °C and 42 °C, for 40–48 hours (SMI ID23). Selective media is needed to enhance the growth of any *Campylobacter* present and suppress other faecal organisms. Modified charcoal cefoperazone deoxycholate agar (mCCDA) is commonly used (SMI ID23) and colonies appear as small, greyish/white, flat and moist in appearance. Cefoperazone is a β-lactam cephalosporin, so strains of *Escherichia coli* which produce extended spectrum β-lactamase (ESBL) are not inhibited by its presence in culture media. It has therefore been suggested that a different selection of antibacterial agents should be considered, especially when culturing environmental water or food samples (Smith *et al.*, 2015). Clearly, the choice needs to be made carefully, taking the sensitivities of *Campylobacter* spp. into account (SMI ID23). Similarly, while 42 °C incubation is optimal for *C. jejuni* and *C. coli* (and reduces the growth of other faecal bacteria), they can grow at 37 °C and there are other clinically relevant species that prefer the lower temperature (SMI ID23). *Camplyobacter* spp. can be isolated in specimens from other sites, such as blood cultures and CSF; in this case, 37 °C is an acceptable incubation temperature due to lack of potential competitor bacteria – and this again enhances the detection of other *Campylobacter* species, which are likely to be associated with more unusual extra-intestinal disease presentations (SMI ID23). Isolates can be examined by Gram stain. They are Gram negative, spiral-shaped or curved bacilli (often orientated in pairs, to resemble seagull wings). The cells do not easily take up safranin counter stain, so carbol (or basic) fuchsin is recommended as an alternative. Where this is not possible, the safranin should be left on the slide for 10 minutes instead of the usual two (SMI ID23).

Given the fastidious nature of the bacteria and the fact that it takes several days to grow, it is perhaps not surprising that immunoassays to detect the bacterial antigen have been developed (e.g., Bessède *et al.*, 2011; Regnath and Ignatius, 2014). Commercially available tests are designed to detect an antigen which is the same in *C. jejuni* and *C. coli* and enzyme immunoassay and immunochromatographic (POCT) assays are on the market. Prospective evaluation studies have found that both EIA and POCT methods perform reasonably well. Gómez-Camarasa *et al.* (2014) tested 300 faecal specimens and found *Campylobacter* in 37 of them by culture, POCT and/or PCR. The POCT test parameters showed a sensitivity of 89% and specificity of 99%. Similarly, when Bessède *et al.* (2011) compared two EIAs and one POCT with culture and PCR they found *Campylobacter* in 37 out of 242 stools samples, with sensitivities and specificities all above 90%; inaccurate results were most often false positives in all three assays. Interestingly, the addition of PCR to these studies highlighted the poor sensitivity of culture methods. Bessède *et al.* (2011) tested all samples with PCR while Gómez-Camarasa *et al.* (2014) only tested those where the results were discrepant, but both teams report finding that immunoassays were more sensitive than culture. Regnath and Ignatius (2014) were unable to use PCR in their investigations of an EIA and POCT so they calculated the positive and negative percentage agreement (PPA and NPA) instead of sensitivity and specificity. They still reached similar conclusions to the previous authors that the parameters of the immunological tests suggested that they could be useful in diagnosis of campylobacteriosis (Regnath and Ignatius, 2014).

An additional feature of some immunoassays is the use of a genus-specific antigen. Regnath and Ignatius (2014) used a POCT test which the manufacturers intended to detect only *C. jejuni* or *C. coli* and found that it also picked up another species, *C. lari*. This warrants further investigation and shows that with a suitable range of antibodies in the solid phase, these assays could be adapted to detect all pathogenic species.

Culture is still beneficial, because species differentiation is important (and this requires a live isolate), although it can be tricky in a routine diagnostic situation. While MALDI-TOF is an effective means of identification of and differentiation between all species of *Campylobacter*, isolates from mCCDA do not work well here and subculture onto another medium is recommended first (SMI ID23). Other available methods include PCR, detecting a number of targets which vary between species, including the *gyrA* gene (Bessède *et al.*, 2011) and the *cdt* gene (Lutful Kabir *et al.*, 2011). In the latter case, the cytolethal distending toxin *(cdt)* gene actually comprises three elements – *cdtA, cdtB* and *cdtC* – and diversity within and between them has been used to distinguish between species. Lutful Kabir *et al.* (2011) tested 325 patient faecal samples from which *Campylobacter* spp. had been isolated in culture. From these, 314 were identified as *C. jejuni* and 11 as *C. coli*, based on a combination of the results from the three *cdt A, B* and *C* gene sequences. In some strains, only one or two of the genes were detected, due to mutations, which illustrated the value of the multiple gene approach (Lutful Kabir *et al.*, 2011). This work was subsequently extended as PCR for the *cdtB* gene follow by restriction fragment length polymorphism (RFLP) of the PCR product to allow detection and differentiation of *C. jejuni, C. coli* and five other less commonly occurring *Campylobacter* spp. associated with human disease (Kamei *et al.*, 2014). Although useful diagnostically, these assays are under development and are mostly the domain of research or reference laboratories. Other methods are also available in specialist laboratories, including pulsed field gel electrophoresis (PFGE), multilocus sequence typing (MLST) and whole genome sequencing (WGS) (SMI ID23).

4.2.5 Treatment and Control

In most cases, patients with campylobacteriosis make a full recovery without medical intervention, although if the diarrhoea is severe and prolonged, management of fluid and electrolytes might be needed (Ketley and van Vliet, 2012). Where antibiotic therapy is required, a range of drugs including macrolides, tetracyclines and chloramphenicol are considered effective (Murray *et al.*, 2013). Fluoroquinolones, such as ciprofloxacin have been widely used, but the prevalence of resistant strains has been increasing since the 1990s (Allos, 2001). Erythromycin is the drug of choice, since it is relatively cheap, safe to give to children and during pregnancy and it causes minimal disruption to normal faecal flora. Also and importantly, resistance to erythromycin is not common amongst *Camplylobacter* spp. isolates (Allos, 2001), although it is reportedly found more often in *C. coli* than *C. jejuni* (e.g., Gaudreau *et al.*, 2014).

The high and increasing incidence of campylobacteriosis in people globally means that it is a public health issue of concern. However, the high prevalence of *Campylobacter* in chickens makes control a challenge. The highest standards of hygiene in poultry farms and meat processing sites are required, along with careful preparation and thorough cooking of chicken meals in the kitchen (Allos, 2001). Achieving both these strands of reduction in transmission essentially involve education to change people's behaviour. Millman *et al.* (2014) investigated the self-reported food preparation activities of 182 subjects who had laboratory confirmed campylobacteriosis in the preceding five years, compared with 185 controls who had not reported any food-associated gastrointestinal disease. They explored a number of variables through a longitudinal questionnaire-based study and culturing of organisms from swabs taken from surfaces in the kitchens of participants' homes. The only significant difference which they found between the two groups was that those who had experienced *Campylobacter* spp. infection were more likely to have said that they washed chicken before cooking and also pre-washed, bagged salads before eating (Millman *et al.*, 2014). Rinsing the meat could lead to spread of bacteria around the kitchen and re-washing the salad increases the risk of contamination from surfaces or utensils (Millman *et al.*, 2014). Washing the meat is contrary to current public health advice and further cleansing of salad is unnecessary. Both these activities suggest a lack of understanding of the rationale behind the advisory information and this highlights the importance for public health professionals of clear communication. MacRitchie *et al.* (2014) also reported that they had raised awareness of campylobacteriosis in a consumer survey conducted in North East Scotland. They proposed a range of possible control measures, including better farming practises, vaccination of chickens, use of chemical wash on the meat and irradiation of the final product. Of the 210 participants, about half (103) lived in a rural area, while the rest were from towns and cities. There was a variation in the self-reported awareness of *Campylobacter* infection at the beginning of the survey. Fewer than half of those questioned had heard of *Campylobacter* and not all of those who did know about the bacteria associated it with chicken. The study found that, regardless of the demographic variable group, very few respondents felt that they would be happy to buy and eat chicken meat which had been treated with chemicals or UV radiation (MacRitchie *et al.*, 2014). They expressed a clear preference to put the onus on the producers and to encourage farmers to use more hygienic methods. However, the high prevalence of the infection in chickens does mean that the consumer needs to take some responsibility for prevention of food poisoning with *Campylobacter* spp.

4.3   *Clostridium difficle*

Classification
Kingdom: Bacteria
Phylum: Firmicutes
Class: Clostridia
Order: Clostridiales
Family: Clostridiaceae
Genus: *Clostridium*
Species: *C. difficile, C. perfringens, C. tetani*

4.3.1   Introduction

Clostridium spp. are Gram positive, spore-forming bacilli. A number of species are known to cause disease in humans, including *C. perfringens, C. tetani, C. sordellii, C. novyi* and *C. difficile* (Riley, 2012). While the sites of infection and clinical presentations vary considerably, pathogenicity is associated with the production of toxins in all cases. *C. difficile* has emerged as a significant cause of healthcare-associated infection (HCAI) throughout the world since the 1970s (Murray *et al.*, 2013). Acquisition of the spores can lead to enteric disease in patients whose immune response is compromised and whose balance of normal flora has been disrupted through use repeated of antibiotics or other drug treatments (Riley, 2012). A particularly virulent new strain of *C. difficile* was recognised in the early 2000s in the United States and Europe, which has now spread to many other countries (Murray *et al.*, 2013). Molecular testing has allowed isolates to be classified into a number of distinct ribotypes which is useful for clinical and epidemiological purposes.

4.3.2   Pathogenesis and Clinical Symptoms

C. difficile is an opportunistic pathogen and infection occurs when the spores are able to develop inside the gut. The main risk factors for developing *C. difficile* infection are prolonged use of antibiotics, extended periods as a patient in hospital (or other healthcare environment) and pre-existing conditions causing depletion of the immune system or affecting the bowel (Sun and Hirota, 2015). A small proportion of people are natural carriers of the bacterium, while others acquire it through coming into contact with the spores (Murray *et al.*, 2013). Disease occurs when the organism is able to exploit a niche caused by reduction in the normal gut flora. The spores are able to germinate and the bacteria can grow and divide. They produce protease enzymes to enable them to penetrate the mucous layer lining the gut epithelium and move along the surface using their flagellae. Bacteria enter intestinal cells using interactions with specific cell surface markers and begin producing toxins. The opportunity for this 'colonisation' is presented by the absence of the usual level of protection afforded by the normal intestinal flora (Murray *et al.*, 2013) and it is usually associated with extended or repeated use of broad-spectrum antibiotics (SMI B10). Patients usually experience an inflammatory diarrhoea ('antibiotic-associated diarrhoea', AAD), which can be relatively mild and self-limiting but may be prolonged; in some cases serious sequelae develop, including toxic megacolon and pseudomembranous colitis, which can be fatal (Riley, 2012; Valiente *et al.*, 2014).

Pathogenicity is associated with the activity of the bacterial toxins (Sun and Hirota, 2015). Toxin A is an enterotoxin, while toxin B is a cytotoxin. Toxin A is known to promote the activation of polymorphonuclear leucocytes and cytokines, leading to inflammation and alteration of the electrolyte and fluid balance across intestinal epithelial cells – thus causing diarrhoea. Toxin B disrupts actin in the cytoskeleton, thus destroying cells (Murray *et al.*, 2013). This cellular damage means that patients' faeces often contains blood and pus. There is some debate about the roles and interactions between the two toxins (Sun and Hirota, 2015). Toxin B may enhance the pro-inflammatory response caused by toxin A, but animal studies suggest that it does not cause inflammation on its own (Sun and Hirota, 2015). It has been found that there are some *C. difficile* strains in which toxin A production cannot be detected, but which still cause disease (including pseudomembranous colitis). Interestingly, no strains which produce toxin A but not toxin B have been detected (Sun and Hirota, 2015). In more virulent strains, such as ribotype 027, a mutation in a regulatory gene means that production of these toxins is not controlled (Valiente *et al.*, 2014). In addition some of these strains produce another toxin called binary toxin; while the function of this protein is not fully understood, its detection is helpful in diagnosis and also prognosis (Valiente *et al.*, 2014; Murray *et al.*, 2013). While only around 10% of *C. difficile* isolated from patients exhibit production of this binary toxin, when present it is associated with a worse clinical outcome (Valiente *et al.*, 2014; Sun and Hirota, 2015).

4.3.3 Epidemiology

While *C. difficile* infection was a recognised nosocomial infection during the twentieth century, incidence increased sharply at the turn of the twenty-first century. Reported cases in United States rose five-fold from 2000 onwards and large outbreaks followed in other countries including the UK (Denève *et al.*, 2009). There was also a rise in serious infections and mortality. These clinical observations and analysis of bacterial isolates indicated that a new virulent, highly transmissable strain of *C. difficile* was in circulation. It was classified as ribotype RT027 (Cairns *et al.*, 2015). This new strain was shown to produce greater quantities of toxins A and B than other strains, along with the binary toxin; it also had a relatively high rate of spore production, contributing to the noted high transmission rate and a changed antibiotic resistance profile (Denève *et al.*, 2009). Incidence of *C. difficile*–associated disease peaked in Europe towards the end of the 2000s; after this, stricter surveillance and tighter control measures appear to have reduced the number of cases considerably. In England the reported incidence almost halved between the 2009–2010 and the 2014–2015 reporting periods from 25,604 confirmed cases to 14,165 (https://www.gov.uk/government/statistics/clostridium-difficile-infection-annual-data).

There has also been a change in the prevalent ribotypes isolated. The proportion of RT027 bacteria in the UK has declined, while RT017 and RT078 have become more prominent (CDRN, 2014). Interestingly, this shift has not been seen in North America (Cairns *et al.*, 2015) and by 2012–2013 RT027 appeared to have become more common in Germany and Eastern Europe than previously (Davies *et al.*, 2014). While heightened awareness and improved control measures are likely to be reducing incidence, it has been suggested that efforts are being hampered by lack of consistency in laboratory diagnostic algorithms and use of sub-optimal test methods (Davies *et al.*, 2014).

Point to consider 4.3: What are the implications of the documented variability in predominance of *C. difficile* ribotypes at different times and in different geographical areas?

4.3.4   Laboratory Diagnosis

The gold standard for detection of toxigenic *C. difficile* remains inoculation of an aliquot of the sample (faeces or isolate from agar plate culture) into monolayer culture of VERO cells and examination for characteristic cytopathic effect (CPE) after 24 hours. The specificity of the CPE is confirmed by pre-incubation of a second aliquot with *C. sordellii* antitoxin (to neutralise any *C. difficile* toxin(s) present) before culture (Swindells *et al.*, 2010). This is not available in most routine laboratories and diagnosis often relies on a combination of immunoassays to detect bacterial glutamate dehydrogenase (GDH) and toxins A and B. Standard enzyme immunoassays and immunochromatographic tests are available to detect these. The GDH is considered a reliable indicator of the presence of a toxigenic *C. difficile*. Evaluations of both the EIA and IHC formats indicate good specificity (Swindells *et al.*, 2010; Beaulieu *et al.*, 2014), meaning that a negative result is reliable. Thus a GDH assay is a good initial screening test. If GDH is detected in a sample, then the next step would be to examine for toxin A and/or B. Again there are EIA and POCT kits available for this. Given that pathogenesis is always associated with toxin B (see above), then some manufacturers have taken the logical step to combine GDH and toxin B into one POCT format. Others use a configuration which will detect both toxins, such as the one shown in Figure 4.1. A coloured line appearing on the left-hand side indicates that the GDH was detected and one on the right hand side signifies the presence of toxin A and/or B in the sample. All available kits provide a fairly quick result; the test shown in Figure 4.1 is designed to an output in about 30 minutes. There are sensitivity issues with all formats of the toxin assay (Swindells *et al.*, 2010; Beaulieu *et al.*, 2014) and it is generally accepted that any sample in which GDH is detected but toxin B is not must be further investigated with PCR for

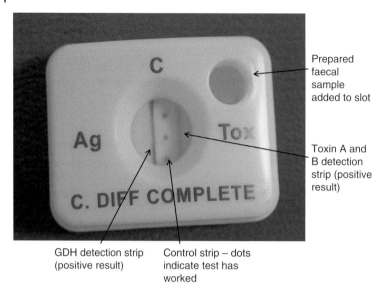

Figure 4.1 Point-of-care test card intended to detect *Clostridium difficile* glutamate dehydrogenase (Ag) and toxins A+ B (Tox). (*See insert for colour representation of the figure.*)

the toxin B gene (*tcd*B). If the PCR does not detect the gene, then it indicates either a false-positive GDH result or that the person is carrying *C. difficile*, but it is not a toxigenic strain – that is, the *tcd*B has been deleted (Swindells *et al.*, 2010). It is important to note that caution should be observed in this situation, as the result may be genuine. There are some reports of *C. difficile* isolates from symptomatic patients which were found to have neither the toxin A nor the toxin B gene, but did have the binary toxin gene (Eckert *et al.*, 2015). These strains of the bacterium seem to be unusual, but this illustrates the value of questioning results and consulting the literature when evaluating unexpected laboratory findings. Since *C. difficile* PCR kits are available commercially, laboratories considering the use of PCR as their routine first line test should be aware of this issue (Beaulieu *et al.*, 2014).

Ribotyping of isolates, using multilocus variable repeat analysis 'MLVA' ('enhanced DNA fingerprinting') is available at reference centres (SMI B10). This can be useful during investigations of outbreaks. It is also important to monitor *C. difficile* strains at regional and national levels given the variability and fluctuations which have been reported (CDRN, 2014; Davies *et al.*, 2014).

4.3.5   Management, Treatment and Control

When a patient has been diagnosed with *C. difficile* infection, their care and treatment must be reviewed. The use of broad-spectrum antibiotics should obviously be stopped. There is some evidence that drugs which are proton pump inhibitors (treatments for GI tract ulcers and acid reflux) exacerbate *C. difficile* symptoms, so discontinuing their use should also be considered (Wilcox, 2013). In cases where specific anti-microbial treatment is indicated, metronidazole or vancomycin are generally used, although some newer antibiotics, such as fidaxomicin are becoming available (Ivarsson *et al.*, 2015).

Mild to moderate disease would usually be treated with metronidazole. Vancomycin is reserved for severe cases – partly due to improved efficacy and also partly to restrict its use, to reduce the risk of drug resistance (Wilcox, 2013). Good supportive therapy (e.g., fluid and electrolyte balance, nutrition) is important and in some situations, bowel surgery is indicated.

Since the disease is associated with antibiotic use, a number of alternatives to this have been tried – with mixed results. Probiotics are an attractive option, with the idea being to augment the gut normal flora. The evidence required to assess efficacy is somewhat lacking, due to insufficient, robust, well-designed studies. A similar problem applies to the issue of whether to use prebiotics as a prophylactic measure (Wilcox, 2013). Bacteriophage therapy is a possible future option but at the time of writing work is at very early stages (Ivarsson *et al.*, 2015). Passive immunisation with pooled immuno-globulin does appear to have some benefit (Wilcox, 2013) and treatments using mono-clonal antibodies against toxins A and B have also been developed (Ivarsson *et al.*, 2015).

In patients who experience severe, recurrent *C. difficile*–associated disease it would be logical to consider replacement of the whole gut flora. To achieve this, a limited number of small trials of faecal transplants have been carried out. This involves taking faeces from a healthy donor, filtering (to remove potential pathogens) mixing with nor-mal saline (to make manipulation easier) and administering it to the patient via an enema, colonoscopy or a nasogastric tube! Evidence suggests that using the faecal preparation in combination with vancomycin can be effective, but not enough data has been collected yet to confirm this (Wilcox, 2013; Ivarsson *et al.*, 2015). As well as safety issues, including the quality of the donated faeces and processing pre-transplant, there are clearly some patient acceptability concerns. Nonetheless, if the patient is debilitated and distressed, they may well think it worth overcoming the natural revulsion to the idea, in order to improve their quality of life.

Highlighting the importance of infection control is an essential tool in reducing the spread of *C. difficile* in health and care settings. People who have contact with patients should always wash their hands thoroughly in soap and water, since spores are not eliminated by alcohol hand gel (Landelle *et al.*, 2014). The microbiology laboratory has a key role to play, since rapid and accurate identification of the causative organism ensures that the appropriate management and treatment are applied to the situation (Swindells *et al.*, 2010; Barbut *et al.*, 2014). A number of other organisms including *C. perfringens*, *Staphylococcus aureus* and *Candida albicans* are known to cause antibi-otic-associated diarrhoea (SMI B10). If the patient has a GI tract infection, but the organism is not *C. difficile*, then a different treatment and management strategy will be necessary; if symptoms are not attributable to an infectious cause at all, then infection control measures are not required. Providing guidelines for prescribers which restrict the use the antibiotics most commonly implicated in *C. difficile* disease (such as cepha-losporins and fluoroquinolones) can also be of value (Feazel *et al.*, 2014). Such 'antibi-otic stewardship' programmes can be planned locally or regionally and implemented with a multidisciplinary team (pharmacists and infection control team, as well as infec-tious diseases and gastroenterology clinicians). Surveillance is also an important tool in control. Schemes have been widely implemented in many countries since the mid-2000s (e.g., CDRN, 2014; Gase *et al.*, 2013) as part of a policy to reduce incidence of the infec-tion. These usually include mandatory reporting of confirmed cases – which therefore involves the laboratory.

4.4   *Cryptosporidium*

Phylum: Apicomplexa
Subclass: Coccidiasina
Family: Cryptosporidiidae
Genus: *Cryptosporidium*
Species *C. parvum, C. hominis*

4.4.1   Introduction

The *Cryptosporidium* genus contains at least 26 species and they are associated with enteric disease in humans and animals. It is a protozoan parasite, which is spread through oocysts; these have thick walls and are resistant to external environmental conditions (http://www.cdc.gov/dpdx/cryptosporidiosis/index.html). In contrast to the biology of other parasites in the Coccidiasina subclass, such as *Toxoplasma gondii*, the oocysts do not require further development outside of the host's body and are infectious straight away (Murray *et al.*, 2013). Transmission is usually via the faecal–oral route, most commonly from contaminated water (Chalmers and Katzer, 2013), although food-borne outbreaks do occur (Chalmers and Katzer, 2013). Interestingly, oocysts have been found in the sputum of immunocompromised patients (Gunn and Pitt, 2012). While *C. hominis* is, as the name implies, considered to be a human pathogen, *C. parvum* is predominantly found in cattle and therefore transmission to people is zoonotic (Chalmers and Katzer, 2013). Water systems often become infected through 'run-off' of faecal waste, for example, associated with flooding.

4.4.2   Pathogenesis and Clinical Symptoms

The incubation time for cryptosporidiosis is about a week, although this does appear to depend on the concentration of oocysts in the contaminated water or food on ingestion (Chalmers and Katzer, 2013). Most healthy adults will experience few symptoms, apart from an inconvenient watery diarrhoea, low-grade fever and slight abdominal pain, which resolves within a few weeks (Murray *et al.*, 2013). Some people may have nausea and vomiting and appetite may be reduced (Chalmers and Katzer, 2013). In young children and the elderly, infection with *Cryptosporidium* spp. can be more severe, mainly due to their immune systems being underdeveloped or waning respectively (SMI B31). Underlying malnourishment can also be a contributory factor (Chalmers and Katzer, 2013; Checkley *et al.*, 2015). In patients who are seriously immunocompromised, such as those with acquired immune deficiency syndrome (AIDS), diarrhoea tends to be severe and chronic and is accompanied by high frequency of stool production, which may continue for months or even years (Murray *et al.*, 2013). This often leads to malabsorption and occasionally other GI tract symptoms including hepatitis (SMI B31). The parasite can also spread to other areas of the body, notably the respiratory tract (Murray *et al.*, 2013).

Inside the host, the oocysts excyst, setting free four sporozoites. These attach to the microvilli of the gut through receptor-ligand interactions and release of digestive enzymes (Gunn and Pitt, 2012). This allows the parasite to start its complex replication cycle, inside a vacuole on the surface of the host's gut epithelium – for details, see Bouzid *et al.* (2013).

Although ordinarily, *Cryptosporidium* does not invade the tissue, there are nevertheless changes to the operation of the gut, causing malabsorption and over-secretion (Bouzid *et al.*, 2013). The mechanisms for this are not established, but possible reasons include specific damage to the epithelial cells and the inflammatory response elicited by the presence of the parasite (Bouzid *et al.*, 2013, Checkley *et al.*, 2015). Particular virulence factors have been postulated, but none have been definitively identified. This is at least in part due to the difficulty of growing the *Cryptosporidium* parasite *in vitro* (Bouzid *et al.*, 2013; Chalmers and Katzer, 2013), but research into pathogenicity is also complicated by the range of host and parasite species involved (Bouzid *et al.*, 2013). The link between the severity of disease and impairment of the host's T-cell response is well-documented, implying that cellular immunity is important in depressing parasite activities. Some studies have suggested that a protective IgG response may develop, particularly after repeated exposure, but that it may not be long lasting (Bouzid *et al.*, 2013).

4.4.3 Epidemiology

Cryptosporidium spp. are an important cause of outbreaks of diarrhoea throughout the world. Infection is usually associated with contaminated water, but the parasite can also be transmitted in food. It has been estimated that up to 25% of children with diarrhoea at any one time have cryptosporidiosis (Checkley *et al.*, 2015). Outbreaks and sporadic cases in humans are mainly caused by *C. parvum* and *C. hominis*, but *C. meleagridis*, *C. felis*, *C. canis* and *C. caniculus* have all been found in human faecal samples (Chalmers and Katzer, 2013). Apart from *C. hominis*, the reservoir hosts for *Cryptosporidium* spp are other animals (e.g., *C. parvum* is found in cattle, *C. felis* in cats and *C. meleagridis* in birds), making for a complicated picture of anthroponotic and zoonotic transmission patterns (Chalmers *et al.*, 2011a). Rigorous testing of potable water sources undoubtedly prevents outbreaks, but the methods employed involve filtration of large volumes of water through membranes and examining those for the presence of *Cryptosporidium* by microscopy (often using immunofluorescence – see below). Since oocysts are small (5–7 μm) and usually present in relatively low concentrations, this approach lacks sensitivity and it also provides no information about species – which might be very important when investigating an outbreak (Chalmers and Katzer, 2013).

There are between 3,000 and 6,000 laboratory confirmed cases annually in the UK; the reported number tends to vary according to whether there was a big outbreak in a particular year. This data is likely to be an underestimate of the total cases, since most healthy adults will not seek medical advice and would therefore only submit faecal samples for laboratory testing if included as part of an investigation of a known outbreak. Studies which analyse isolates suggest that the proportion of them which are *C. hominis* has been increasing during the twenty-first century (Chalmers *et al.*, 2011a). Outbreaks involving this species are more likely to occur in urban areas, such as through contamination of mains water or public swimming pools (Chalmers *et al.*, 2011a; Deshpande *et al.*, 2015). By contrast *C. parvum* is mostly implicated in cases from rural places in Wales, South West England and parts of North, West and East Scotland (Chalmers *et al.*, 2011a; Deshpande *et al.*, 2015).

Point to consider 4.4: Why do you think there is a difference in prevalence of *C. hominis* and *C. parvum* in urban and rural areas?

As well as large outbreaks associated with contaminated water, food sources can harbour oocysts and their spread is facilitated by their stability in the external environment. For instance, in 2012, there was an outbreak of cryptosporidiosis in Finland associated with consumption of frisée salad leaves (Åberg *et al.*, 2015). Customers who ate lunch in one of five restaurants across four cities during a two-week period in October 2012 were affected. Epidemiological investigations traced the source of infection to a supplier in the Netherlands (Åberg *et al.*, 2015). Transport through Europe and processing at each site did not appear to have significantly affected the viability of the oocysts, since over 250 people developed symptoms after eating the frisée salad. Faecal samples were obtained from only a small number of patients, but where *Crytosporidium* was isolated, genetic analysis showed that they were the same, rare subtype of *C. parvum* (Åberg *et al.*, 2015). Another potential route of infection is through direct contact with animals, including on farms or petting zoos (Chalmers *et al.*, 2011a; Deshpande *et al.*, 2015). Utsi *et al.* (2015) describe an outbreak amongst visitors to a farm where handling of the animals was encouraged. In this case, 46 people reported relevant clinical symptoms and 22 faecal specimens were obtained for genetic analysis. All these samples, plus faeces from a lamb, were found to have contained a single subtype of *C. parvum* (Utsi *et al.*, 2015).

4.4.4 Laboratory Diagnosis

Since the clinical picture is often non-specific, laboratory tests are important. An audit of *Cryptosporidium* diagnosis in microbiology laboratories in England and Wales conducted across 9 months starting in August 2013 (Chalmers *et al.*, 2015), found that routine testing for the parasite in samples from patients with diarrhoea was only done in about one third of laboratories. Other departments used protocols based on specific details to screen out samples, such as stool consistency, patient age and reported clinical details, resulting in only a small proportion being investigated for *Cryptosporidium* (Chalmers *et al.*, 2015). All such criteria may be flawed and accurate diagnosis for the individual is needed, regardless of whether anti-protozoal treatment is indicated. Also, this more constrained approach can hamper epidemiological studies, the tracking transmission of species or particular strains, as well as investigation of outbreaks because isolates need to be identified and, where possible, fully characterised. Surveys suggest that most diagnostic laboratories use light microscopy as their initial test (Chalmers *et al.* 2015; Manser *et al.*, 2013). This involves concentrating the faeces using a procedure such as the modified formol-ether method. A saline smear is prepared, which is then air-dried and fixed in methanol. Staining is most commonly carried out using the modified Ziehl-Neelson stain (mZ-N) or auramine phenol (AP). Oocysts are 5-7 μm in diameter and often present in low concentrations (due to being shed intermittently), so these bright stains are needed (Gunn and Pitt, 2012). Fluorescent-labelled anti-oocyst antibodies are also available (Murray *et al.*, 2013), again intended to enhance the sensitivity of microscopy, although at low concentrations of oocysts, labelled antibodies may not be more effective than mZ-N (Manser *et al.*, 2013). Enzyme immunoassays to detect cryptosporidial antigens are also used (Chalmers *et al.*, 2015) and these are reported to have better sensitivity than microscopy, particularly compared to the mZ-N method (SMI B31; Chalmers *et al.*, 2015). Commercially available IF and EIA kits usually combine testing for *Cryptosporodium* with *Giardia duodenalis* (Chalmers *et al.*,

2011b; Chalmers and Katzer, 2013), since these are the most common protozoal causes of diarrhoea.

The EIA format has been developed into immunochromatographic assays (rapid diagnostic tests, RDTs) and again they are often formatted to detect both *Cryptosporidium* and *Giardia* in one test card (Chalmers *et al.*, 2011b, Manser *et al.*, 2013). These assays all use antibodies raised against *Cryptosporodium* oocyst antigen, which is most likely to be derived from *C. parvum*. This is a potential concern given the prevalence of *C. hominis* and also the need to detect the less common species of *Cryptosporidium*. Evaluations to date suggest that assays based on antigen detection have good sensitivity when challenged with *C. hominis* (e.g., Van den Bossche *et al.*, 2015), but other species may be missed (Chalmers and Katzer, 2013). The RDTs are reported to have specificities of up to 100%, but sensitivities may be around 85% when compared to real-time PCR (Chalmers *et al.*, 2011b; Alexander *et al.*, 2013; Van den Bossche *et al.*, 2015). This may be related to the issue of species variation. Chalmers *et al.* (2011b) tested 259 samples, 152 of which were found to be 'true positives'; *Cryptosporidium* was detected in only 129 of these by RDT, but the species identification is not stated. Similarly, Alexander *et al.* (2013) assayed 200 faecal samples and found oocysts by AP microscopy in 105 of them. These results were confirmed by PCR, thus also allowing identification of the species, which revealed that while the majority of isolates (89) were *C. parvum*, there were 14 *C. hominis* and one each of *C. meleagridis* and *C. ubiquitum*. The RDT only detected the parasite in 92 of the samples, but analysis of the species found in the false-negative samples is not given. In contrast, Van den Bossche *et al.* (2015) compared four RDT kits by testing 160 samples, 15 of which contained *Cryptosporidium*, as confirmed by PCR. They reported 100% sensitivity for all four assays, but molecular results showed that isolates were all either *C. parvum* or *C. hominis*. This limitation should be borne in mind when using RDTs or EIAs for diagnosis, particularly if that is the only test being used. For maximum sensitivity and specificity (including species confirmation and strain identification), PCR of parasite DNA is necessary. The targets are the small subunit ribosomal RNA and the heat shock protein hsp70 genes; sequences from this combination allow species identification (Chalmers and Katzer, 2013). Ideally, PCR would be the diagnostic assay of choice for *Cryptosporidium*, due to the accuracy and detailed information that it provides (Chalmers *et al.*, 2015; Manser *et al.*, 2013).

4.4.5 Treatment and Control

Most cases of cryptosporidiosis are relatively mild and self-limiting. The individual should be advised to keep themselves hydrated and rest. Oocysts can be shed for several weeks, even after symptoms have cleared. The patient should be advised of this so that they can be careful when in contact with immunocompromised people (Murray *et al.*, 2013). Where anti-protozoal treatment is indicated, the current options are rather limited. A few anti-microbial drugs, including spiramycin and paromomycin have been shown to have some efficacy in children over the age of one year with chronic diarrhoea caused by *Cryptosporidium* spp., especially those who are malnourished (Sparks *et al.*, 2015; Checkley *et al.*, 2015). While the anti-protozoal nitazoxanide is also licenced in the United States for this purpose (Checkley *et al.*, 2015), it has not been made widely available in Europe. Spiramycin may be helpful in patients who are human immunodeficiency virus (HIV) positive, but it is not particularly effective in those with AIDS, in whom treatment is most needed. Clinical management is based on reviewing the HAART (see Chapter 2) medication to increase the CD4+ count and reduce HIV viral load (Sparks *et al.*, 2015), along with supportive therapy. Current research is focussed in developing specific anti-cryptosporidial drugs, which exploit the parasite's unique enzymes (Sparks *et al.*, 2015). Control during outbreaks can be difficult unless the source is known, but where public drinking water sources are affected, supply can be curtailed and/or people can be advised to boil water before use, which should denature the parasite.

4.5 Norovirus

Family: Caliciviridae
Genus: *Norovirus*
Species *Norwalk virus*

4.5.1 Introduction

The virus was first described after being identified as the cause of an outbreak of diarrhoea and vomiting in a school in Norwalk, Ohio in the late 1960s (Gray and Desselberger, 2009). Subsequently, the 27 nm Norwalk virus was commonly associated with episodes of gastroenteritis spreading through groups of people in close proximity. It is particularly found in institutional settings such as hospitals, care homes and educational establishments (Lopman *et al.*, 2012) and it is a recognised hazard of travel on cruise ships (Bert *et al.*, 2014). Norovirus is a positive sense single-stranded RNA virus and under electron microscopy, the surface glycoprotein conformation gives the appearance of cup-like structures; this explains the family name, since '*calix*' is the Latin for 'cup'. Although Norwalk virus is the type species, the genus designation Norovirus is now preferred over the species name. There is another genus of human pathogens in the *Caliciviridae* family, namely the sapoviruses, which appear similar microscopically and are also associated with gastrointestinal infection (Gray and Desselberger, 2009). Genetic analysis has shown that there are at least seven genogroups of norovirus (GI to GVII), which are further subdivided into genotypes. Noroviruses isolated from humans fall into genogroups I, II or IV; within each genotype there is considerable strain variation (Vinjé, 2015). Most water and food-borne outbreaks of norovirus are associated with GI genotypes particularly GI.1 and GI.6 (Vinjé, 2015; Barclay *et al.*, 2014), while most healthcare-associated infections are caused by strains of GII.4 (Vinjé, 2015; Morter *et al.*, 2011).

4.5.2 Pathogenesis and Clinical Symptoms

After ingestion of the virus, the incubation time can range from half a day to two days. The virus infects the villi of the small intestine and disrupts the usual enzyme activity; this leads to diarrhoea, combined with abdominal pain, nausea and often uncontrollable projectile vomiting. The outcome is malabsorption and sometimes dehydration, if the vomiting is severe (Gray and Desselberger, 2009). Immune response is characterised by serum antibody production, along with increased levels of IFN-γ and IL-2. Some studies have found that the specific antibodies are cross reactive against viruses within the same genogroup, but these do not necessarily have a protective effect (Gray and Desselberger, 2009). In healthy adults, symptoms generally last for one to two days and although unpleasant, the illness is usually self-limiting. By contrast, in very young, very old or immunocompromised patients, the infection can last longer and can be badly debilitating or even fatal (Gray and Desselberger, 2009). Excretion of the virus in faeces can continue for several weeks after the symptoms have subsided, with the length of time prolonged in those who have higher viral loads during the acute illness; genogroup GII viruses are associated with higher viral loads (Vinjé, 2015).

Although pre-existing anti-norovirus antibody does not appear to protect against infection, during an outbreak there will nevertheless be some people who do not seem to have been affected. Some of these may, of course, have very mild symptoms which they do not notice, but there is an interesting cohort of people who appear to have natural resistance to the invasion of their gut epithelial cells by the virus. The histo-blood group antigens (HBGAs) – ABO (H) and the Lewis antigen (Le) – are formed by the addition of specific carbohydrate molecules to cell membrane gly-colipids and the N- and O-linked terminal glycans of glycolipids (Rydell *et al.*, 2011; Gray and Desselberger, 2009). These HBGAs are found in saliva and on epithelial cells, as well as erythrocytes. Gut epithelium antigens contain a number of different oligosaccharide combinations, but they all contain an α 1, 2–linked fructose (Rydell *et al.*, 2011; Gray and Desselberger, 2009), which has been found to be important in forming a receptor for norovirus. The production of this is controlled by the FUT2 gene product (α 1,2 fucosyltransferase). About 20% of Caucasians have a specific mutation in this gene which means that they do not have these HBGAs on their cell membranes ('non-secretors') and therefore have some resistance to norovirus infec-tion (Rydell *et al.*, 2011; Gray and Desselberger, 2009), although this phenomenon may not afford total protection. It appears that variations in two other glycopro-teins – the products of the FUT3 and the ABO genes – also contribute to blocking viral binding (Ruvoën-Clouet *et al.*, 2013). There is some evidence to suggest that different strains of the virus have different affinities for combinations of variants of these three gene products (Ruvoën-Clouet *et al.*, 2013). A major drawback in research is lack of an *in vitro* culture system for the virus in which to explore the virus-antigen interactions. This means that most investigations rely on use of human volunteers, so are either time limited and involving small numbers or are epidemiological studies (Rydell *et al.*, 2011; Ruvoën-Clouet *et al.*, 2013). Nevertheless it is an intriguing observation, which may yield useful leads for treatment or vaccine development.

4.5.3   Epidemiology

Transmission of norovirus between human hosts is incredibly efficient and it is one of the most important causes of outbreaks of gastrointestinal disease worldwide (Lopman *et al.*, 2012; Barclay *et al.*, 2014). In temperate climates, notable outbreaks are seasonal, leading to the term 'winter vomiting disease' for the illness. Its infectivity is partly due to the large numbers of progeny virus (10^5 to 10^9 per gram of faeces/vomit) excreted during the acute phase of the disease (Barclay *et al.*, 2014) and the low infectious dose required, which is estimated at between 10 and 100 particles (Rondy *et al.*, 2011). The faecal-oral route of infection is effective by itself and can lead to contamination of food and water (Barclay *et al.*, 2014); in norovirus, transmission is enhanced by the projectile vomiting, which creates aerosols full of virus particles (Gray and Desselberger, 2009). The virus remains viable on fomites for several weeks and is difficult to remove from surfaces without strong disinfectant (hypochlorite at 1,000 parts per million is recommended) or steam cleaning (Barclay *et al.*, 2014). It has been shown to persist in water for more than two months (Barclay *et al.*, 2014). Due to this ease of spread and persistence, outbreaks of norovirus are very common (Lopman *et al.*, 2012; Barclay *et al.*, 2014), and can be difficult to control, particularly in enclosed environments. In the healthcare setting, the physical proximity and immunological vulnerability of people mean that outbreaks can be prolonged. Also, more than one strain of norovirus can be circulating in the community – or indeed a single hospital (e.g., Morter *et al.*, 2011)! Since antibodies do not provide much cross protection (Gray and Desselberger, 2009), it is theoretically possible to recover from the illness and then contract another through infection with a different strain, within a short time period during the outbreak 'season'. It is important to remember that 'diarrhoea' can be caused by a range of infections, as well as non-infectious agents, especially in hospital patients. Therefore laboratory diagnosis is very important in confirming norovirus as the cause of an outbreak of gastroenteritis and in identifying which individuals are affected, as well as monitoring strain variation.

In healthcare settings, outbreaks of norovirus can be 'suspected' or 'confirmed'; in either case, there is an obligation to report it in order to set up outbreak control measures (Norovirus Working Party, 2012). For individuals, norovirus is suspected when they report vomiting at least twice within 24 hours, or have two or more bouts of diarrhoea within 24 hours or a combination of both (assuming that these symptoms cannot be explained by other causes, such as medication). Where two or more people on the same ward or department have suspected norovirus infection, this constitutes a 'suspected outbreak'. Once this is supported by laboratory diagnosis of norovirus, the outbreak is then confirmed and in the UK this is logged onto a central database (Public Health England: (http://bioinformatics.phe.org.uk/noroOBK/); definition of an outbreak: http://bioinformatics.phe.org.uk/noroOBK/outbreak.html). An important part of controlling outbreaks in healthcare settings is 'cohort nursing'. This includes putting patients who are known to be infected in single rooms, the same bay or a specific area of the ward (Norovirus Working Party, 2012). Staff with gastrointestinal symptoms are also required to stay at home; meanwhile the ward, section or sometimes the whole hospital, is closed to visitors (Barclay *et al.*, 2014; Norovirus Working Party, 2012; Morter *et al.*, 2011). This shows why laboratory diagnosis is needed: while it is sensible to exclude staff with gastrointestinal symptoms regardless of the cause, a patient who has diarrhoea due to a different infection, should not be put in close proximity to those who have norovirus and thus be put at risk of a co-infection. The best efforts of restricting movement of people, careful hand washing, cleaning of walls, furniture and fomites and monitoring staff and patients for relevant symptoms can be confounded by the virus itself. Morter *et al.* (2011) monitored the norovirus isolates found in an acute hospital during one winter and noted that while they were all genotype GII.4, there were

56 distinct strains! At some points in the season, more than one strain was circulating within the hospital. The authors point out that this puts a significant onus on the infection control team to ensure that measures are both effective and implemented rapidly (Morter *et al.*, 2011).

The value of laboratory diagnosis is also highlighted in food-related outbreaks of norovirus. In this situation, the people affected are generally healthy (so may have few or no symptoms) and may have no direct, regular connection with the venue of infection or others involved in the outbreak. Also, when there are reports of gastrointestinal symptoms after eating out, the main suspect will usually be bacterial food poisoning. This is illustrated by the following case: In early 2009, there were 240 identified cases of food-related illness in seven weeks amongst people who had dined at one restaurant in England (Smith *et al.*, 2011). Investigations eventually led to the conclusion that the cause was norovirus. Although a limited number of faecal and food samples were available for testing, the virological picture was complicated. Viruses from genogroup I and several genotypes in group II were found in diners, staff and food (Smith *et al.*, 2011). It eventually became clear that while infection from shellfish (razor clams and oysters) had occurred, the outbreak was protracted through person to person contact; since there was a range of virus types, there was also probable introduction and spread of strains of virus from outside of the establishment. Some staff had not kept to the guidelines regarding not attending work while symptomatic, or for two days after symptoms subside; this could have contributed to either direct contact with customers if they were waiting staff or to continued food infection if they were kitchen staff. Assumptions were made about the probable cause of the 'food poisoning' which diners were reporting and this meant that the problem was not notified (to the Health Protection Agency) until well into the seventh week (Smith *et al.*, 2011). In addition, the enhanced hygiene and cleaning measures necessary to contain norovirus were not implemented until it was known that this virus was the causative agent – that is, on the basis of the laboratory identification.

Point to consider 4.5: Norovirus infection also known as 'winter vomiting disease' due to the increased incidence in the winter. What factors might be contributing to this?

4.5.4   Laboratory Diagnosis

The original mainstay of diagnosis was identification of the virus particles in faeces and vomit by electron microscopy (EM). The particles are approximately 27 nm in diameter and exhibit the distinctive morphology of the Caliciviridae family. As Figure 4.2 shows the cup-like appearance of the viral surface glycoproteins is visible after negative staining. However, there are some important drawbacks to EM in this context, including sensitivity (the limit of detection is 10^6 particles per mL), specificity (virus particles are small and there are a number of different viruses with similar glycoprotein structure), turnaround time for results (only one sample can be examined at a time), the requirement for highly skilled operators and the high cost of purchasing and maintaining an electron microscope (Jeffrey and Aarons, 2009). The virus cannot be cultured *in vitro*, but EIAs to detect the viral antigen in faecal samples were developed during the 1990s. These allowed processing of a large number of samples simultaneously. As well as the issues

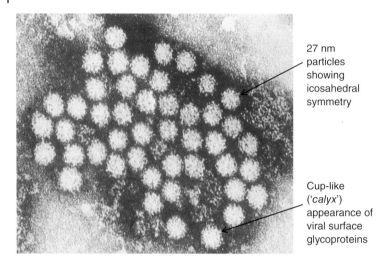

27 nm
particles
showing
icosahedral
symmetry

Cup-like
('*calyx*')
appearance of
viral surface
glycoproteins

Figure 4.2 Norovirus particles in a faecal sample detected by electron microscopy. Source: Courtesy of Mr I. Phillips.

with preparation of faeces and vomit to release viral antigen in a detectable form, there were problems caused by using antibody derived from a single norovirus genotype (e.g., Morillo *et al.*, 2011). The iterations of norovirus EIA kits available at the time of writing use combinations of antibodies, in order to maximise capture of genotypes and strains. While the specificity of these assays is generally good (over 90%), sensitivities can be as low as 70% (Vinjé, 2015), which restricts their usefulness.

Reverse transcriptase PCR and real-time quantitative PCR methods have largely super-seded EIA as the test procedure used in the main laboratory. The target is usually a highly conserved region across the ORF1-ORF2 junction of the RNA polymerase gene (Vinjé, 2015). There are a number of commercial kits available for detection of norovirus by PCR (Dunbar *et al.*, 2014; Vinjé, 2015) and they are generally reported to perform well. The increased sensitivity compared to other detection methods means that virus can be detected in faecal samples for up to three weeks after clinical symptoms have resolved (Vinjé, 2015). This does mean that care needs to be taken when interpreting results from individuals with low viral loads, but when investigating outbreaks, it can be useful in identifying people who might have been affected in early stages of the event. Since there are a large range of geno-types and norovirus is an RNA virus, it is important to be vigilant for false-negatives due to genome variation. Commercially available assays should have inbuilt controls to account for this (Vinje, 2015) and the PCR target is considered to be genetically stable.

Many diagnostic laboratories use multiplex PCR formats, to allow testing for a range of gastroenteritis viruses (e.g., van Maarseveen *et al.*, 2010). Assays which include probes for a range of viral, bacterial and protozoan causes of acute diarrhoeal disease have also been developed (e.g., McAuliffe *et al.*, 2013). Both of these are reported to perform as well as the individual PCRs and potentially reduce time taken to identify the pathogen in a particular patient (van Maarseveen *et al.* 2010, McAuliffe *et al.*, 2013). These have obvious advantages, including determining the exact cause of an outbreak of gastroenteritis quickly, rather than working through a list of possible causes one by one. Another benefit diagnostically is that people with relevant symptoms who happen to be in the vicinity of a norovirus outbreak, but are actually infected with a different organism can be identified (and managed appropriately). This needs to be balanced against the expense of running a multiplex system and the issue of interpretation of results where a patient has detectable levels of more than one gastrointestinal tract pathogen.

Most outbreaks of norovirus involving person-to-person transmission appear to be caused by strains of the GII.4 genotype (Gray and Desselberger, 2009; Barclay *et al.*, 2014). However, more than one can be circulating simultaneously within a particular setting (e.g., Morter *et al.*, 2011) and it can be important to distinguish between them for epidemiological reasons. Thus, sequencing of isolates is a key technique in norovirus monitoring and control, which highlights the contribution of laboratory scientists. An interesting example of the value of sequencing is provided by the account of an outbreak of norovirus in the Netherlands reported by Rondy *et al.* (2011). Residents and staff from a psychiatric care institution contracted diarrhoea and vomiting after returning from a pilgrimage to the Catholic site of Lourdes in Southern France. Some people later confirmed as 'cases' were symptomatic on the journey home, while others became ill more than three days after the return. Sequencing of virus isolated from faecal samples provided three key pieces of epidemiological information. Firstly that the people who were thought to have been infected while in France had a strain of norovirus GII.4 identical to one which was found in Lourdes at that time (Rondy *et al.*, 2011). This strain was also found in other Dutch people who had been part of the group, but who were not associated with the psychiatric unit and also pilgrims who had returned home to other countries (Rondy *et al.*, 2011). During the time of the outbreak at the psychiatric institution, norovirus infection also occurred in a nearby hospital. There was a possibility of contact with a traveller to Lourdes at this second institution, but it also raised the question of whether the infections contracted in the Netherlands were attributable to a strain circulating locally. The second finding from the sequencing was that the virus brought back from Lourdes subsequently spread around the psychiatric institution to residents and staff who had not travelled (Rondy *et al.*, 2011). Thirdly, it was shown that cases associated with the other hospital were carrying a different GII.4 strain of the virus. The laboratory results therefore allowed the infection control team to track the transmission of the virus through the institution and for the analysis to be clear – not distracted by the possibility of a second source of the infection.

While PCR is sensitive and specific, with protocols that allow for the result to be provided within one working day if required, a rapid point-of-care test (POCT) could be useful in circumstances where cohorting is indicated, such as on a hospital ward. There are a number of immunochromatographic kits available such as the one shown in Figure 4.3. The prepared faecal sample is loaded at one end and a result which is easy to interpret is usually provided in less than 30 minutes. As is visible in Figure 4.3, a coloured 'control' line appears if the test has worked and a second 'test' line is visible if viral antigen is detected in the sample. POCT tests for norovirus are generally reported to have very high specificities (up to 100%), but sensitivity can be below 50% in some settings (Bruggink *et al.*, 2015). This means that while positive results can be considered reliable, negative specimens will need to be referred to the main laboratory for PCR. It is important to find an assay which can detect genotype GI and GII noroviruses, because although most outbreaks in healthcare settings are GII.4 strains, it cannot be assumed that all of them will be. Bruggink *et al.* (2015) conducted an evaluation of a POCT kit which involved testing 100 stool samples known to contain norovirus,

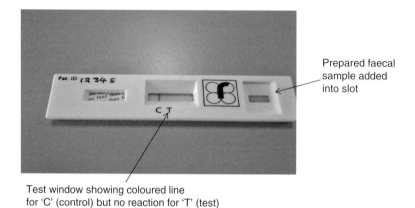

Prepared faecal sample added into slot

Test window showing coloured line for 'C' (control) but no reaction for 'T' (test)

Figure 4.3 Immunochromatographic assay for norovirus in a faecal sample (virus not detected). (*See insert for colour representation of the figure.*)

100 classified as not containing any faecal pathogens and 12 in which other viruses had been found. They found the test easy to perform and it gave a result within 20 minutes (Bruggink *et al.*, 2015). Using PCR as the 'gold standard', it was confirmed the samples contained genogroup I and II noroviruses and that there was no cross-reactivity with other faecal viruses. There were three false-positives, giving a specificity of 97%, but sensitivity was 87%; the false-negatives included GI and GII strains, including a GII.4 isolate (Bruggink *et al.*, 2015). Similarly, Bruins *et al.* (2010) evaluated a POCT kit by testing 537 samples taken from symptomatic patients, with real-time PCR as the gold standard and reported specificity of 99%, but a sensitivity of 57%. From the laboratory scientist's point of view, results of published evaluations of such tests need to be considered carefully. If specimens from all patients with negative POCT results will have to be checked by PCR (Bruins *et al.*, 2010), it may not make logistical or financial sense to use the immunochromatographic assay at all. It would therefore be essential to present data from the literature (and, if possible, a local evaluation) to the infection control team as part of any decision-making process.

4.6   *Helicobacter pylori*

Phylum Proteobacteria,
Class Epsilonproteobacteria,
Order Campylobacterales,
Family Helicobacteraceae,
Genus *Helicobacter*
Species *H. pylori, H. cinaedi, H. fennelliae*

4.6.1   Introduction

Some discoveries in microbiology are stories of serendipity, but *Helicobacter pylori* was found and linked with specific disease through systematic and determined science by two doctors in Australia. In the early 1980s, Marshall (a general practitioner) and

Warren (a pathologist) noticed that biopsy samples from patients with peptic ulcers and stomach cancers all contained the same unusual bacterium (Ahmed, 2005). The prevailing medical opinion at the time was that ulcers were caused by a stressful and unhealthy lifestyle, leading to build up of acid in the stomach and duodenum, which damaged the mucosal lining. It took Marshall and Warren some time to convince others that antibiotics were a better treatment for ulcers than antacids. Isolating the organism *in vitro* was not straightforward, as it is slow growing; because it does not grow in non-primates, the animal model used to prove Koch's postulates was Marshall himself (Ahmed, 2005)! He ingested a preparation containing *H. pylori* and developed a mild gastritis; subsequent studies on symptomatic patients from whose samples the bacterium was isolated indicated that the symptoms could be alleviated with antibiotics and bismuth sulphate (Kusters *et al.*, 2006). Warren and Marshall were awarded the Nobel Prize for their work in 2005 (Ahmed, 2005).

Point to consider 4.6: Marshall and Warren found it difficult to overturn the 'received wisdom' that ulcers were caused by lifestyle rather than an infectious agent. How did they use scientific methods to convince their colleagues that their theory had merit?

It is now known that *H. pylori* is a Gram negative bacillus related to *Campylobacter* spp., but in a separate family. Morphologically the organism is found as helical, curved or straight forms and it has several unipolar flagellae which affords it motility. It is estimated that a large proportion of the world's population is infected with *H. pylori*, rates varying from 40% to 50% in developed countries to 70% to 100% in developing countries (Murray *et al.*, 2013). The bacterium colonises the stomach mucosa and initial infection leads to acute gastritis and dyspepsia (Ketley and van Vliet, 2012). This can develop into chronic gastritis, which is usually asymptomatic, with the stomach tissue appearing morphologically normal on investigation (McColl, 2010; Murray *et al.*, 2013). A small number of people (less than 20% of those infected) experience long-term sequelae including stomach and duodenal ulcers, non-cardia gastric carcinoma and more rarely, B-cell mucosa–associated lymphoid tissue (MALT) lymphoma (McColl, 2010; Plummer *et al.*, 2015).

4.6.2 Pathogenesis and Clinical Symptoms

H. pylori colonises the gastric mucosa specifically (Ketley and van Vliet, 2012; Sheh and Fox, 2013). It survives the low pH in the stomach by producing a highly potent form of urease enzyme that converts the urea into NH_4 and HCO_3^{2-}. In this less acidic environment, the bacteria migrate to the epithelium and attach to host cells by specific adhesins (Sheh and Fox, 2013). Here they produce pathogenic proteins, particularly vacuolating cytotoxin A (Vac A) and the product of the cytotoxin-associated gene A (Gag A), which detrimentally alter the epithelial cells' function. The bacteria evade the host's non-specific immune response, through catalase and superoxide dismutase, to neutralise reactive oxygen and nitrogen species (Kusters *et al.*, 2006; Sheh and Fox, 2013). A pro-inflammatory T-cell response occurs, involving T_H cells, IL-2, IL-17, IL-22 and IFN-γ but there is T_{reg} activity which moderates the immune-mediated damage (Sheh and Fox, 2013). In addition, the bacterial lipopolysaccharide and flagellin are not very immunogenic

(Kusters *et al.*, 2006), which limits the extent and effectiveness of the B-cell function. These conditions are favourable to the *H. pylori* setting up chronic infection and research indicates that the interaction between the pathogen and the host immune system can worsen the tissue damage (Sheh and Fox, 2013).

Up to 10% of people chronically infected with *H. pylori* will develop duodenal or gastric ulcers (McColl, 2010). These appear to develop when inflammation occurs in the stomach antrum (which does not normally produce gastric acid) and stimulates the production of excess gastrin (McColl, 2010). This in turn induces the other areas of the organ to over-secrete acid, which damages the cells in both the stomach itself and the duodenum. *H. pylori* can invade these ulcerated areas, thus exacerbating the condition (McColl, 2010). The acid can also adversely affect areas higher up the GI tract, for example causing gastro-oesophageal reflux disease (GORD, also written as 'GERD') and Barrett's oesophagus. Gastric cancer develops in up to 3% of cases (McColl, 2010). It arises when the inflammatory response causes atrophy in the stomach mucosa, but metaplasia in the intestine (McColl, 2010). The virulence factor VacA is associated with the atrophy as it seems to induce apoptosis (Sibony and Jones, 2012); the proliferation of cells is thought to be mediated by Cag A, which seems to enhance conditions for upregulation of the cell cycle, while also downregulating the expression of p53, which is a tumour suppressor (Sibony and Jones, 2012). There is some evidence that the Cag A protein may protect against Barrett's oesophagus (e.g., Fischbach, *et al.*, 2012), although not reflux disease. Treatment of the *H. pylori* infection may relieve the symptoms of GORD/GERD (e.g., Saad *et al.*, 2012). The reasons and mechanisms for these observations require further investigation.

It has also been suggested that interaction between *H. pylori* and the commensal gut bacteria could affect the development of gastric cancer, although the nature of the pathogenesis and the optimal species make up to reduce risk is not at all clear at present (Sibony and Jones, 2012; Sheh and Fox, 2013). Interestingly, co-infection with parasitic helminths may afford some protection against tissue damage and atrophy, due to the alterations that the worms cause to the host immune response (e.g., Du *et al.*, 2006). Also there is some evidence that infection with *H. pylori* early in life – leading to a situation where host and bacterium are more tolerant of each other – can protect against asthma and inflammatory bowel disease (Sibony and Jones, 2012). Research has indicated an inverse relationship between *H. pylori* infection and coeliac disease (CD) (Lebwohl *et al.*, 2013). Over 100,000 patients were investigated by endoscopy over the course of four and a half years and nearly 2,700 of them were found to have CD. Amongst these 4.4% had *H. pylori* infection, compared to the 8.8% prevalence of the bacterium in those without CD. Allowing for the large discrepancy between the numbers of people in the two groups, this still suggests an interesting area for research. The authors speculate that the bacterium might alter host immune response to gluten (Lebwohl *et al.*, 2013).

4.6.3   Epidemiology

The route of transmission of *H. pylori* is still uncertain, but it is now known that it is a very common infection and is probably present in about 50% of the global population – with the geographical variations mentioned above. It has been estimated that about 6% of all recorded cancers are attributable to chronic *H. pylori* infection (Plummer *et al.*, 2015) making it a good example of an infectious agent which is recognised as a carcinogen (McColl, 2010). Factors affecting the likelihood of a chronically infected individual

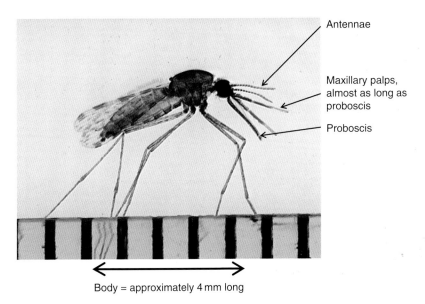

Antennae

Maxillary palps, almost as long as proboscis

Proboscis

Body = approximately 4 mm long

Figure 2.1 Female *Anopheles* spp. mosquito. Source: Courtesy of Dr A Gunn.

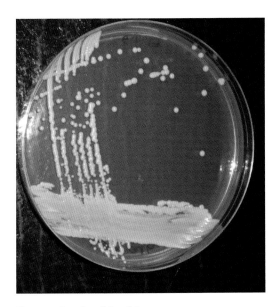

Figure 2.2a *Candida albicans* grown on Sabouraud dextrose agar.

Clinical Microbiology for Diagnostic Laboratory Scientists, First Edition. Sarah J. Pitt.
© 2018 John Wiley & Sons Ltd. Published 2018 by John Wiley & Sons Ltd.

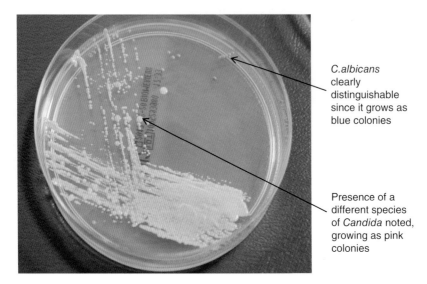

C.albicans clearly distinguishable since it grows as blue colonies

Presence of a different species of *Candida* noted, growing as pink colonies

Figure 2.2b Mixed growth of *Candida* spp. on chromogenic agar.

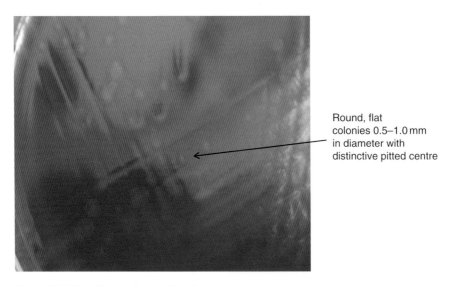

Round, flat colonies 0.5–1.0 mm in diameter with distinctive pitted centre

Figure 2.3 *Eikenella corrodens* on blood agar.

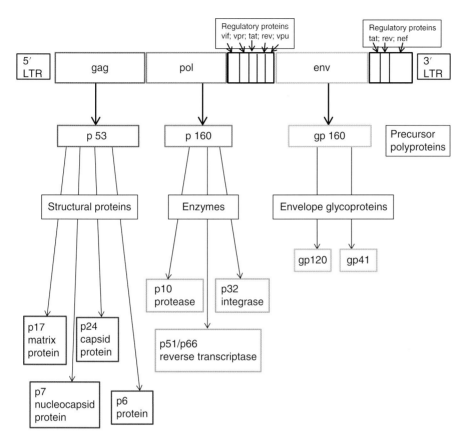

Figure 2.4 HIV-1 genome and gene products.

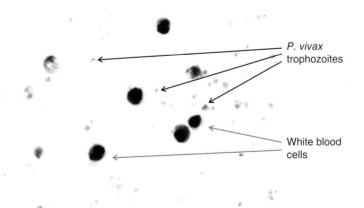

Figure 2.5a Thick blood film stained with Giemsa stain showing early trophozoites of *Plasmodium vivax*.

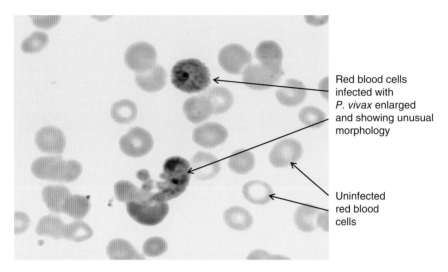

Red blood cells
infected with
P. vivax enlarged
and showing unusual
morphology

Uninfected
red blood
cells

Figure 2.5b Thin blood film stained with Giemsa stain showing late trophozoites of *Plasmodium vivax*.

Figure 2.6 Crow (*Corvus cornix*) in a rural habitat. Source: Courtesy of Dr A Gunn.

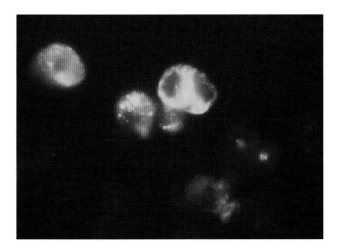

Figure 3.2 Slide prepared from respiratory sample stained with fluorescent-labelled anti-RSV antibody and viewed under UV microscope showing cells infected with the virus. Source: Courtesy of Mr I. Phillips.

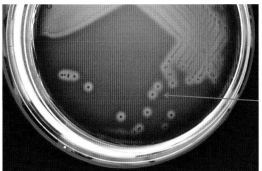

Clear zones of complete red blood cell lysis around bacterial colonies

Figure 3.3a *Streptococcus* spp. grown on blood agar exhibiting beta-haemolysis.

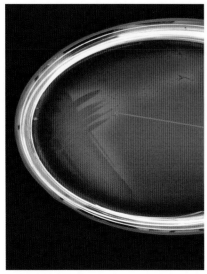

Greenish zones around areas of bacterial growth due to incomplete lysis of red blood cells

Figure 3.3b *Streptococcus* spp. grown on blood agar exhibiting alpha-haemolysis.

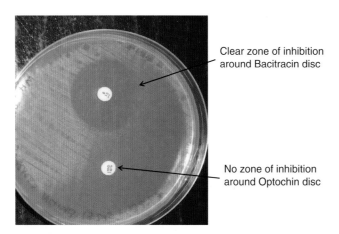

Clear zone of inhibition around Bacitracin disc

No zone of inhibition around Optochin disc

Figure 3.3c *Streptococcus pyogenes* grown on Mueller-Hinton agar and tested for susceptibility to bacitracin.

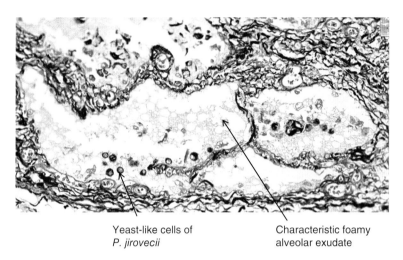

Yeast-like cells of
P. jirovecii

Characteristic foamy
alveolar exudate

Figure 3.4 Section of lung biopsy stained with Grocott's methenamine silver stain, showing pneumocystis. Source: Courtesy of Ms L. Dixon; photos and interpretive annotations by Dr A. Gal.

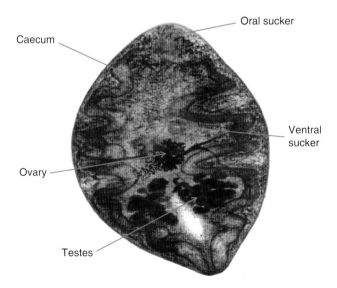

Oral sucker

Caecum

Ventral
sucker

Ovary

Testes

Figure 3.5a Mounted preparation of adult fluke of *Paragonimus* spp. Source: Courtesy of Dr A Gunn.

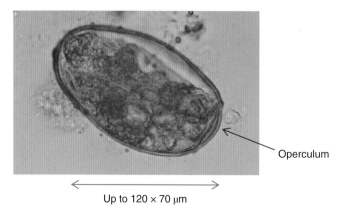

Operculum

Up to 120 × 70 μm

Figure 3.5b Egg of *Paragonimus* spp. Source: Courtesy of Dr A Gunn.

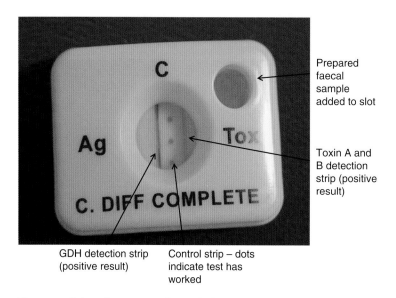

Prepared faecal sample added to slot

Toxin A and B detection strip (positive result)

GDH detection strip (positive result)

Control strip – dots indicate test has worked

Figure 4.1 Point-of-care test card intended to detect *Clostridium difficile* glutamate dehydrogenase (Ag) and toxins A+ B (Tox).

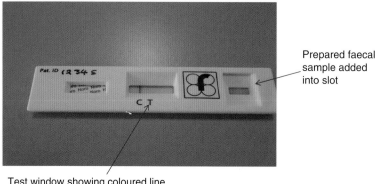

Prepared faecal sample added into slot

Test window showing coloured line for 'C' (control) but no reaction for 'T' (test)

Figure 4.3 Immunochromatographic assay for norovirus in a faecal sample (virus not detected).

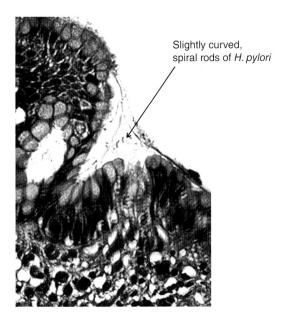

Slightly curved,
spiral rods of *H. pylori*

Figure 4.4a Section of stomach biopsy stained with
Giemsa stain showing *Helicobacter pylori* within gastric
mucosa. Source: Courtesy of Ms L Dixon; photo and
interpretive annotation Dr A Gal.

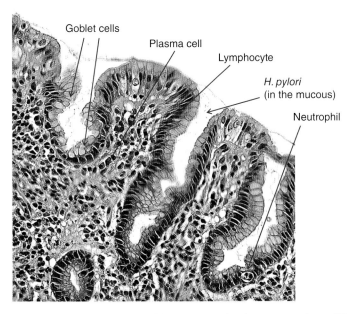

Goblet cells

Plasma cell

Lymphocyte

H. pylori
(in the mucous)

Neutrophil

Figure 4.4b Section of stomach biopsy stained with Haematoxylin and Eosin stain showing
Helicobacter pylori and signs of inflammation and metaplasia within the gastric tissue. Source:
Courtesy of Ms L Dixon; photo and interpretive annotation Dr A Gal.

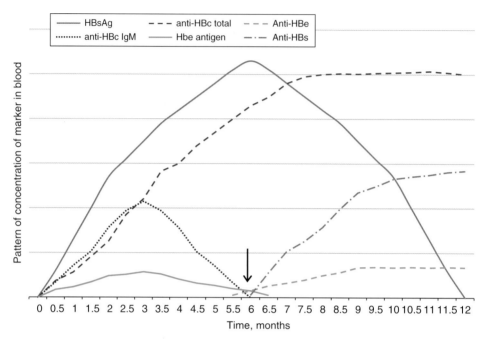

Figure 4.5a Patterns of serological markers during an acute HBV infection.

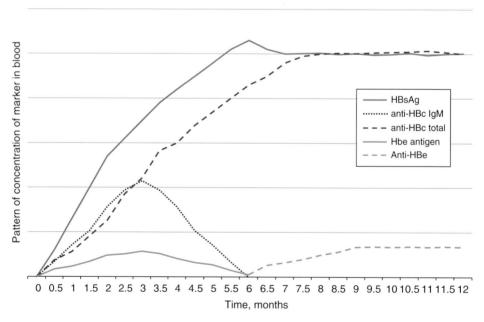

Figure 4.5b Patterns of serological markers during a chronic HBV infection where the patient is anti-'e' antigen-positive.

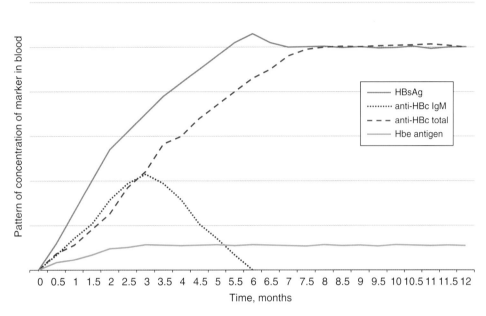

Figure 4.5c Patterns of serological markers during a chronic HBV infection where the patient is 'e' antigen-positive.

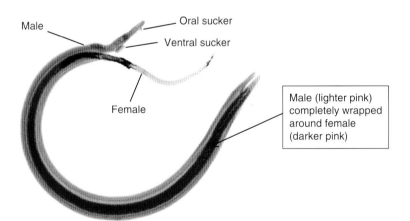

Figure 4.6 Stained and mounted preparation of a pair of adult *Schistosoma mansoni* helminths. Source: Courtesy of Dr A Gunn.

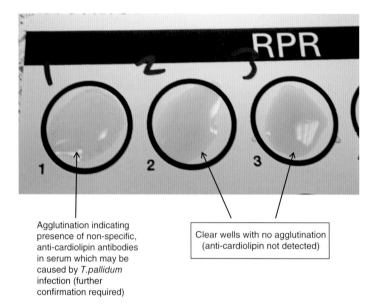

Agglutination indicating presence of non-specific, anti-cardiolipin antibodies in serum which may be caused by *T.pallidum* infection (further confirmation required)

Clear wells with no agglutination (anti-cardiolipin not detected)

Figure 5.2 Rapid plasma reagin (RPR) agglutination card for non-treponemal (cardiolipin) antibodies which may indicate *Treponema pallidum* infection.

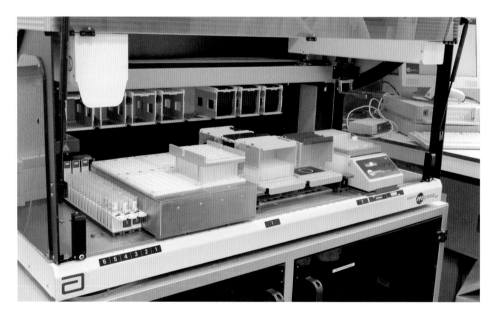

Figure 6.1 Automated laboratory system used to prepare samples for nucleic acid amplification analyses, including polymerase chain reaction.

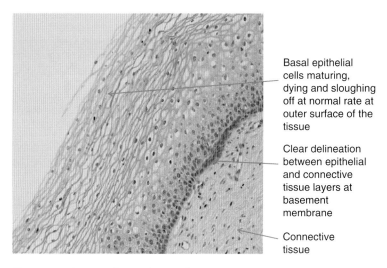

Basal epithelial cells maturing, dying and sloughing off at normal rate at outer surface of the tissue

Clear delineation between epithelial and connective tissue layers at basement membrane

Connective tissue

Figure 6.2a Section of normal cervical squamous epithelium stained with Papanicolaou stain. Source: Courtesy of Ms L. Dixon.

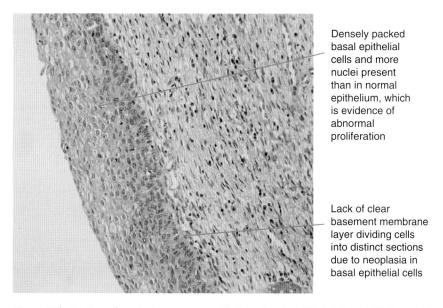

Densely packed basal epithelial cells and more nuclei present than in normal epithelium, which is evidence of abnormal proliferation

Lack of clear basement membrane layer dividing cells into distinct sections due to neoplasia in basal epithelial cells

Figure 6.2b Section of cervical squamous epithelium showing CIN-2 stained with Papanicolaou stain. Source: Courtesy of Ms L. Dixon.

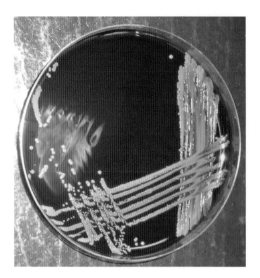

Figure 7.1a *Staphylococcus aureus* growing on blood agar as golden yellow colonies.

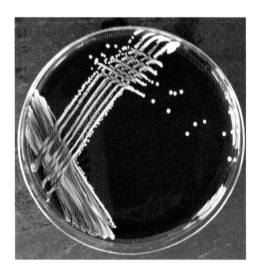

Figure 7.1b *Staphylococcus epidermidis* growing on blood agar as white colonies.

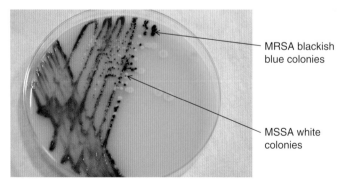

Figure 7.2 Growth of methicillin-resistant *Staphylococcus aureus* (MRSA) and methicillin-sensitive *Staphylococcus aureus* (MSSA) on chromogenic agar.

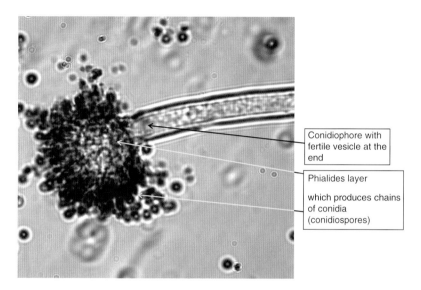

Figure 7.3a Conidiophore *Aspergillus niger* prepared in lactophenol cotton blue.

Figure 7.3b *Aspergillus niger* grown on Sabouraud dextrose agar.

Figure 7.4a Pigeons (*Columba livia*) in an urban setting in the United Kingdom. Source: Courtesy of Dr A Gunn.

Figure 7.4b Eucalyptus tree (Eucalyptus spp.) in a suburban setting in Australia.
Source: Courtesy of Mrs S. Weeks.

developing ulcers and cancer appear to include genetic variations in cytokine sequences (Kusters *et al.*, 2006; Sibony and Jones, 2012) and gender (males have double the risk) (Sheh and Fox, 2013). Geographical location determines the chance of being infected (see above) and the higher rates of cancer reported from East Asia, Eastern Europe and South America (Sheh and Fox, 2013) are probably a reflection of the prevalence of the infection. The picture is a little more complicated, because the incidence of gastric cancer is relatively low in Africa and South Asia (Suzuki *et al.*, 2012; Sheh and Fox, 2013). Analysis of isolates indicates that this might be due to bacterial strain variation, leading to differences in the *cag*A and *vac*A gene sequences which affect the virulence of the proteins (Suzuki *et al.*, 2012).

4.6.4   Laboratory Diagnosis

There are several different approaches to diagnosis and the choice between histology or microbiology depends largely on the clinical situation. Where an ulcer or cancer is suspected, then endoscopic investigation would be indicated and this would include collection of biopsy tissue. A preliminary result can be obtained with the rapid urease test, which is a POCT and can be carried out in the endoscopy clinic (SMI B55). A piece of biopsy tissue is exposed to a solution of urea, which also contains a coloured pH indicator. In the presence of urease, ammonia will be produced, increasing the pH. It may take up to 6 hours for the colour change to become detectable and a test would not be considered negative until 24 hours after it was set up. Thus it is not a particularly rapid test, but it is fast compared to histology or culture, which may take a week or more. A positive urease test result is not definite confirmation of *H. pylori* infection, since there may be other sources for the urease (including other bacterial species – although not many are likely to be found in gastric mucosa!) and it should be followed up with a laboratory-based test. It is important to note that drugs which a patient with an ulcer might be taking could interfere with the bacterial activity leading to false negative results. These include proton-pump inhibitors, H_2 receptor antagonists and antimicrobials (McColl, 2010) which should all be stopped before biopsy collection.

The biopsy material can be sent to the microbiology laboratory for culture or PCR if available, but currently the standard investigation would be routine histology. Bacteria should be detectable in sections stained with haematoxylin and eosin (H&E) or Giemsa stain (McColl, 2010; SMI B55) and this has the added advantage that the nature and extent of any damage to the gastric mucosa can also be assessed. As Figure 4.4 shows, *H. pylori* is usually found as slightly curved, spiral rods (Figure 4.4a). In this example, the bacteria can be seen within the mucous in the gastric glands but have not invaded the epithelial cells. The signs of inflammation are clear in the section stained with H&E (Figure 4.4b). There are neutrophils which are associated with acute inflammation, as well as plasma cells and lymphocytes which are markers of a chronic inflammatory response. In addition, signs of metaplasia are visible in the formation of the goblet cells (Figure 4.4b). The bacteria have been noted to alter their morphology from the curved/spiral form to a coccoidal appearance, particularly within damaged tissue or after antimicrobial therapy (Kusters *et al.*, 2006). It is uncertain whether these coccoids are viable (Kusters *et al.*, 2006), but their formation in a biopsy sample may affect the sensitivity and specificity of histological examination.

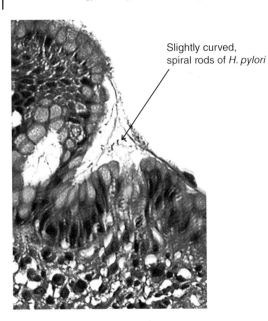

Slightly curved,
spiral rods of *H. pylori*

Figure 4.4a Section of stomach biopsy stained with Giemsa stain showing *Helicobacter pylori* within gastric mucosa. Source: Courtesy of Ms L Dixon; photos and interpretive annotation by Dr A Gal. (*See insert for colour representation of the figure.*)

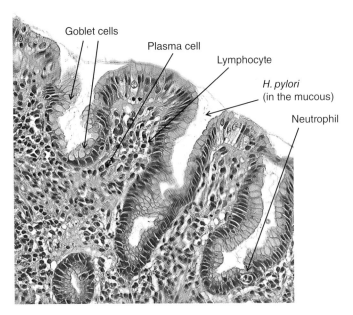

Goblet cells

Plasma cell

Lymphocyte

H. pylori (in the mucous)

Neutrophil

Figure 4.4b Section of stomach biopsy stained with Haematoxylin and Eosin stain showing *Helicobacter pylori* and signs of inflammation and metaplasia within the gastric tissue. Source: Courtesy of Ms L. Dixon; photos and interpretive annotations by Dr A. Gal. (*See insert for colour representation of the figure.*)

Microbiological culture of the biopsy material could be used to complement the histology, particularly when an isolate is required for antimicrobial susceptibility testing or to check there are no viable organisms after a course of treatment (SMI B55). It is important to ensure that biopsy tissue intended from the microbiology laboratory is put into appropriate transport medium (SMI B55) and not formalin! Depending on the clinical situation, it may be appropriate to attempt to isolate the organism from faecal samples and blood cultures (SMI ID26). A blood agar–based medium is recommended, with antifungals (and carefully selected antibiotics) added, as selective agents (SMI ID26). Media designed to isolate *Neisseria gonorrhoeae*, such GC agar, can also be useful, although some strains of *H. pylori* are sensitive to colistin and/or polymyxin. Thus it is appropriate to include a non-selective medium as well (SMI B55). Plates should be incubated at 37 °C under microaerophilic conditions for up to seven days. Colonies are usually 1–2 mm in diameter and can be identified by looking for the characteristic seagull morphology under Gram stain; helical forms are not usually obtained from plate cultures (SMI ID26). For best results, it is recommended that dilute carbol fuchsin is used as the counter stain (SMI B55), or acridine orange if the primary specimen was a blood culture (SMI ID26).

Where the patient has dyspepsia, but no symptoms suggesting an ulcer or cancer, then some less-invasive tests are available. The urea breath test (UBT) is a method for detecting urease which requires the patient to drink a solution containing ^{13}C- or ^{14}C-labelled urea. Any HCO_3 resulting from the enzyme activity will thus be labelled. This is subsequently converted into CO_2 in the lungs and then can be measured using mass spectrometry of a breath sample (McColl, 2010; SMI B55). This is currently the gold standard non-invasive test, but collecting samples requires skill and the relevant analysers are not universally available (SMI B55). There are a number of commercial kits available for detection of the '*H. pylori* stool antigen' (HpSA), which might be a suitable alternative to biopsy to test the effectiveness of course of treatment. There are both plate EIA and immunochromatographic lateral flow formats currently on the market and while the reported specificities tend to be reasonably good (around 90%), some have rather poor sensitivities. In a study, five kits were evaluated using stool samples from 198 patients undergoing endoscopic investigation for 'dyspeptic symptoms' (Korkmaz *et al.*, 2013). Amongst these, 90 were found to have *H. pylori* infection by histology and rapid urease test – which indicates the value of laboratory diagnosis. The best performing test was an EIA, which had a sensitivity of 92% and specificity of 94% (Korkmaz *et al.*, 2013). The three rapid immunochromatographic tests which were evaluated had sensitivities of 69%, 79% and 89%. These results suggest that at present, the HpSA detection test could be used in conjunction with others, but the limitations need to be considered.

Serological tests for IgG are available, but their diagnostic usefulness is limited, as they can only indicate chronic (or previous, but eradicated) *H. pylori* infection, which does not help in deciding the cause of clinical symptoms or whether a course of treatment has been effective. Also as discussed above, *H. pylori* antigens are not very immunogenic, which means that the sensitivity and specificity of most available assays are relatively low. Burucoa *et al.* (2013) conducted a systematic evaluation of a set of 29 commercially available IgG kits comprising 17 standard laboratory–based EIAs and 12 POCT rapid serological tests, comparing the results to histology, the biopsy urease test and the urea breath test. Testing sera taken from 45 patients with confirmed *H. pylori* infection and 47 in whom it had not been detected, they found that most of the EIA assays gave sensitivities of over 90%, with three at 100% – although one was only 58%; specificities were lower, ranging from 57% to 98% (Burucoa *et al.*, 2013). This had a consequent effect on the PPVs, one of which was just under 70%. The POCTs used a variety of formats (immunochromatographic, agglutination, EIA) and their performance characteristics were generally less satisfactory than those for the laboratory-based tests. For all of the assays, the 'antigen' used to detect the antibody was not disclosed, which also reduces the diagnostic value to some extent. One way to improve the test parameters might be to use multiple antigens. Formichella *et al.* (2013) reported the development and evaluation of an immunochromatographic assay involving purified CagA, VacA and four other proteins. They reported 98% sensitivity and 96% specificity compared with histology, which is an improvement on many of the existing commercial IgG assays. However, all the evidence suggests that serology is only useful as one of a range of tests for the infection.

4.6.5   Treatment and Management

About 80% of people with duodenal ulcers will be cured by antimicrobial treatment; the others are likely to have non-*H. pylori* ulcers due to prolonged use of NSAIDs (McColl, 2010). Treatment regimens are based on the combination of a proton pump inhibitor (PPI) with one or more antibiotics (Malfertheiner *et al.*, 2012). The most widely used formula is a PPI and clarithromycin plus either amoxicillin or metronidazole (Malfertheiner *et al.*, 2012). The efficacy of this has reportedly reduced in the twenty-first century, due to the development of clarithromycin-resistant strains of *H. pylori*. If the treatment does fail, then a 'bismuth-containing quadruple therapy' is recommended; this comprises bismuth salts, tetracycline, metronidazole and a PPI. In regions where clarithromycin resistance is known to be high, then the quadruple therapy should be the first-line treatment.

Eradication of the *H. pylori* infection may reverse the atrophy associated with gastric cancer, but evidence for this is limited (Malfertheiner *et al.*, 2012) and the effect is probably only seen in the early stages of the pathological process. Unfortunately, removal of the bacteria does not alter the course of the intestinal metaplasia (McColl, 2010; Malfertheiner *et al.*, 2012). Therefore, treatment and management of *H. pylori*–associated cancers involves appropriate oncology and palliative care.

Point to consider 4.7: Would it be cost-effective to screen asymptomatic people for *H. pylori* infection? If so should that be either areas of high prevalence, low prevalence or both? Which laboratory test would you recommend for a screening programme and why?

4.7   Liver Infections

The clinical condition of hepatitis is caused by the body's response to infection of the liver. While there are some specifically hepatotropic microorganisms, liver disease can also be a complication of other infections. Hepatitis is one of the recognised presenting features of severe Q fever (Murray *et al.*, 2013), a characteristic of yellow fever and can also occur as a late symptom of infection with some of the viruses in the *Herpesviridae* family, particularly CMV and EBV (SMI G5; Johannessen and Ogilvie, 2012).

In acute hepatitis, patients experience a range of symptoms from nothing at all, through non-specific, vague feelings of being unwell, to obvious hepatitis. Symptomatic infection usually starts with a "flu-like illness", which in the case of hepatitis B can resemble a 'serum sickness', with arthralgia and a characteristic rash (Harrison *et al.*,

2009; SMI G5). After a few days, the abdominal discomfort becomes considerable; urine may be darker than usual and faeces paler. There is an aversion to food and certain smells. Jaundice ('icterus') then becomes apparent (though not in all patients), with a yellow tinge visible on the skin and in the whites of the eyes (SMI G5). The symptoms are caused by a combination of viral activity and the immune response to it. Thus, while the clinical disease is unpleasant and full recovery takes several weeks, it is a sign that the body has reacted appropriately to the presence of the virus. In the case of hepatitis B and C, symptomatic acute infection makes progression to chronic infection less likely. Of course there can be an overreaction, leading to fulminant hepatitis, which is often fatal (Harrison *et al.*, 2009; SMI G5). Chronic hepatitis B and C infections arise because the virus is not cleared from the body after the acute phase. Viral replication continues and eventually there can be an inflammatory immune reaction, which leads to tissue damage (Harrison *et al.*, 2009; SMI G5).

Laboratory investigations often start with liver function tests (LFTs), particularly when there are no symptoms of patent hepatitis. A profile of raised serum bilirubin, alkaline phosphatase and gamma-glutamyltranspeptidase (indicating disruption of normal liver function), along with high alanine aminotransferase (ALT) and aspartate aminotransferase (AST), which are markers of liver damage (SMI G5) would be expected. In chronic infection, protein production by the liver is significantly reduced, leading to abnormally low serum albumin concentrations and a prolonged prothrombin time (SMI G5). Serology for virus specific antibodies and antigens would then be indicated. Where chronic hepatitis is suspected, histology of liver biopsies to determine the nature and extent of damage would be appropriate (SMI G5).

Point to consider 4.8: Investigate liver function tests. Can the biochemical profile of a set of 'abnormal liver function tests' be indicative of the probable cause of the pathology?

4.7.1 Hepatitis Viruses

The hepatitis viruses are so-called due to their tropism for the liver and the clinical condition they are associated with. Hepatitis A, hepatitis B and hepatitis D viruses were isolated and characterised in the 1970s. This facilitated laboratory investigation of viral hepatitis. It was nevertheless recognised that there were a significant number of patients with both acute and chronic illnesses which fitted the clinical and biochemical criteria for 'viral hepatitis', but who did not appear to be infected with any of these known viruses. Thus, the term 'non-A, non-B hepatitis' was used to categorise them. Thanks to more sophisticated molecular techniques, Hepatitis C and E viruses were discovered in the 1980s and they accounted for the majority of 'non-A non-B' infections. In the 1990s, hepatitis G was added to the list of hepatitis viruses which are human pathogens. They will be considered here in alphabetical order.

4.7.1.1 Hepatitis A Virus
Hepatitis A virus (HAV) is in the family *Picornaviridae* and is the type species in the genus *Hepatovirus*. The virus particles are small (27 nm), icosahedral and with a positive sense single stranded genome. There is only one serotype (Murray *et al.*, 2013) but six known genotypes (I-IV) of HAV (Harrison *et al.*, 2009)

The virus is spread via the faecal-oral route and hepatitis A is a common, though often asymptomatic, childhood infection in many parts of the world. Estimates suggest that in some countries up to 90% of adults have serological evidence of previous infection (Harrison *et al.*, 2009). In Western Europe and the United States, acquisition in childhood is now relatively rare, due to improved sanitary conditions, leaving adults vulnerable to infection. Surveys suggest that in this setting between 5% and 20% of young adults have antibodies to the virus (Harrison *et al.*, 2009). A study in the southwest of England found 45% of blood donors had anti-HAV IgG (Dalton *et al.*, 2008a), but the mean age for patients was 45 years and the result does not take into account vaccination history (see below). Since it is endemic in many countries, people travelling from non-endemic areas can acquire hepatitis A infection through contaminated food and water. A common route is through consumption of shellfish, particularly while on holiday (Wasley *et al.*, 2006; Harrison *et al.*, 2009).

Hepatitis A is an acute disease, with an incubation time of three to five weeks. While clinical signs are not common in children, at least 70% of adults infected with HAV will experience symptoms including jaundice (Harrison *et al.*, 2009) and these may last for several weeks. The disease can be debilitating but is usually self-limiting; rare complications include acute fulminating hepatitis and chronic autoimmune hepatitis (Harrison *et al.*, 2009). The pathogenesis of hepatitis A is not well understood. Although the virus is known to replicate in hepatocytes, it is believed that HAV-activated CD8+ T lymphocytes contribute to the liver damage (Harrison *et al.*, 2009).

The virus can be detected in stool samples during the incubation period, but once the patient becomes symptomatic, shedding appears to stop. Therefore the best method for diagnosing current or recent infection in the laboratory is an EIA assay for serum anti-HAV IgM, with the appropriate checks for non-specific reactions due to rheumatoid factor or other acute infections (SMI V27). It should be noted that IgM may be raised after recent vaccination, so this must be ascertained when taking the patient's history. There is some evidence that HAV might be detectable in saliva (Mahboobi *et al.*, 2012) and so collection of oral fluid can be considered as an alternative to peripheral blood when following up contacts of cases (https://www.gov.uk/government/publications/hepatitis-a-oral-fluid-testing-for-household-contacts). Samples should be taken to the laboratory and tested for the presence of HAV RNA by PCR. Past infection or successful vaccination would be confirmed by looking for serum anti-HAV IgG.

Incidence of hepatitis A in the UK is low. Dalton *et al.* (2008a) identified 20 cases out of over 4,000 patients tested over a five-year period in Devon and Cornwall. There were 289 laboratory confirmed cases across England and Wales during 2014 (https://www.gov.uk/government/publications/laboratory-reports-of-hepatitis-a-and-c-2014). A high proportion of cases (31%) were in the 15-year-old to 34-year-old age group, but the over 65 year olds also accounted for a significant amount (16%). Where information was collected, the majority (60%) of people reported a history of travel abroad – although possible sources of infection were not noted. Genotype analysis revealed that all isolates were either IA, IB or IIIA, with no particular pattern reflecting the presumed geographical location of acquisition (https://www.gov.uk/government/publications/laboratory-reports-of-hepatitis-a-and-c-2014). As it is spread faecal-orally, good hygienic practises should prevent hepatitis A infection, although the virus can be found in unexpected sources. For instance, there have been a number of outbreaks in Europe and elsewhere associated with consumption of imported sun-dried tomatoes (e.g., Carvalho *et al.*, 2012; Donnan *et al.*, 2012).

Specific anti-viral treatment is not usually needed, but vaccines are available for travellers to endemic areas. Both live attenuated and killed HAV vaccines have been developed (Harrison *et al.*, 2009), but those licensed in the UK are the latter type. Two doses are given, 6 to 18 months apart (SMI G5) and despite the inactivated formulation, protection is expected to last for at least 10 years (https://www.nathnac.org/pro/factsheets/hep_a.htm). Hepatitis A vaccine is now part of the routine childhood immunisation in some countries, such as the United States and parts of Spain and Italy (Wasley *et al.*, 2006) and this is reported to have reduced incidence considerably, but it seems unlikely that the UK policy of offering targeted vaccination will change.

Point to consider 4.9: would you advocate universal childhood vaccination against hepatitis A in the UK? Justify your answer.

4.7.1.2 Hepatitis B Virus

Hepatitis B virus (HBV) is classified in the *Hepadnaviridae* family and is the type species of the *Orthohepadnavirus* genus. It was discovered in 1965 by Blumberg, who was an immunologist investigating interactions between serum proteins from different individuals. He noted an interaction between serum from two patients, one from Australia and the other from the United States who seemed unlikely to have a genetic connection. It subsequently transpired that what had occurred was an antigen-antibody reaction between the virus particle in the Australian patient's serum and a specific antibody carried by the American (due to a previous HBV infection). This is an interesting example of how following up an unexpected observation can lead to a key scientific discovery. For more information, see Blumberg (2002). HBV is a relatively complicated virus, with an incompletely double-stranded DNA genome and thus a replication cycle involving RNA intermediates (Harrison *et al.*, 2009; SMI G5). The whole 42 nm virus (Dane particle), comprises a 27 nm core surrounded by surface antigen. Viral replication produces excess of this outer surface antigen (HBsAg) which is formed into separate 22 nm particles (Harrison *et al.*, 2009). Both of these forms are present in blood of an infected patient, along with various other markers, which are used as an aid to determining the stage of infection, infectivity of the individual and in some situations, the patient's prognosis. Figure 4.5a shows the pattern of antigens and antibodies detectable in a patient's serum during acute HBV infection and Figures 4.5b and 4.5c indicate the two possible outcomes of a chronic infection. In all scenarios, as well as the HBsAg, antibody to the HBV core antigen (HBc) will be detectable, along with the HBV 'e' antigen (HBe) or the anti-HBe antibody. Anti-core antibody tests measure total antibody or IgM. Assays tend to be set up to determine the levels of HBe and anti-HBe simultaneously, since they are not usually present together (see below). In the early stages of an HBV infection, the concentration of HBsAg rises, along with anti-core IgM and HBe (Figure 4.5a). Full resolution of an acute infection takes about six months and involves production of antibody against HBsAg (anti-HBS) and anti-HBe and elimination of the respective antigens. Although anti-HBc IgM wanes within three to six months, total anti-HBc levels rise (Figure 4.5a). The definition of chronic HBV infection is persistence of HBsAg for six months or longer (SMI G5). Anti-HBS antibody is not produced to detectable levels (Figures 4.5b and 4.5c), as the immune response to the infection is unable to clear the

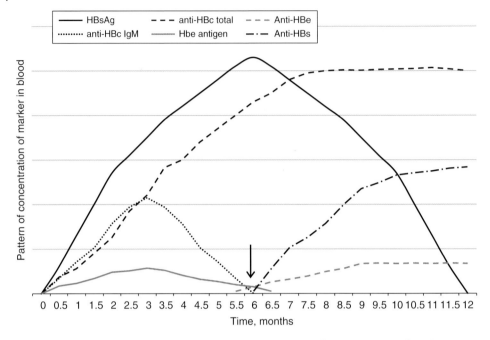

Figure 4.5a Patterns of serological markers during an acute HBV infection. (*See insert for colour representation of the figure.*)

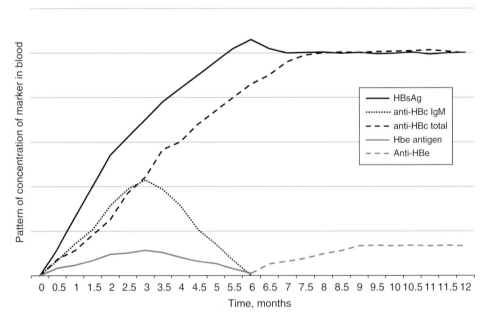

Figure 4.5b Patterns of serological markers during a chronic HBV infection where the patient is anti-'e' antigen-positive. (*See insert for colour representation of the figure.*)

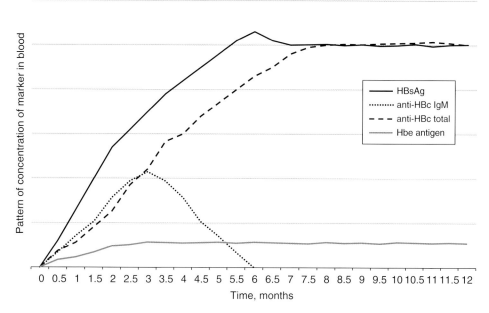

Figure 4.5c Patterns of serological markers during a chronic HBV infection where the patient is 'e' antigen-positive. (*See insert for colour representation of the figure.*)

virus. In some cases, HBe antigen levels decrease and anti-HBe is produced (Figure 4.5b), but in other patients HBe persists (Figure 4.5c). While all patients with chronic HBV infection are likely to experience long-term liver damage and pose an infection risk to others, the possibility for both is much higher when the patient remains HBe antigen positive (Harrison *et al.*, 2009).

There are thought to be at least 10 genotypes of HBV (A to J), based on variations in the overall viral gene sequence, but in particular within the S gene (which codes for the surface antigen). There are also four main serological subtypes ('serotypes'), classified according to the combination of antigens expressed on the HBsAg. The *a* antigen is common to all (and indeed anti-*a* antibody appears to be the protective element raised in response to vaccination). There are also two other sets of antigens, which are found in four possible combinations namely *adw, adr, ayw* and *ayr* (Kramvis *et al.*, 2005). It should be noted that the genotype and serotype are different formats used for grouping isolates, so serotypes are not subdivisions within each genotype. As an example, strains of *adw* serotype have been found in all 8 genotypes (Kravmis *et al.*, 2005). The four serotypes are now broken down further, as finer discrimination between the antigens has been recognised. This is exemplified by the finding that *awy*1, *awy*2, *awy*3 and *awy*4 can now be distinguished and they are considered to be discrete serotypes. Both the genotypes and serotypes vary considerably in their geographical distribution. As an illustration of this, genotype A isolates are reported from many parts of the world including Europe, Africa, Southeast Asia and North and South America, while genotype G is mostly found in France, Germany and the United States (Kravmis *et al.*, 2005). The serotype *adw* is most prevalent in Northern Europe, the United States and Australia and is also found in some parts of Southeast Asia (e.g., Indonesia and Thailand). The *ayr* subtype is mostly found in eastern Europe, Middle East, Central and South Asia and Africa. In Southeast Asia, *adr* is the main subtype. The *ayr* combination appears to be rare (Kramvis *et al.*, 2005; Harrison *et al.*, 2009).

The virus infects parenchymal liver cells and replication can take place for a month or more after primary infection before the patient experiences any symptoms (Murray *et al.*, 2013). There is an increase in Kupffer and endothelial cells in the affected area, along with thickening of the hepatic venules and accumulation of bile in the bile canaliculi (Tong, 2012). It is known that HBV core and HBV 'e' antigens elicit a T- and B-cell response (Tong, 2012; Harrison *et al.*, 2009) which results in a range of immunological outcomes, including lysis of infected hepatocytes through cytotoxic T-cell activity. Thus, the damage is thought to be mainly caused by the host's immune response rather than the virus itself. It does not appear to correlate with viral load (Harrison *et al.*, 2009), but a weak immune response is likely to result in chronic infection. Tissue necrosis usually occurs in foci. The repeated stimulus of the immune system by the presence of the virus leads to a destructive autoimmune response, which causes fibrosis and eventually cirrhosis and in some cases carcinoma (Tong, 2012; Harrison *et al.*, 2009).

Up to one third of people across world have a current or evidence of a past infection with HBV. The World Health Organisation estimates that 240 million people worldwide have chronic HBV infection and that over 780,000 people die each year from complications including cirrhosis and hepatocellular carcinoma (WHO, 2015). There are marked geographical variations in prevalence. It is < 2% in Western Europe and North America, rising to 2% to 8% in Southern and Eastern Europe, Middle East, Central Asia, Central and South America and is > 8% in Southeast Asia and sub-Saharan Africa (Ott *et al.*, 2012). Rates can be much higher than these averages in local areas and in specific populations. In the UK, the chronic carriage is reported at < 1% overall. Hepatitis B is a notifiable disease, which means that a 'case' requires laboratory confirmation (SMI G5). There are around 700 to 800 new cases of acute hepatitis B reported each year in England and Wales (PHE 4). HBV infection is most commonly found in people between 15 and 45, with peak incidence amongst those in their late twenties and early thirties (https://www.gov.uk/government/collections/hepatitis-b-guidance-data-and-analysis).

The initial test would be serology for HBsAg. If this is detected, then it must first be confirmed, through re-testing of an aliquot from the original blood sample. Where possible, this should involve the use of both the same assay and an alternative method (SMI V4). This finding can also be corroborated through testing for total antibody to hepatitis B core antigen (anti-HBc), which should be positive regardless of whether the person has acute or chronic infection (SMI V4; Figures 4.5a, 4.5b and 4.5c).

In a new diagnosis, it is important to work out the stage of infection, but this may not be possible without follow-up tests some months later, to see whether the infection is resolving or not. The HBe/ anti-HBe status may help, since as an acute infection resolves, serum levels of HBe fall and anti-HBe titre consequently rises. Occasionally the blood sample is collected at a time near the crossover point (arrow on Figure 4.5a) and both antigen and antibody can be detected, albeit at low concentrations. While this might initially be puzzling in terms of interpretation of the results, it is good news for the patient, because it indicates that the body has mounted an effective immune response to the virus. It should be noted that chronic carriers can be either HBe or anti-HBe positive (Figure 4.5b and 4.5c), so the presence of the anti-HBe does not preclude a diagnosis of chronic carriage. In all situations, when a person is diagnosed with hepatitis B, contact tracing is essential even when they are anti-HBe positive. It has been estimated that during active viral replication, there may be up to 10^8 virions mL^{-1} of blood (SMI G5). The virus has also been found in most other body fluids including saliva, vaginal and seminal fluid, breast milk and wound exudates (Harrison *et al.*, 2009). Once a patient has been confirmed as a chronic HBV carrier, they should be monitored regularly through measurement of HBV DNA viral load in blood, since serological assays are essentially qualitative (SMI G5).

It is worth bearing in mind that HBV biology is relatively complex and this can lead to apparently inexplicable sets of laboratory results, which need to be thought about carefully. There are some so-called 'escape mutant' genotypes which are hypothesised to have arisen as result of selection pressure by a number of factors, including the antiviral lamivudine (Clements *et al.*, 2010). The difference in the composition of their HBsAg proteins means that they are not detected by most standard EIA assays (e.g., Larralde *et al.*, 2013) and, perhaps more worryingly it has been reported that the anti-HBS antibody produced through vaccination is not protective against some of these variants (Clements *et al.*, 2010). Cases have also been reported where serum HBsAg falls to undetectable levels while other markers, including HBV DNA, are present and indicate active HBV infection. Again, sequencing has shown that the Hepatitis B viruses associated with these 'occult' infections have significant genetic mutations. The replication of the virus can cause liver damage and there is a particular risk that the effects of Hepatitis C Virus can be exacerbated in a dual infection with an occult HBV (Harrison *et al.*, 2009).

Acute symptomatic HBV infection should resolve, so specific antiviral therapy should not be required (Harrison *et al.*, 2009; Tong, 2012). If fulminant hepatitis develops, this is life-threatening. There is some, limited, evidence that lamivudine may be helpful in this situation, but it inhibits viral replication rather than the immune response (Harrison *et al.*, 2009). The only realistic options are supportive therapy followed by liver transplant (Harrison *et al.*, 2009; Tong, 2012). There are a number of effective drug therapies for chronic hepatitis B, but as these target viral replication, they are most effective in earlier stages of the course of the infection – before the damage due to fibrosis and cirrhosis is too extensive (Harrison *et al.*, 2009) The first option is usually IFN-α (or pegylated IFN- α which is considered more efficacious), which is administered for prolonged periods of 24 to 48 weeks (Tong, 2012). The nucleoside analogue lamivudine has been used extensively for a number of years, but resistant strains of HBV have emerged. Other nucleoside (e.g., adefovir) and nucleotide analogues (e.g., tenofovir) are available and may be used in combination with each other or with pegylated IFN-α (Harrison *et al.*, 2009; Tong, 2012; Murray *et al.* 2013). The regime is adjusted according to the circumstances, but the patient may be on treatment for years and only a small proportion actually clear the virus (Tong, 2012).

Control of HBV infection is an important focus of public health policy. A safe, effective vaccine has been available since the early 1990s (Harrison *et al.*, 2009). It is administered as three doses over six months. Since the younger a person is at the age of acquisition of the virus, the more likely they are to become and chronic carrier, a strategy of screening pregnant women to prevent perinatal infection is widely adopted. In the UK, if the mother is a carrier with the HBsAg, HBe profile, then the neonate is given immunoglobulin and started on a course of vaccine at birth; if the mother's serological results show HBsAg, anti-HBe, then the baby is given the vaccine only, due to the lower transmission risk (SMI G5). Many countries have incorporated the HBV vaccine into their universal childhood immunisation programmes (Locarnini *et al.*, 2015) and this appears to have been very effective in reducing incidence of acute Hepatitis B infection, along with the prevalence of chronic infection and its sequelae (Locarnini *et al.*, 2015; Ott *et al.*, 2012). Several countries in Southeast Asia have reported that after 10 years of universal childhood HBV vaccination, the number of reported cases of hepatocellular carcinoma (HCC) have decreased dramatically. In Taiwan, incidence of HCC has been reduced by 80% with mortality rates down by more than 90% (Locarnini *et al.*, 2015). The majority of countries in Europe have moved to a policy of vaccinating all babies or children (Lernout *et al.*, 2014). The UK introduced HBV vaccination into its routine childhood immunization schedule in the Autumn of 2017. The vaccine is also given to healthcare and other public sector workers and offered to those at increased risk of exposure through their life style.

4.7.1.3 Hepatitis C Virus

Hepatitis C virus (HCV) is single-stranded positive sense RNA virus, classified in the family *Flaviviridae* within its own genus of *Hepacivirus*. It was discovered in the late 1980s, in an intriguing set of experiments which took isolating and characterising proteins in blood samples from patients with 'non-A, non-B hepatitis' as the starting point (Harrison *et al.*, 2009; Murray *et al.*, 2013). For an account of this intriguing work; see, for example, Alter (1999). Once an assay was available to test for HCV, evidence of infection with this virus was found in many people diagnosed clinically with chronic hepatitis (Harrison *et al.*, 2009). The virus is transmitted through blood and body fluids and it is estimated that up to 3% of people across the world are infected with it (Tapper and Afdhal, 2013).

Some of those people had unfortunately acquired HCV through blood transfusion or blood products, given to treat serious conditions such as haemophilia (Harrison *et al.*, 2009) before screening for the virus was possible. It was also spread widely through sharing of needles when this was not considered as a risk for transmission of blood-borne infections. There is a link between the helminth parasite *Schistosoma mansoni* and HCV, particularly in Egypt for this reason (see below).

About three quarters of patients in the acute stage of HCV infection experience mild, non-specific symptoms. Sometimes patients are ill enough to seek medical advice and they occasionally have jaundice – thus precipitating investigations of LFTs and viral serology (Harrison *et al.*, 2009). In a small proportion of cases, the infection resolves, but up to 80% of people will develop chronic HCV, which may take 10 years or more to become clinically apparent (Murray *et al.*, 2013). Although subclinical, the virus is active during this time and stimulates an inflammatory response. The pathological mechanisms are not well understood, but they are thought to involve inflammatory cytokines and interaction between the HCV and TLRs (e.g., Li *et al.*, 2012). The extent of the damage and the chances of this leading to cirrhosis or HCC appear to be enhanced by co-infection with other pathogens, particularly HBV and HIV (Lacombe and Rockstroh, 2012). There is some evidence of a link between chronic HCV infection and a number of metabolic disorders. These include steatosis, which is reported to be more likely to occur with genotype 3 infection (Roingeard, 2013) and insulin resistance, which may be alleviated with anti-HCV treatment (Conjeevaram *et al.*, 2011). Pathological mechanisms leading to some autoimmune and lymphoproliferative disorders have also been associated with HCV infection (Zignego, 2012).

Initial laboratory diagnosis usually involves detection of anti-HCV antibody in serum, using an EIA; combined antigen-antibody assays are also available and these should enhance sensitivity (SMI V5). In some groups, such as HIV-positive patients, RT PCR is a better first-line test, but the EIA for HCV core antigen may be a suitable alternative (e.g., Cresswell *et al.*, 2014). It is recommended that positive results in new patients are confirmed by a repeat anti-HCV antibody test, RT PCR and/or HCV core antigen EIA on a follow-up blood sample. When acute HCV is suspected, it is important to bear in mind the date of (possible) exposure when selecting the test method, since it may take three months for antibody levels to rise, whereas the virus should be detectable by PCR within four weeks (SMI V5). In chronic infection, regular laboratory monitoring of viral load is carried out to determine the effectiveness of anti-HCV treatment (SMI V5; Murray *et al.*, 2013). This can be achieved with PCR but the HCV core antigen test is also quantifiable. It is useful to ascertain the genotype where possible, for epidemiological purposes and also to inform treatment plans. There are 11 known genotypes of HCV, with several subtypes within each (WHO: http://www.who.int/csr/disease/hepatitis/whocdscsrlyo2003/en/index2.html). Genotypes 1, 2 and 3 are found throughout the world, while genotype 4 predominates in the Middle East, genotype 5 in Southern Africa and genotypes 6-11 in Asia. (WHO: http://www.who.int/csr/disease/hepatitis/whocdscsrlyo2003/en/index2.html).

In the UK, it was estimated that at least 214,000 people had chronic HCV infection in 2014 (Harris, 2014) and that the most common route of acquisition was intravenous drug use (IVDU). About 90% of the isolates sequenced were found to be genotype 1 or 3, in roughly equal proportions (Harris, 2014). The annual rate of hospital admissions for HCV-related liver disease and the number of deaths has quadrupled since the late 1990s (Harris, 2014), although some of this reported rise is likely to be attributable to improvements in the accuracy of the virological diagnoses.

Identification of the genotype is important in deciding the treatment regime. IFN-α (or pegylated IFN-α) in combination with Ribavirin is effective in reducing viral activity if the HCV isolate is genotype 2 or 3; strains of the former are almost always completely susceptible (Tapper and Afdhal, 2013). Genotype 1 viruses are often refractory to interferon-based therapies (Harrison *et al.*, 2009; Tapper and Afdhal, 2013), although they may respond to a longer course of treatment. If not, a protease inhibitor, such as boceprevir or telaprevir (SMI G5) is recommended, but the IFN-α and Ribavirin regime must be continued, to mitigate against drug resistance (SMI G5; Tapper and Afdhal, 2013). It has been suggested that treatment of genotype 3 may be harder than previously assumed and that specifically designed direct acting antivirals (DDAs) are needed (Tapper and Afdhal, 2013).

Blood samples from patients on treatment should be monitored regularly for HCV viral load to evaluate the effectiveness of the anti-viral therapy. The aim is to clear the virus and this is taken to have occurred when there is no detectable virus for at least six months after the course of treatment has finished (SMI G5). This is referred to as the 'sustained virological response' (SVR). Being able to trace the molecular evolution of a particular strain of virus is useful in tracking outbreaks, including in forensic cases. Following an outbreak in Spain, fairly extensive investigations comparing sequences in the NS5B and E1 and E2 regions of the genome of isolates proved that an anaesthetist transmitted HCV to at least 275 of his patients through effectively 'sharing' needles with them; he did this by injecting himself with small aliquots of morphine before administering the rest of the dose to the person that it had been prescribed for (Gonzalez-Candelas *et al.*, 2013).

Point to consider 4.10: Since most patients with HCV infection do not present until they are experiencing hepatic damage, do you think screening of asymptomatic people would be a good idea? If so, who would you target in a screening programme?

4.7.1.4   Hepatitis D (Delta) Virus

Hepatitis Delta virus (HDV) is a defective agent comprising a small piece of circular RNA surrounded by hepatitis B surface antigen (Harrison *et al.*, 2009; Hughes *et al.*, 2011). It requires HBV as a 'helper virus' in order to replicate. It can therefore only set up an infection in a new host if it is co-transmitted with its 'helper' or if the person has a pre-existing HBV infection (Murray *et al.*, 2013). The prevalence of Delta is not fully documented, partly due to the difficulty of testing for such an unusual entity and partly because detection of HBsAg might lead to an unquestioned diagnosis of HBV infection. It is thought to have a global distribution and is estimated to infect 15 to 20 million people (Noureddin and Gish, 2014), with an overall prevalence of HDV in 5% of HBV carriers (Harrison *et al.*, 2009; SMI G5).

Infection with the HDV on top of HBV exacerbates the liver damage and increases the likelihood of the patient developing fulminating hepatitis (Murray *et al.*, 2013) or the long-term sequelae of cirrhosis and HCC (Noureddin and Gish, 2014). Although HDV does appear to replicate inside individual hepatocytes, it requires the HBsAg in order to leave and infect another cell. Thus it seems likely that the immune response to the HDV makes a significant contribution to the hepatic pathology (Harrison *et al.*, 2009; Hughes *et al.*, 2011).

HDV can be diagnosed through detection of anti-HD IgM and IgG antibodies, as well as RNA PCR, which is the most sensitive in the acute phase (SMI G5). Results should be interpreted with caution, since IgM is known to persist (so it cannot be a reliable marker of acute infection), while viraemia tends to be short-lived (Nourreddin and Gish, 2014). It should also be noted that HDV interferes with expression of HBV markers, which may lead to an unusual profile for the latter (Harrison *et al.*, 2009). There is some evidence that chronic HDV infection can respond to treatment with IFN-α, but results from clinical trials have been rather variable (Hughes *et al.*, 2011; Noureddin and Gish, 2014).

4.7.1.5   Hepatitis E Virus

Hepatitis E virus is a small (32–34 nm) positive sense single-stranded RNA virus (Harrison *et al.*, 2009). It was first isolated in the early 1980s (Kamar *et al.*, 2012) and was originally classified in the *Caliciviridae*, but is now assigned to the family *Hepeviridae*, genus *Orthohepevirus* (SMI G5). There is one species known to infect mammals and it belongs to a single serotype, but isolates from human infections fall into four distinct genotypes (1–4). Genotypes 1 and 2 are thought to be solely human pathogens, while genotypes 3 and 4 appear to be swine viruses, which people acquire zoonotically (Dalton *et al.*, 2008b; Kamar *et al.*, 2012).

In the 1990s, HEV was recognised as the cause of outbreaks of hepatitis thought to be spread via the faecal-oral route, but not attributable to HAV – most notably in South

Asia (Harrison *et al.*, 2009). It is now known to be endemic in developing countries and associated with contaminated food and water (Kamar *et al.*, 2012). Unlike hepatitis A, it is an infection of young adults rather than children and the patients tend to be symptomatic (Dalton *et al.*, 2008b). Although HEV is usually self-limiting, it can be extremely dangerous in pregnancy; up to 20% of woman infected in the third trimester contract fatal, fulminant hepatitis and there is also some evidence for vertical transmission (SMI G5). There is a viraemic phase and there have been some incidences of transmission through blood transfusion (Dalton *et al.*, 2008b).

In Europe it was originally considered to be a hazard of travel abroad, but once laboratory diagnostic tests were available, studies in the UK and elsewhere found evidence that it is a common cause of sporadic acute hepatitis in people with no history of overseas trips, particularly males over the age of 55 (SMI G5; Dalton *et al.*, 2008a).

The prevailing genotype in Europe is type 3, which means that people acquire it from pigs. Thus the risk factors are close contact with these animals (e.g., working as farmers or vets) or consumption of processed pork meat products, particularly when they have been inadequately cooked. So perhaps middle-aged men should avoid the barbequed sausages at summer parties! Hepatitis E does appear to have the ability to persist, particularly in immunosuppressed patients (Kamar *et al.*, 2012), leading to chronic infection. It is also suggested that infection exacerbates underlying liver damage caused by conditions such as a pre-existing infection or alcohol related liver disease (Dalton, 2008b; Kamar *et al.*, 2012).

The incidence of laboratory confirmed HEV infection in England and Wales was 869 in 2014 (https://www.gov.uk/government/uploads/system/uploads/attachment_data/file/404107/hpr0515_zoos.pdf). This is three times that reported for HAV (see above) and adds weight to the suggestion that HEV may now be a more common cause of acute hepatitis in the UK and should always be considered as part of laboratory diagnosis (Dalton *et al.*, 2008a). One study revealed a rate of HEV viraemia in units of donated blood in the UK of just over 1 in 2800 (Hewitt *et al.*, 2014). The donors were asymptomatic at the time of donation and unaware of the infection. This leads to the question of whether blood donors should be routinely screened for HEV. This was introduced for UK donations in April 2017 (https://www.nhsbt.nhs.uk/media/news/change-to-nhsbt-pricing-of-products-in-201718-and-introduction-of-universal-screening-for-hepatitis-e/).

The optimal method for laboratory diagnosis of acute HEV is RT-PCR of stool or blood samples. There are serological assays for IgM, which may be more suitable in some situations, but results need to be interpreted with caution for two reasons: the first is that IgM is never found at detectable levels in some people and the second is that the kits are usually manufactured using isolates of genotype 1 or 2, so may have reduced specificity for genotype 3 (SMI G5). Similar problems with sensitivity and specificity have been reported for HEV- IgG assays (Kamar *et al.*, 2012).

There is no treatment available for acute hepatitis E, though supportive therapy may be required. In situations where the person already had pre-existing liver damage, the outcome may be serious enough to warrant liver transplant (Dalton *et al.*, 2008b). There is a vaccine in development but it is not yet available (SMI G5).

Point to consider 4.11: In which clinical situations should hepatitis E infection be considered as a possible diagnosis?

4.7.1.6 Hepatitis G Virus (GB-Virus C)

The GBV-C virus is a flavivirus which possibly originated in monkeys and does not appear to have a particularly high prevalence in humans. It may cause symptoms in immunocompromised patients, although this has not been definitively proven (Harrison *et al.*, 2009; SMI G5). It is blood-borne, so can be found in co-infection with HCV and HIV. Interestingly (unlike Delta agent) GBV-C actually seems to delay the course of disease with these two viruses, reducing viral loads and enhancing responses to treatment (Sahni *et al.*, 2014; SMI G5). Laboratory diagnosis is usually through detection of GBV-C viraemia using RT- PCR (Murray *et al.*, 2013).

4.7.2 Liver Coinfections

From the laboratory scientist's point of view it is important to always bear in mind that a person can be infected with more than one hepatotropic microorganism at a time. Having a chronic hepatitis B or C infection does not preclude someone from contracting acute hepatitis, attributable to another cause. In particular, due to the common transmission routes, a person newly diagnosed with hepatitis B infection should always be tested for co-infection with HCV (and *vice versa*) and HIV status should also be investigated. Co-infection can affect the patient's care (including choice of anti-viral drugs) and prognosis (Lacombe and Rockstroh, 2012). Some combinations of microorganisms are particularly known to alter immune responses, including antibody production; this can mean that laboratory results may need careful interpretation.

Renal failure and hepatitis can occur as a result of infection with the spirochaete *Leptospira interrogans*, the cause of Weil's disease (Bharti *et al.*, 2003). Humans acquire the bacterium through contact with soil or water contaminated with urine from infected reservoir animals, usually rodents. The bacteria are motile, which affords dissemination to sites throughout the body. They also appear to elicit a strong (but poorly understood) immune response, which contributes to the pathology (Bharti *et al.*, 2003). Icterus and abnormal LFTs are found in patients with Weil's disease, along with haemorrhage, renal impairment, respiratory disease and cardiac damage. The jaundice is actually the least of the patient's worries and if they survive, liver function is usually recovered. Interestingly, there have been some reports from India of cases where patients were acutely infected with *L. interrogans*, hepatitis E virus and dengue virus simultaneously (e.g., Singh *et al.*, 2014). Although it is probably impossible to ascertain the order of infection with the three organisms, it is likely that interaction between the organisms and the immune responses which they trigger caused serious systemic symptoms, including hepatic injury.

Chronic infection with some parasites can lead to liver damage. For example, pairs of adults of the helminth *Schistosoma mansoni* live in the inferior mesenteric veins, associated with the large intestine (Gunn and Pitt, 2012). As Figure 4.6 shows, females are longer and thinner than males; they are surrounded and held by the males. After copulation, females produce characteristic ova, which are 150 x 60 μm with a lateral spine (http://www.cdc.gov/dpdx/schistosomiasis/index.html). Eggs induce an immune response in the host, which facilitates their passage through the intestinal wall, to be excreted in faeces. However, some eggs get caught in the blood vessels and are transported to ectopic sites, most notably the liver. Here they induce a Th2 mediated, Il-4 promoted immune response, resulting in the formation of granulomas (Greenwood, 2012) within the

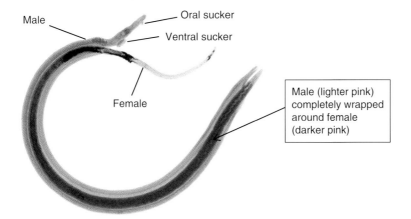

Figure 4.6 Stained and mounted preparation of a pair of adult *Schistosoma mansoni* helminths. Source: Courtesy of Dr A Gunn. (*See insert for colour representation of the figure.*)

hepatic tissue. Similarly, trophozoites (the growing and dividing stage) of the gut protozoan *Entamoeba histolyica* usually encyst before falling into the lumen ready for excretion (http://www.cdc.gov/dpdx/amebiasis/index.html). In some cases, this morphological change does not occur and the parasite can migrate via the intestinal circulation to the liver; the granulomatous reaction around the trophozoites leads to the development of liver abscesses (Gunn and Pitt, 2012). By a very unfortunate quirk of epidemiological fate, co-infection with *S. mansoni* and HCV is a serious public health problem in some countries, most notably Egypt. Mass treatment for schistosomiasis in the 1960s and 1970s involved delivering tartar emetic intravenously; as needles were re-used, HCV was unwittingly transmitted to large cohorts of people, who are now in middle and old age (SMI G5; Sanghvi *et al.*, 2013). Infection with the helminth appears to predispose to higher HCV viral loads – possibly due to the down regulation of the Th-1 immune response (CD4+ and CD8+ cells) – thus exacerbating the liver damage (Strickland, 2006; Sanghvi *et al.*, 2013).

Point to consider 4.12: How important is it for a patient diagnosed with a liver infection to be counselled to avoid activities which might cause further hepatitis damage (e.g., acquisition of another hepatotrophic organism, consumption of alcohol)?

4.8 Exercises

4.1 **A** Construct an algorithm for testing of faecal specimens from patients with diarrhoea, but restrict it to a particular group (e.g., paediatric in-patients, immunocompromised adults, healthy adults in the community).

B Compare this with the appropriate PHE SMI or your laboratory's SOP.

C Reflect on whether there were any important causes of diarrhoea in your chosen group that your algorithm missed out?

D Did the published protocol omit something which you think should be included?

4.2 **A** Select a commercially available POCT test kit for any faecal microorganism and read the most up to date literature on its evaluation

B Decide whether and under which circumstances you would recommend its use.

4.3 **A** Find out about a multiplex PCR system for faecal pathogens (the one in use in your laboratory or a commercially available system)

B Critically discuss the selection of pathogens included in the assay.

4.9 Case Studies

4.1 Mr DE is a 47 year old who has presented with usual tiredness, dyspepsia, upper abdominal pain, bloating and nausea. He has a stressful job which includes regular overseas travel and a fairly unhealthy diet.

a) What do these symptoms suggest and which investigations should be carried out (not only microbiology laboratory tests)?

Histological examination of the gastric biopsy stained with haematoxylin and eosin stain indicated the presence of Gram negative, flagellate coccobacilli, which were identified as *Helicobacter pylori*.

b) Would this be enough to confirm the clinical diagnosis? If not, which further tests could be performed on the tissue sample?

c) Assuming that the diagnosis is confirmed, which treatment would be recommended for Mr DE?

4.2 Mr WA is a 27-year-old known intravenous drug abuser. He has presented with pyrexia, nausea, headache, right-sided abdominal pain and jaundice. Blood is collected and sent to the biochemistry laboratory. Results indicate that the patient has abnormal liver function tests, consistent with acute viral hepatitis.

The virological laboratory results are as follows:

Hepatitis B surface antigen: NOT detected
Anti-Hepatitis B core IgM antibody: NOT detected
Anti-Hepatitis B core total antibody: NOT detected
Anti- Hepatitis B surface antibody: DETECTED
Anti-Hepatitis C antibody: equivocal
Anti- Hepatitis A IgM: DETECTED
Anti- Hepatitis A IgG: equivocal

a) What do these results indicate?

b) Which follow-up tests should be carried out and what time interval would be appropriate?

References

Åberg R *et al.* (2015). *Cryptosporidium parvum* caused by a large outbreak linked to frisée salad in Finland, 2012. *Zoonoses and Public Health*, **62**: 618–624. doi: 10.1111/zph.12190

Ahmed N (2005). 23 years of the discovery of *Helicobacter pylori*: Is the debate over? *Annals of Clinical Microbiology and Antimicrobials*, **4**: 17. doi: 10.1186/1476-0711-4-17

Alexander CL, Niebel M and Jones B (2013). The rapid detection of *Cryptosporidium* and *Giardia* species in clinical stools using the Quik Chek immunoassay. *Parasitology International*, **62**: 552–553. doi: 10.1016/j.parint.2013.08.008

Allos BM (2001). *Campylobacter jejuni* Infections: Update on Emerging Issues and Trends. *Clinical Infectious Diseases*, **32**: 1201–1206. doi: 10.1086/319760

Alter H (1999). Discovery of non-A, non-B hepatitis and identification of its etiology. *The American Journal of Medicine*, **107**: 16–20. doi: 10.1016/S0002-9343(99)00375-7

Atchison CJ *et al.* (2016). Rapid declines in age group–specific rotavirus infection and acute gastroenteritis among vaccinated and unvaccinated individuals within 1 year of rotavirus vaccine introduction in England and Wales. *Journal of Infectious Diseases*, **213**: 243–249. doi: 10.1093/infdis/jiv398

Barbut F *et al.* (2014). Does a rapid diagnosis of *Clostridium difficile* infection impact on quality of patient management? *Clinical Microbiology and Infection*, **20**: 136–144. doi: 10.1111/1469-0691.12221

Barclay L *et al.* (2014). Infection control for norovirus. *Clinical Microbiology and Infection* **20**: 731–740. doi: 10.1111/1469-0691.12674

Beaulieu C *et al.* (2014). Clinical characteristics and outcome of patients with *Clostridium difficile* infection diagnosed by PCR versus a three-step algorithm. *Clinical Microbiology and Infection*, **20**: 1067–1073. doi: 10.1111/1469-0691.12676

van den Berg B *et al.* (2014). Guillain-Barre syndrome: Pathogenesis, diagnosis, treatment and prognosis. *Nature Reviews Neurology*, **10**: 469–482. doi: 10.1038/nrneurol.2014.121

Bert F *et al.* (2014). Norovirus outbreaks on commercial cruise ships: A systematic review and new targets for the public health agenda. *Food and Environmental Virology*, **6**: 67–74. doi: 10.1007/s12560-014-9145-5

Bessède E *et al.* (2011). New methods for detection of campylobacters in stool samples in comparison to culture. *Journal of Clinical Microbiology*, **49**: 941–944. doi: 10.1128/JCM.01489-10

Bharti AR *et al.* (2003). Leptospirosis: A zoonotic disease of global importance. *Lancet Infectious Diseases*, **3**(12): 757–771. doi: 10.1016/S1473-3099 (03) 00830-2

Blumberg BS (2002). The discovery of the hepatitis B virus and the invention of the vaccine: A scientific memoir. *Journal of Gastroenterology and Hepatology*, **17**: S502–S503. doi: 10.1046/j.1440-1746.17.s4.19.x

Bouzid M *et al.* (2013). *Cryptosporidium* pathogenicity and virulence. *Clinical Microbiology Reviews*, **26**: 115–134. doi: 10.1128/CMR.00076-12

Bruggink L *et al.* (2015). Evaluation of the updated RIDA® QUICK (Version N1402) immunochromatographic assay for the detection of norovirus in clinical specimens. *Journal of Virological Methods*, **223**: 82–87. doi: 10.1016/jviromet.2015.07.019

Bruins MJ *et al.* (2010). Evaluation of a rapid immunochromatographic test for the detection of norovirus in stool samples. *European Journal of Microbiology and Infectious Disease*, **29**: 741–743. doi: 10.1007/s10096-010-0911-5

Burucoa C *et al.* (2013). Comparative evaluation of 29 commercial *Helicobacter pylori* serological kits. *Helicobacter* **18**: 169–179. doi: 10.1111/hel.12030

Cairns MD *et al.* (2015). Genomic epidemiology of a protracted hospital outbreak caused by a toxin a–negative *Clostridium difficile* sublineage PCR riboytpe 017 strain in London, England. *Journal of Clinical Microbiology*, **53**: 3141–3147. doi: 10.1128/JCM.006-48-15

Carvalho C *et al.* (2012). A possible outbreak of hepatitis A associated with semi-dried tomatoes, England, July–November 2011. *Eurosurveillance*, **17(6):** pii = 20083. http://www.eurosurveillance.org/ViewArticle.aspx?ArticleId=20083

CDRN (2014). *Clostridium difficile* Ribotyping Network 2011–2013 Report. https://www.gov.uk/government/uploads/system/uploads/attachment_data/file/329156/C_difficile_ribotyping_network_CDRN_report.pdf

Chalmers RM *et al.* (2011a) Epidemiology of anthroponotic and zoonotic human cryptosporidiosis in England and Wales, 2004-2006. *Epidemiology and Infection*, **139**: 700–712. doi: 10.1017/S0950268810001688

Chalmers RM *et al.* (2011b) Comparison of diagnostic sensitivity and specificity of seven *Cryptosporidium* assays used in the UK. *Journal of Medical Microbiology*, **60**: 1598–1604. doi: 10.1099/jmm.0.034181-0

Chalmers RM and Katzer F (2013). Looking for *Cryptosporidium*: The application of advances in detection and diagnosis. *Trends in Parasitology*, **29**: 237–251 doi: 10.1016/j.pt.2013.03.001

Chalmers RM *et al.* (2015). An audit of the laboratory diagnosis of cryptosporidiosis in England and Wales, *Journal of Medical Microbiology*, **64**: 688–693. doi: 10.1099/jmm.0.000089

Checkley W *et al.* (2015). A review of the global burden, novel diagnostics, therapeutics and vaccine targets for cryptosporidium. *Lancet Infectious Diseases*, **15**: 85–94 doi: 10.1016/S1473-3099(14)70772-8

Clemente JC *et al.* (2012) The impact of the gut microbiota on human health: An integrative view. *Cell*, **146**: 1258–1270. doi: 10.1016/j.cell.2012.01.035

Clements CJ *et al.* (2010). Global control of hepatitis B virus: Does treatment-induced antigenic change affect immunization? *Bulletin of the World Health Organisation*, **88**: 66–73. doi: 10.2471/BLT.08.065722

Cody AJ *et al.* (2015). Wild bird-associated *Campylobacter jejuni* isolates are a consistent source of human disease in Oxfordshire, United Kingdom. *Environmental Microbiology Reports*, 7: 782–788. doi: 10.1111/1758-2229.12314

Conjeevaram HS *et al.* (2011). Changes in insulin sensitivity and body weight during and after peginterferon and ribavirin therapy for hepatitis C. *Gastroenterology*, **140**: 469–477. doi: 10.1053/j.gastro.2010.11.002

Cresswell FV *et al.* (2014). Hepatitis C core antigen testing: A reliable, quick and potentially cost-effective alternative to hepatitis C polymerase chain reaction in diagnosing acute hepatitis C virus infection. *Clinical Infectious Diseases*, **60**: 263–266. doi: 10.1093/cid/ciu782

Dalton HR *et al.* (2008a). Autochthonous hepatitis E in Southwest England: A comparison with hepatitis A. *European Journal of Clinical Microbiology & Infectious Diseases*, **27**: 579–558. doi: 10.1007/s10096-008-0480-z

Dalton HR et al. (2008b) Hepatitis E: an emerging infection in developed countries. *The Lancet Infectious Diseases*, **8**: 698–709. doi: 10.1016/S1473-3099(08)70255-X

Davies KA *et al.* (2014). Underdiagnosis of *Clostridium difficile* across Europe: The European, multicentre, prospective, biannual, point-prevalence study of *Clostridium difficile* infection in hospitalised patients with diarrhoea (EUCLID). *Lancet Infectious Diseases*, **14**: 1208–1219. doi: 10.1016/S1473-3099(14)70991-0

Deneve C *et al.* (2009). New trends in *Clostridium difficile* virulence and pathogenesis. *International Journal of Antimicrobial Agents*, **33(S1):** S24–S28. doi: 10.1016/S0924-8579(09)70012-3

Deshpande AP *et al.* (2015). Molecular characterisation of *Cryptosporidium parvum* isolates from human cryptosporidiosis cases in Scotland. *Parasitology*, **142**: 318–325. doi: 10.1017/S0031182014001346

Donnan EJ *et al.* (2012). A multistate outbreak of hepatitis A associated with semidried tomatoes in Australia, 2009. *Clinical Infectious Diseases*, **54**: 775–781. doi: 10.1093/cid/cir949

Du Y *et al.* (2006). *Helicobacter pylori* and *Schistosoma japonicum* co-infection in a Chinese population: Helminth infection alters humoral responses to *H. pylori* and serum pepsinogen I/II ratio. *Microbes and Infection*, **8**: 52–60. doi: 10.1016/j.micinf.2005.05.017

Dunbar NL *et al.* (2014). Evaluation of the RIDAGENE real-time PCR assay for the detection of GI and GII norovirus. *Diagnostic Microbiology and Infectious Disease*, **79**: 317–321. doi: 10.1016/j.diagmicrobio.2014.03.017

Eckert C *et al.* (2015). Prevalence and pathogenicity of binary toxin–positive *Clostridium difficile* strains that do not produce toxins A and B. *New Microbes and New Infections*, **3**: 12–17. doi: 10.1016/j.nmni.2014.10.003

Feazel LM *et al.* (2014). Effect of antibiotic stewardship programmes on *Clostridium difficile* incidence: A systematic review and meta-analysis. *Journal of Antimicrobial Chemotherapy*, **69**: 1748–1754. doi: 10.1093/jac/dku046

Fischbach, LA *et al.* (2012). The association between Barrett's esophagus and *Helicobacter pylorii*: A meta-analysis. *Helicobacter*, **17**: 163–175. doi: 10.1111/j.1523-5378.2011.00931.x

Formichella L *et al.* (2013). A novel line immunoassay based on recombinant virulence factors enables highly specific and sensitive serologic diagnosis of *Helicobacter pylori* infection. *Clinical and Vaccine Immunology*, **20**: 1703–1710. doi: 10.1128/CVI.00433-13

Gase KA *et al.* (2013) Comparison of 2 *Clostridium difficile* surveillance methods national healthcare safely network's laboratory-identified event reporting module versus clinical infection surveillance. *Infection Control and Hospital Epidemiology*, **34**: 284–290. doi: 10.1086/669509

Gaudreau C *et al.* (2014) Antimicrobial susceptibility of *Campylobacter jejuni* and *Campylobacter coli* isolates obtained in Montreal, Quebec, Canada, from 2002 to 2013. *Journal of Clinical Microbiology*, **52**: 2644–2646. doi: 10.1128/JCM.00362-14

Gómez-Camarasa C *et al.* (2014). Evaluation of the rapid RIDAQUICK Campylobacter® test in a general hospital. *Diagnostic Microbiology and Infectious Disease*, **78**: 101–104. doi: 10.1016/j.diagmicrobio.2013.11.009

González-Candelas F *et al.* (2013). Molecular evolution in court: analysis of a large hepatitis C virus outbreak from an evolving source. *BMC Biology*, **11**: 76. doi: 10.1186/1741-7007-11-76

Gray JJ and Desselberger U (2009). Viruses other than rotaviruses associated with acute diarrhoeal disease. In: *Principles and Practice of Clinical Virology*, *6th edn*. Zuckerman AJ, Banatvala JE, Schoub BD, Griffiths PD and Mortimer P, eds. Chichester: Wiley-Blackwell.

Greenwood D (2012). Protozoa. In: *Medical Microbiology: A Guide to Microbial Infections, 18th edn*. Greenwood D, Barer M, Slack R and Irving W, eds. London: Churchill Livingstone – Elsevier.

Gunn A and Pitt SJ (2012). *Parasitology: An Integrated Approach*. Chichester: Wiley-Blackwell.

Harris H (2014). PHE Hepatitis C in the UK: 2014 report https://www.gov.uk/government/uploads/system/uploads/attachment_data/file/337115/HCV_in_the_UK_2014_24_July.pdf

Harrison TJ, Dusheiko GM and Zuckerman AJ (2009). *Hepatitis Viruses*. In: *Principles and Practice of Clinical Virology, 6th edn*. Zuckerman AJ, Banatvala JE, Schoub BD, Griffiths PD and Mortimer P, eds. Chichester: Wiley-Blackwell.

Hewitt PE *et al.* (2014). Hepatitis E virus in blood components: A prevalence and transmission study in southeast England. *Lancet*, **384(9956):** 1766–1773. doi:10.1016/S0140-6736(14)61034-5

Hughes SA, Wedemeyer H and Harrison P (2011). Hepatitis delta virus. *Lancet*, **378**(9785): 73–85. doi: 10.1016/S0140-6736(10)61031-9

Huizinga R *et al.* (2015). Innate immunity to *Campylobacter jejuni* in Guillain-Barré syndrome. *Annals of Neurology*, **78**: 343–354. doi: 10.1002/ana.24442

Humphries RM and Linscott AJ (2015). Laboratory diagnosis of bacterial gastroenteritis. *Clinical Microbiology Reviews*, **28**: 3–31. doi: 10.1128/CMR.00073-14

Ivarsson ME, Leroux J-C and Castagner B (2015). Investigational new treatments for *Clostridium difficile infection. Drug Discovery Today*, **20**: 602–608. doi: 10.1016/j.drudis.2014.12.003

Jeffrey K and Aarons E (2009). Diagnostic approaches. In: *Principles and Practice of Clinical Virology, 6th edn*. Zuckerman A J, Banatvala JE, Schoub BD, Griffiths PD and Mortimer P, eds. Chichester: Wiley-Blackwell.

Johannessen I and Olgilvie MM (2012). Herpesviruses. In: *Medical Microbiology: A Guide to Microbial Infections, 18th edn*. Greenwood D, Barer MR, Slack RCB and Irving W, eds. London: Churchill Livingstone-Elsevier.

Kaakoush N *et al.* (2015). Global epidemiology of *Campylobacter* infection. *Clinical Microbiology Reviews*, **28**: 687–720. doi: 10.1128/CMR.00006-15

Kamar N *et al.* (2012). Hepatitis E. *Lancet*, **379(9835):** 2477–2488. doi: 10.1016/S140-6736(11)61849-7

Kamei K *et al.* (2014) A PCR-RFLP assay for the detection and differentiation of *Campylobacter jejuni, C. coli, C. fetus, C. hyointestinalis, C. lari, C. helveticus* and *C. upsaliensis. Journal of Medical Microbiology*, **63**: 659–666. doi: 10.1099/jmm.0.071498-0

Ketley JM and van Vliet ANM (2012). Campylobacter and helicobacter. In: *Medical Microbiology: A Guide to Microbial Infections, 18th edn*. Greenwood D, Barer MR, Slack RCB and Irving W, eds. London: Churchill Livingstone – Elsevier.

Korkmaz H *et al.* (2013). Comparison of the diagnostic accuracy of five different stool antigen tests for the diagnosis of *Helicobacter pylori* infection. *Helicobacter*, **18**: 384–391. doi: 10.1111/hel.12053

Kotloff KL *et al.* (2013). Burden and aetiology of diarrhoeal disease in infants and young children in developing countries (the Global Enteric Multicenter Study, GEMS):

A prospective, case-control study. *The Lancet*, **382**(9888)**:** 209–222. doi: 10.1016/S0140-6736(13)60844-2

Kramvis A, Kew M and François G. (2005) Hepatitis B virus genotypes. *Vaccine*, **23**: 2409–2423. doi: 10.1016/j.vaccine.2004.10.045

Kusters JG, van Vliet, AHM and Kuipers EJ (2006). Pathogenesis of *Helicobacter pylori* Infection. *Clinical Microbiology Reviews*, **19**: 449–490. doi: 10.1128/CMR.00054-05

Lacombe K and Rockstroh J (2012). HIV and viral hepatitis coinfections: Advances and challenges. *Gut*, **61**: 147–158. doi: 10.1136/gutjnl-2012-302062

Landelle C *et al*. (2014) Contamination of healthcare workers' hands with *Clostridium difficile* spores after caring for patients with *C. difficile* infection. *Infection Control and Hospital Epidemiology*, **35**: 10–15. doi: 10.1086/674396

Larralde O *et al*. (2013). Hepatitis B escape mutants in Scottish blood donors. *Medical Microbiology and Immunology*, **202**: 207–214. doi: 10.1007/s00430-012-0283-9

Lebwohl B *et al*. (2013). Decreased risk of celiac disease in patients with *Helicobacter pylori* colonization. *American Journal of Epidemiology*, **178**: 1721–1730. doi: 10.1093/aje/kwt234

Lernout T *et al*. (2014). A cohesive European policy for hepatitis B vaccination: Are we there yet? *Clinical Microbiology and Infection*, **20(s5):** 19–24. doi: 10.1111/1469-0691.12535

Leslie JH and Young VB (2015). The rest of the story: The microbiome and gastrointestinal infections. *Current Opinion in Microbiology*, **23**: 121–125. doi: 10.1016/j.mib.2014.11.010

Li K *et al*. (2012).Activation of chemokine and inflammatory cytokine response in hepatitis C virus–infected hepatocytes depends on toll-like receptor 3 sensing of hepatitis C virus double-stranded RNA intermediates. *Hepatology*, **55**: 666–675. doi: 10.1002/hep.24763

Locarnini S *et al*. (2015). Strategies to control hepatitis B: Public policy, epidemiology, vaccine and drugs. *Journal of Hepatology*, **62**: s76–s86. doi: 10.1016/j.antiviral.2015.06.008

Lopman B *et al*. (2012). Environmental transmission of norovirus gastroenteritis. *Current Opinion in Virology*, **2**: 96–102. doi: 10.1016/j.coviro.2001.11.005

Lutful Kabir SM *et al*. (2011). Evaluation of a Cytolethal distending toxin (*cdt*) gene-based species-specific multiplex PCR assay for the identification of *Campylobacter* strains isolated from diarrheal patients in Japan. *Japanese Journal of Infectious Diseases*, **64**: 19–27.

McAuliffe GN *et al*. (2013). Systematic application of multiplex PCR enhances the detection of bacteria, parasites, and viruses in stool samples. *Journal of Infection*, **67**: 122–129. doi: 10.1016/j.jinf.2013.04.009

McColl, KEL (2010). *Helicobacter pylori* infection. *New England Journal of Medicine*, **362**: 1597–1604. doi: 10.1056/NEJMcp1001110

van Maarseveen NM *et al*. (2010). Diagnosis of viral gastroenteritis by simultaneous detection of adenovirus group F, astrovirus, rotavirus group A, norovirus genogroups I and II, and sapovirus in two internally controlled multiplex real-time PCR assays. *Journal of Clinical Virology*, **49**: 205–201. doi: 0.1016/j.jcv.2010.07.019

MacRitchie LA, Hunter CJ and Strachan NJC (2014). Consumer acceptability of interventions to reduce Campylobacter in the poultry food chain. *Food Control*, **35**: 260–266. doi: 10.1016/j.foodcont.2013.06.005

Mahboobi N *et al*. (2012). Oral fluid and hepatitis A, B and C: A literature review. *Journal of Oral Pathology and Medicine*, **41**: 505–516. doi: 10.1111/j.1600-0714.2011.01123.x

Malfertheiner P *et al*. (2012). Management of Helicobacter pylori infection—The Maastricht IV/Florence consensus report. *Gut*, **61**: 646–664. doi: 10.1136/gutjnl-2012-302084

Manser M *et al*. (2013). Detection of *Cryptosporidium* and *Giardia* in clinical laboratories in Europe – A comparative study. *Clinical Microbiology and Infection*, **20**: O65–O71. doi: 10.1111/1469-0691.12297

Millman C *et al*. (2014). Perceptions, behaviours and kitchen hygiene of people who have and have not suffered camplobacteriosis: A case control study. *Food Control*, **41**: 82–90. doi: 10.1016/j.foodcont.2014.01.002

Morillo SG *et al*. (2011). Norovirus 3rd generation kit: An improvement for rapid diagnosis of sporadic gastroenteritis cases and valuable for outbreak detection. *Journal of Virological Methods*, **173**: 13–16. doi: 10.1016/jviromet.2010.12.017

Morter S *et al*. (2011). Norovirus in the hospital setting: Virus introduction and spread within the hospital environment. *Journal of Hospital Infection*, **77**: 106–112. doi: 10.1016/j.jhin.2010.09.035

Murray PR, Rosenthal, KS and Pfaller MA (2013). *Medical Microbiology*, 7th edn. Philadelphia: Elsevier-Saunders.

National Travel Health Network and Centre. (2017) https://www.nathnac.org/pro/factsheets/hep_a.htm

Norovirus Working Party (2012). *Guidelines for the management of norovirus outbreaks in acute and community health and social care settings*. London: HPA Publications. http://www.his.org.uk/files/9113/7398/0999/Guidelines_for_the_management_of_norovirus_outbreaks_in_acute_and_community_health_and_social_care_settings.pdf

Noureddin M and Gish R (2014). Hepatitis delta: Epidemiology, diagnosis and management 36 years after discovery. *Current Gastroenterology Reports*, **16**: 365. doi: 10.1007/s/11894-013-0365-x

Ott JJ *et al*. (2012). Global epidemiology of Hepatitis B infection: New estimates of age-specific HBsAG seroprevalence and endemicity. *Vaccine*, **30**: 2212–2219. doi: 10.1016/j.vaccine.2011.12.116

Plummer M *et al*. (2015). Global burden of gastric cancer attributable to *Helicobacter pylori*. *International Journal of Cancer*, **136**: 487–490. doi: 10.1002/ijc.28999

Public Health England: http://bioinformatics.phe.org.uk/noroOBK/ Definition of an outbreak: http://bioinformatics.phe.org.uk/noroOBK/outbreak.html

Public Health England. (2015). Identification of *Campylobacter* species. UK Standards for Microbiology Investigations. ID23 Issue 3. https://www.gov.uk/uk-standards-for-microbiology-investigations-smi-quality-and-consistency-in-clinical-laboratories

Public Health England. (2014). Investigation of Gastric Biopsies for *Helicobacter pylori*. UK Standards for Microbiology Investigations. B55 Issue 5.2. http://www.hpa.org.uk/SMI/pdf

Public Health England (2014). Identification of *Helicobacter* species. UK Standards for Microbiology Investigations. ID26 Issue 2.2. http://hpa.org.uk/SMI/pdf

Public Health England. (2014). Processing of Faeces for Clostridium difficile. UK Standards for Microbiology Investigations. B10 Issue 1.5. http://www.hpa.org.uk/SMI/pdf

Public Health England. (2014). Investigations of Faecal Specimens for Enteric Pathogens. UK Standards for Microbiology Investigations. B30 Issue 8.1. http://www.hpa.org.uk/SMI/pdf

Public Health England. (2014). Investigation of Specimens other than Blood for Parasites. UK Standards for Microbiology Investigations. B31 Issue 4.1. http://www.hpa.org.uk/ SMI/pdf

Public Health England. (2013). Gastroenteritis and Diarrhoea. UK Standards for Microbiology Investigations. S7 Issue 1. http://www.hpa.org.uk/SMI/pdf

Public Health England. (2014). Hepatitis B Diagnostic Serology in the Immunocompetent (including Hepatitis B in Pregnancy). UK Standards for Microbiology Investigations. V4 Issue 5.3. http://www.hpa.org.uk/SMI/pdf

PHE 1: Infection report. (2014). Hepatitis A: Oral fluid testing for household contacts. https://www.gov.uk/government/publications/ hepatitis-a-oral-fluid-testing-for-household-contacts

PHE 2: Infection report. (2014). Laboratory reports of hepatitis A and C. https://www.gov. uk/government/publications/laboratory-reports-of-hepatitis-a-and-c-2014

PHE 3: Infection report. (2015). Common animal associated infections quarterly report (England and Wales). https://www.gov.uk/government/uploads/system/uploads/ attachment_data/file/404107/hpr0515_zoos.pdf

PHE 4: Infection report. (2015). Hepatitis B guidance data and analysis. https://www.gov. uk/government/collections/hepatitis-b-guidance-data-and-analysis

SMI G5: Investigation of Hepatitis. (2015). UK Standards for Microbiology Investigations. G5 Issue 1.2 http://www.hpa.org.SMI/pdf

SMI V27: Public Health England. (2014). Hepatitis A Virus Acute Infection Serology. UK Standards for Microbiology Investigations. V27 Issue 3. http://www.hpa.org.uk/SMI/pdf

SMI V5: Public Health England. (2014). Investigation of Hepatitis C Infection by Antibody Testing or Combined Antigen/Antibody Assay. UK Standards for Microbiology Investigations. V5 Issue 6.2. http://www.hpa.org.uk/SMI/pdf

Regnath T and Ignatius R (2014). Accurate detection of *Campylobacter* spp. antigens by immunochromatography and enzyme immunoassay in routine microbiological laboratory. *European Journal of Microbiology and Immunology*, **4**: 156–158. doi: 10.1556/ EUJMI-D-14-00018

Riley TV (2012). *Clostridium*. In: Greenwood D, Barer MR, Slack RCB and Irving W, eds. *Medical Microbiology: A Guide to Microbial Infections*. London: Churchill Livingstone –Elsevier.

Roingeard P (2013). Hepatitis C virus diversity and hepatic steatosis. *Journal of Viral Hepatitis*, **20**: 77–84. doi: 10.1111/jvh.12035

Rondy M *et al.* (2011). Norovirus disease associated with excess mortality and use of statins: A retrospective cohort study of an outbreak following a pilgrimage to Lourdes. *Epidemiology and Infection*, **139**: 453–463. doi: 10.1017/S0950268810000993

Ruvoën-Clouet N *et al.* (2013). Norovirus and histo-blood groups: The impact of common host genetic polymorphisms on virus transmission and evolution. *Reviews in Medical Virology*, **23**: 355–366. doi: 10.1001/rmv.1757

Rydell GE *et al.* (2011). Susceptibility to winter vomiting disease: A sweet matter. *Reviews in Medical Virology*, **21**: 370–382. doi: 10.1002/rmv.704

Saad AM, Choudhary A and Bechtold ML (2012). Effect of Helicobacter pylori treatment on gastroesophageal reflux disease (GERD): Meta-analysis of randomized controlled trials. *Scandinavian Journal of Gastroenterology*, **47**: 129–135. doi: 10.3109/00365521.2011.648955

Sahni H *et al*. (2014). GBV-C viraemia and clinical events in advanced HIV infection. *Journal of Medical Virology*, **86**: 426–432. doi: 10.1002/jmv.23845

Sanghvi MM, Hotez PJ and Fenwick A (2013). Neglected tropical diseases as a cause of chronic liver disease: The case of Schistosomiasis and Hepatitis C Co-infections in Egypt. *Liver International*, **33**: 165–168. doi: 10.1111/liv.12052

Schielke A *et al*. (2014). Epidemiology of campylobacteriosis in Germany – Insights from 10 years of surveillance. *BMC Infectious Diseases*, **14**: 30. doi: 10.1186/1471-2334-14-30

Sheh A and Fox JG (2013). The role of gastrointestinal microbiome in *Helicobacter pylori* pathogenesis. *Gut Microbes* **4**: 505–531. doi: 10.4161/gmic.26205

Sibony M and Jones NL (2012). Recent advances in *Helicobacter pylori* pathogenesis. *Current Opinion in Gastroenterology*, **28**: 30–35. doi: 10.1097/MOG0b013e32834dda51

Singh RK *et al*. (2014). Comparison between three rare cases of co-infection with dengue, leptospira and hepatitis E: Is early endothelial involvement the culprit mortality? *Annals of Medical Health Science Research*, **4(S1)**: S32–S34. doi: 10.4103/2141-9248.131707

Smith AJ *et al*. (2011). A large foodborne outbreak of norovirus in diners at a restaurant in England between January and February 2009. *Epidemiology and Infection*, **140**: 1695–1701. doi: 10.1017/S0950268811002305

Smith S *et al*. (2015). Restoring the selectivity of modified charcoal cefoperazone deoxycholate agar for the isolation of *Campylobacter* species using tazobactam, a β-lactamase inhibitor. *International Journal of Food Microbiology*, **210**: 131–135. doi: 10.1016/j.ijfoodmicro.2015.06.014

Sparks H *et al*. (2015). Treatment of cryptosporidium: What we know, gaps, and the way forward. *Current Tropical Medicine Reports*, **2**: 181–187. doi: 10.1007/s40475-015-0056-9

Strickland GT (2006). Liver disease in Egypt: Hepatitis C superseded schistosomiasis as a result of iatrogenic and biological factors. *Hepatology*, **43**: 915–922. doi: 10.1002/hep.21173

Sun X and Hirota SA (2015). The roles of host and pathogen factors and the innate immune response in the pathogenesis of *Clostridium difficile* infection. *Molecular Immunology*, **63**: 193–202. doi: 10.1016/j.molimm.2014.09.005

Suzuki R, Shiota S and Yamaoka Y (2012). Molecular epidemiology, population genetics, and pathogenic role of *Helicobacter pylori*. *Infection, Genetics and Evolution*, **12**: 203–213. doi: 10.1016/j.meegid.2011.12.002

Swindells J *et al*. (2010). Evaluation of diagnostic tests for *Clostridium difficile* infection. *Journal of Clinical Microbiology*, **48**: 606–608. doi: 10.1128/JCM.01579-09

Tam CC *et al*. (2012). Longitudinal study of infectious intestinal disease in the UK (IID2 study): Incidence in the community and presenting to general practice. *Gut*, **61**: 69–77. doi: 10.1136/gut.2011.238386

Tapper EB and Afdhal NH (2013). Is 3 the new 1: Perspectives on virology, natural history and treatment for hepatitis C genotype 3. *Journal of Viral Hepatitis*, **20**: 669–677. doi: 10.1111/jvh.12168

Tong CYW (2012). *Hepadnaviruses*. In: *Medical Microbiology: A Guide to Microbial Infections*, 18th edn. Greenwood D, Barer MR, Slack RCB and Irving W, eds. London: Churchill Livingstone – Elsevier.

Utsi L *et al*. (2015). Cryptosporidiosis outbreak in visitors of a UK industry-compliant petting farm caused by a rare *Cryptosporidium parvum* subtype: A case-control study. *Epidemiology and Infection*, **144(5)**: 1000–1009. doi: 10.1017/S0950268815002319

Valiente E, Cairns MD and Wren BW (2014). The *Clostridium difficile* PCR ribotype 027 lineage: A pathogen on the move. *Clinical Microbiology and Infection*, **20**: 396–404. doi: 10.1111/1469-0691.12619

Van den Bossche D *et al.* (2015). Comparison of four rapid diagnostic tests, ELISA, microscopy and PCR for the detection of *Giardia lamblia*, *Cryptosporidium* spp. and Entamoeba histolytica in feces. *Journal of Microbiological Methods*, **110**: 78–84. doi: 10.1016/j.mimet.2015.01.016

Vinjé J (2015). Advances in laboratory methods for detection and typing of norovirus. *Journal of Clinical Microbiology*, **53**: 373–381. doi: 10.1128/JCM.01535-14

Wasley A *et al.* (2006). Hepatitis A in the era of vaccination. *Epidemiologic Reviews*, **28**: 101–111. doi: 10.1093/epirev/mxj012

WHO (2016). http://www.who.int/topics/rotavirus_infections/en/

WHO (2004). http://www.who.int/csr/disease/hepatitis/whocdscsrlyo2003/en/index2.html

WHO (2015). Fact sheet 204. http://www.who.int/mediacentre/factsheets/fs204/en

Wilcox MH (2013). Updated Guidance on the Management and Treatment of *Clostridium difficile* Infection. London: Public Health England.

Zignego AL *et al.* (2012). The hepatitis C virus infection as a systemic disease. *Internal and Emergency Medicine*, **7** (**Suppl 3**): S201–S208. doi: 10.1007/s11739-012-0825-6

5

Congenital, Perinatal and Neonatal Infections

5.1   Introduction

The intimate psychological and physiological relationship between mother and child that occurs during pregnancy and childbirth is mirrored by the microbiological interactions. Microorganisms that infect a woman during pregnancy can adversely affect both her and her unborn child. In some situations, the health of the mother is put in jeopardy and this obviously has consequences for the foetus she is carrying. Other infections can be mild or asymptomatic in the mother, while the organism is capable of infecting and causing damaging pathology in the baby – either *in utero* or during the perinatal period. Neonates are especially vulnerable to infection and can acquire both normal flora and potential pathogens from their mother. Therefore the aim of antenatal and perinatal care is to ensure that the mother has a safe and uneventful pregnancy and is delivered, at full term, of a healthy child as far as this is possible. Clearly there are situations where this does not happen and the role of the microbiology laboratory is to contribute to the prevention and diagnosis of infectious causes of adverse events during pregnancy, delivery and the immediate postnatal phase of the baby's life. In this chapter infections will be considered under four categories, namely, infections which adversely affect the mother during pregnancy, congenital infections, perinatal infections and neonatal infections.

5.2   Antenatal Screening

Microbiological screening of women during pregnancy is conducted to ensure the welfare of the mother, as well as to protect her baby from potential infection. Being pregnant is evidence of putting oneself at risk of contracting a sexually transmitted infection (STI) – however remote and unlikely that may seem to a particular individual – so as well as management of the patient (e.g., Allstaff and Wilson, 2012) there may also be public health issues to be addressed. Some infections pose a risk to the mother during the time of her pregnancy, while others can be asymptomatic in the adult but extremely dangerous to the foetus. Diagnosis of symptomatic infections as well as detecting subclinical ones is therefore important. Screening policies always represent a balance between what is technically feasible, how the results can be interpreted and whether

Clinical Microbiology for Diagnostic Laboratory Scientists, First Edition. Sarah J. Pitt.
© 2018 John Wiley & Sons Ltd. Published 2018 by John Wiley & Sons Ltd.

Table 5.1 Microorganisms and markers tested for as part of antenatal screening in UK.

Infection	Specimen and marker	Implication of result
Aymptomatic bacteriuria	Urine for culture and sensitivity	If significant UTI pathogen isolated consider monitoring/antibiotic treatment to prevent pyelonephritis
Hepatitis B	Blood/serum for HBsAg	If HBsAg detected, risk of perinatal infection in baby; test for further markers[1]
Human immuno-deficiency virus	Blood/serum for p24/andti-gp41	If HIV detected, risk of perinatal infection in baby; test for viral load and consider HAART[2]
Rubella virus	Anti-rubella IgG*	If no antibody detected or titre <10 mIU/ml then risk of primary rubella[3] which may result in congenital infection in baby
Syphilis	Anti-treponemal IgG	If antibody detected, then confirm; if *T. pallidum* infection then risk of congenital infection in baby;[4] consider antibiotic treatment

[1] see perinatal infections below;
[2] see Chapter 2;
[3] see congenital infections below;
[4] see congenital infections below;
* routine screening discontinued in England, Wales and Scotland in 2016.
Source: Information taken from (NICE, 2016).

testing everyone is cost effective. Table 5.1 lists the microorganisms and markers which are usually screened for routinely as part of antenatal care in the UK.

Additional tests may be carried out according to individual circumstances – for instance, a woman with known high risk of hepatitis C infection might have investigations for this virus. The range of infections screened for also varies a little between countries depending on epidemiology and priorities. In Austria, France and Italy, it is national policy that women are regularly tested for serological evidence of primary toxoplasmosis throughout pregnancy (Sagel *et al.*, 2012), while U.S. guidelines recommend collection of vaginal swabs to screen for group B streptococci (Verani *et al.*, 2010). Neither of these measures are currently considered necessary in the UK. Similarly, in a report on best practise for antenatal care in Africa, activities to prevent malaria infection during pregnancy are emphasised, while Rubella screening is not mentioned (Lawn and Kerber, 2006).

Point to consider 5.1: What do you think of the UK antenatal screening policy? In the light of evidence from other countries do you think it is justifiable or should the list of organisms tested for be different (i.e., are there tests which you think should be included or discontinued)?

5.3 Infections Which Adversely Affect the Mother During Pregnancy

In the 1950s, Sir Peter Medawar proposed the hypothesis that since the foetus is immunologically a 'semi-allograft' (i.e., half the antigens expressed will be paternal and therefore immunogenic to the mother), then some kind of immunosuppression must occur in order to avoid rejection of the foetus by the mother (Mor and Cardenas, 2010; Kraus *et al.*, 2012). Subsequent research has found evidence to both support and dispute this theory (Kraus *et al.*, 2012), but what is not in doubt is that the immune response is altered during pregnancy. Thus, the consequences of infection can be different from those usually expected in otherwise healthy adult women. Current thinking suggests that it is helpful to think of pregnancy as a situation where the immune system is modified rather than suppressed (Mor and Cardenas, 2010; Kraus *et al.*, 2012). The observable immune response is due to interactions between the maternal immune system and the placenta and foetus, which both carry paternal antigens (Mor and Cardenas, 2010). Also, rather than producing a uniform anti-inflammatory response throughout the pregnancy, it seems that there are three distinct immunological phases, which coincide with the three trimesters. During the early stages of pregnancy, the fertilised embryo must penetrate the uterine epithelium and weaken the endometrium, in order to be implanted and begin growing and dividing. The immune response at this stage is characterised by pro-inflammatory reactions, in order to repair this damage quickly. In the second period of gestation, the mother, placenta and foetus have adapted to each other, enhanced by an anti-inflammatory response. In the third trimester, the pro-inflammatory response begins to return, in preparation for ejection of the foetus and placenta during delivery (Mors and Cardenas, 2010).

Studies of particular components of the immune system have demonstrated interesting and sometimes puzzling results. For example, embryonic trophoblast cells have been shown to secrete β – defensins 1 and 3 (Mor and Cardenas, 2010). Also, levels of α – 1-3 defensins in the blood of pregnant women have been found to be higher than in those same women 6 months after the birth (Kraus *et al.*, 2012). Defensins are non-specific anti-microbial peptides with known activity against viruses, bacteria and fungi and yet pregnant women are observed to be more susceptible to infection. In contrast, levels of natural killer (NK) cells – which are important in protection against viral infections – were noted by Kraus *et al.* (2012) to be raised during early pregnancy (attributable to their role in trophoblast invasion) but to decline markedly after the first trimester. This would account for the greater severity and poorer outcomes of some infections. It is clear that overall the immune response during pregnancy is complex, but still not yet well understood and more research is needed in order to allow better management and treatment of patients with infectious diseases (Mors and Cardenas, 2010; Kraus *et al.*, 2012).

When the mother does become ill due to infection, this can have adverse consequences for the foetus. A strong maternal immune reaction can disrupt the maternal/placental/foetal interface and a premature or stillbirth (Kraus *et al*, 2012). An altered susceptibility to a microorganism may cause the mother to become very seriously ill or to die herself, which clearly often results in the loss of the foetus as well. It is important to bear in mind that this is quite a different scenario to congenital infection (see below) where the mother may have mild or sub-clinical illness, while the infectious agent is able to cross the placenta, with a severe detrimental effect on the foetus. Examples of organisms which can cause serious outcomes for the mother during pregnancy include varicella zoster virus, cytomegalovirus, *Listeria monocytogenes*, *Coxiella burnetii*, severe acute respiratory syndrome coronavirus (SARS-CoV), *Leishmania spp.*, *Plasmodium spp.*, influenza A virus and *Chlamydia trachomatis*. The last three will be discussed here.

5.3.1 Malaria

Kingdom: Protista
Phylum: Apicomplexa
Genus: *Plasmodium*
Species: *P. falciparum, P. vivax, P. ovale, P. malariae, P. knowlesi*

5.3.1.1 Pathogenesis and Clinical Symptoms

Infections with both *Plasmodium falciparum* and *Plasmodium vivax* are known to be associated with poor outcomes for the mother during pregnancy (ter Kuile and Rogerson, 2008; Hviid, 2011). (For full details of the life cycle, see https://www.cdc.gov/dpdx/malaria/). When *P. falciparum* infects a red blood cell, it causes the production and expression of parasite cell surface antigens, including *Plasmodium falciparum* erythrocyte membrane protein 1 (PfEMP1). This makes the infected red blood cells 'sticky' and leads to sequestration in smaller blood vessels, through clumping together and binding to endothelial walls; the consequence of this can be serious pathology such as cerebral malaria. These infected red blood cells also accumulate in the placenta. Studies have shown that there is a specific interaction between PfEMP1 and chondroitin sulphate A in the placental tissue (Hviid, 2011; Morley and Taylor-Robinson, 2012). Although the mechanism is not well understood, the outcomes of this pregnancy-associated malaria (PAM) have been well documented (e.g., Desai *et al.*, 2007). Along with the characteristic periodic pyrexia, the most notable feature is severe (and in some cases life-threatening) anaemia in the mother (Desai *et al.*, 2007; ter Kuile and Rogerson, 2008; Hviid, 2011). The risk of mortality from cerebral malaria is increased during pregnancy, as is the likelihood of developing complications such as hypoglycaemia, organ damage, pulmonary oedema and septic shock (Gitau and Eldred, 2005; Morley and Taylor-Robinson, 2012). Not surprisingly therefore, patients with *Plasmodium falciparum* PAM are more likely to experience complications during delivery and to have a premature, low birth weight or even stillborn baby (Desai *et al.*, 2007). The increased risk of morbidity and mortality is particularly marked when it is the first pregnancy for a women living in an endemic area (Desai *et al.*, 2007) or for a traveller to a malarious area who is immunologically naïve (Gitau and Eldred, 2005). If the mother survives the first pregnancy, she will develop antibodies against PfEMP1 (Hviid, 2011; Morley and Taylor-Robinson, 2012), which mitigates against the most serious sequelae in subsequent pregnancies. A detailed understanding of this may contribute to the development of a vaccine to reduce disease during pregnancy (Hviid, 2011). In contrast, *P. vivax* preferentially infects reticulocytes rather than mature red blood cells and does not induce the expression of parasite surface antigens. This means that it does not cause the accumulation of infected red cells in the same way as *P. falciparum*. A pregnant woman infected with *P. vivax* is also at more risk of developing severe anaemia than someone who does not have malaria and her baby is more likely to be of low birth weight (Desai *et al.*, 2007; ter Kuile and Rogerson, 2008). Interestingly, in *P. vivax* infection the adverse effect on the baby is more likely in second and later pregnancies (ter Kuile and Rogerson, 2008).

Co-infection with human immunodeficiency virus (HIV) (see Chapter 2) increases the hazard to the mother from malarial infection during pregnancy. Most work has been reported on *P. falciparum* and HIV co-infection in sub-Saharan Africa (Desai *et al.*, 2007), which is due to the fact that the prevalence of both microorganisms is highest there. As well as a having greater risk of PAM during their first pregnancy, women with HIV infection do not appear to be able to mount an effective antibody response to PfEMP1 and therefore are not protected from the adverse effects during subsequent pregnancies (Desai *et al.*, 2007). There is also some evidence that vertical transmission of HIV (see Section 5.5.2 below) is enhanced by the accumulation of *Plasmodium*-infected red cells in the placenta (Jaworowski *et al.*, 2009), though the research on this is limited and the mechanism is poorly understood (Desai *et al.*, 2007).

Rather counter-intuitively, the opposite may be true for a mixed malaria infection of *P. falciparum* and *P. vivax*. The pathogenicity of *P. falciparum* is reported to be reduced in the presence of *P. vivax* (Mayxay *et al.*, 2004; Desai *et al.*, 2007), One study has reported a significant reduction in the risk of severe malaria for pregnant women infected with both species compared to those with *P. falciparum* only (Desai *et al.*, 2007). A lack of data about the true extent of dual infections in areas of co-transmission has hampered investigations into the extent of and reasons for this observation (Mayxay *et al.*, 2004).

In sub-Saharan Africa, where malaria is endemic, it is estimated that *P. falciparum* infection occurs in at least 25% of all pregnancies (Desai *et al.*, 2007) and the outcomes are noted to be worse in primigravidae. In areas where transmission rates are lower and/or seasonal, reported infection rates vary between 6 and 14% (Desai *et al.* 2007) and previous pregnancies do not appear to have such a marked protective effect. Reliable information about the rates of *P. malariae, P. knowlesi* or *P. ovale* infection during pregnancy are not available (Kattenburg *et al.* 2011), so conclusions about relationships between gravidity and pathology cannot be made.

Point to consider 5.2: Should pregnant women living in non-endemic areas always avoid travel to places where malaria is transmitted?

5.3.1.2   Laboratory Diagnosis

The mainstay of laboratory diagnosis of *Plasmodium* infection, particularly in endemic areas, remains microscopic examination of thick and thin blood films, stained with a suitable Romanowsky stain (Greenwood, 2012). In pregnancy, peripheral parasitaemia levels are frequently lower than otherwise expected – particularly in *P. falciparum* – due to the sequestering of infected erythrocytes within the placenta (see above). The 'gold standard' test in this situation is therefore histological examination of placental tissue, although looking at blood films prepared from placental blood is also valuable (Kattenburg *et al.*, 2011). Unfortunately neither of these procedures can be performed until after delivery. There is some evidence that rapid diagnostic tests (RDTs) can be more sensitive than light microscopy in detection of parasitaemia in peripheral blood during pregnancy (Kattenburg *et al.*, 2011). The nature of the tests mean that some can only identify *P. falciparum* specifically and they may not show dual infections (Gunn and Pitt, 2012). The best alternative to placental histology is reported to be PCR (Desai *et al.*, 2007; Kattenburg *et al.*, 2011), which has the advantage of both good sensitivity, and with the correct primers, the ability to accurately detect all species present in the sample. Availability of this method is limited in rural endemic areas (where the necessity for accurate diagnosis is probably greatest), although work is being undertaken to refine the methods to allow their use in resource poor, tropical climate areas (Gunn and Pitt, 2012).

5.3.1.3 Management and Treatment

Malaria infection during pregnancy warrants very careful monitoring of both mother and baby. Supportive therapy, including balance of fluid and electrolytes and monitoring of platelet count, blood glucose and clotting, as well as parasitaemia levels are all required to enhance the chances of maternal and foetal survival (Lalloo *et al.*, 2016). It is safe to administer anti-malarial drugs, although the optimal treatment depends on the severity of the patient's presentation, the stage in pregnancy and the species of *Plasmodium* (Lalloo *et al.*, 2016). Quinine with clindamycin can be given at any stage in pregnancy and in both severe and uncomplicated infections. During the second or third trimesters, alternative treatments are recommended as having greater efficacy. For uncomplicated malaria, artemether-lumefantrine can be used and in severe infections, artesunate should be considered (Lalloo *et al.*, 2016).

5.3.2 Influenza A

Family: *Orthomyxoviridae*
Genus: *Influenzavirus A*
Species: *Influenza A virus*

5.3.2.1 Pathogenesis and Clinical Symptoms

Three species of influenza virus are known to infection humans: types A, B and C (Gambon and Potter, 2009). While influenza C infection is usually trivial, both influenza A and B are associated with 'seasonal flu', which can be serious, resulting in complications and secondary infections. Influenza A is responsible for epidemic and pandemic disease and it has long been recognised that infection with this virus poses substantial and increased risk to the health of the mother and the foetus during pregnancy (Gambon and Potter, 2009). During the 1918 to 1920 'Spanish flu' pandemic of H1N1 there is estimated to have been a 27% case fatality rate amongst pregnant women (Memoli *et al.* 2013), as compared to 2% overall. This effect was dramatically highlighted more recently during the 2009 'Mexican flu' H1N1 pandemic. Analysis of the data collected across the world suggests that pregnant women with the infection were significantly more likely to be admitted to hospital than non-pregnant women in the childbearing age groups and that they were twice as likely to die (Van Kerkhove *et al.*, 2011). Patients who were pregnant accounted for up to 20% of all patients with influenza who were hospitalised and the majority of them were in their third trimester (Van Kerkhove *et al.*, 2011). Seasonal influenza is also associated with higher than otherwise expected morbidity and mortality during pregnancy in otherwise healthy adult women (Memoli *et al.*, 2013).

The risk of an adverse outcome for the mother due to influenza infection rises with the stage in pregnancy (Memoli *et al.*, 2013) and is also greater if she has a pre-existing cardiac or respiratory condition. The foetus is also affected, with increased likelihood of miscarriage, stillbirth, pre-term delivery, low birth weight and neonatal mortality (Memoli *et al.*, 2013). The reasons for these effects are not fully elucidated, but both physiological and immunological factors appear to contribute. The growth of the uterus during pregnancy increases the pressure in the abdominal cavity and requires the position of the diaphragm to be moved upwards, thus reducing the lung capacity. This means that any respiratory infection can have more serious consequences than otherwise expected, with complications including pulmonary oedema and respiratory collapse (Memoli *et al.*, 2013). The later in the pregnancy that the infection occurs, the more these physical changes will contribute to the physiological problems. The immunological changes which occur during pregnancy (see above) seem to have an effect on the body's ability to respond adequately to the activity of the influenza virus, as evidenced by the higher viraemia observed in some studies (Memoli *et al.*, 2013). As noted for influenza infection in general, the mechanisms of the pathogenesis are currently not clear (Gambon and Potter, 2009; Memoli *et al.*, 2012).

5.3.2.2 Laboratory Diagnosis

Influenza virus can be detected in respiratory samples, such as throat swabs, sputum samples and bronchoalveolar lavages (SMI G8, 2014). Real-time, reverse-transcriptase PCR is the most commonly used test method. Assays are usually set up to detect conserved regions such as the nucleoprotein or matrix genes (SMI G8, 2014). Most routine diagnostic laboratories use multiplex methods for a range of respiratory viruses (e.g., Anderson *et al.*, 2013) and since all respiratory infections can produce severe symptoms during pregnancy, that approach would be recommended (see Figure 3.1). Even during an influenza season and with strong clinical indications that the patient is infected with this virus, the laboratory scientist should always bear in mind that there could be a different causative organism. Immunochromatographic POCT kits might be considered a useful option when a severely ill pregnant patient presents at the accident and emergency department. Comparisons of commercially available tests consistently show that their specificity for influenza A, in adult patients, including pandemic strains such as H1N1, is reported to be at least 98%, but sensitivity can be as low as 50% (Chartrand *et al.*, 2012). There are also bench-top molecular analysers which are intended to be used at the point of care. These can have relatively quick turnaround times (1–2 hours) and evaluation studies report high sensitivities (above 95%) with respect to the main laboratory PCR (e.g., Brendish *et al.*, 2015). The advantages of this must be weighed against the use of a more accurate, multiplex assay in the laboratory where the turnaround time could be 2–4 hours (Anderson *et al.*, 2013). In the case of severe respiratory infection during pregnancy the virological result is important in informing patient treatment, management and prognosis, so the optimal method must be employed.

Point to consider 5.3: Would POCT tests have a role in diagnosis of influenza infection in pregnant women during an epidemic (or pandemic)?

5.3.2.3 Management and Treatment

The approach to the management and treatment of influenza in a pregnant woman depends on the severity of the disease and the trimester (as mentioned above, these two are linked). Early in the pregnancy and in the course of the infection, antiviral chemotherapy would be safe and efficacious. Neuraminidase inhibitors such as oseltamivir or zanamivir are recommended (e.g., Rasmussen and Jamieson, 2014). When the patient presents with severe symptoms and is in the third trimester, supportive therapy is key; this can include intensive care treatment to maintain respiratory function. However, follow-up of patients treated during the 2009 H1N1 pandemic suggested the use of antivirals may also be helpful in this situation (Rasmussen and Jamieson, 2014).

Vaccination of pregnant women to prevent infection during pandemics or seasonal epidemics can be considered. Influenza A vaccine is generally deemed to be safe to administer during pregnancy (Fell *et al.*, 2013; Trotta *et al.*, 2014), as it is a killed vaccine. Until the mid-2000s clinicians generally tended to err on the side of caution and only considered giving the vaccine to women with underlying cardio-respiratory conditions or who would be in their third trimester during a serious outbreak (Fell *et al.*, 2013). It should be noted that the United States is an exception to this, as the influenza vaccine has been administered to pregnant women there since the 1960s (Fell *et al.*, 2013). The evidence gathered in the twenty-first century and in particular during the 2009 H1N1 pandemic indicates that vaccinees are not at greater risk of adverse outcomes to the pregnancy such as spontaneous abortion, stillbirth, pre-term delivery, the baby being small for dates or of low birth weight (Fell *et al.*, 2013). A limitation of the available data is that it is not usual to give the vaccine before the second trimester, so there is little evidence of the effect of the vaccine if given in early pregnancy. A study from Italy of over 86,000 pregnant women during the 2009 H1N1 pandemic, showed that for the more than 6,000 in receipt of the vaccine, there was no increased risk to the baby or greater likelihood of the mother being admitted to intensive care (Trotta *et al.*, 2014). There was a slightly higher rate of pre-eclampsia and diabetes in the women. This suggests that the overall progress of the pregnancy and health of the mother should be taken into account when deciding whether to vaccinate, but that it is a clinically acceptable option.

5.3.3 *Chlamydia*

Domain: Bacteria
Phylum: Chlamydiae
Class: Chlamydiae
Order: Chlamydiales
Family: *Chlamydiaceae*
Genus: *Chlamydia*
Species: *Chlamydia trachomatis*

5.3.3.1 Pathogenesis and Clinical Symptoms

Chlamydia trachomatis is an obligate intracellular bacterium, with a life cycle comprising two stages – the elementary body (EB) and the reticulate body (RB). The EB is the infectious stage (Murray *et al.*, 2013). Its cell wall contains a lipopolysaccharide (LPS), related to, but less toxigenic than Gram negative bacterial LPS. Attachment to host epithelial cells is mediated via the chlamydial major outer membrane protein (MOMP) which varies between strains and is immunogenic (Murray *et al.*, 2013). Variants are grouped into two 'biovars': the LGV biovar which contains organisms responsible for lymphogranuloma venereum (serovars L1, L2, L3) and the trachoma biovar which includes serovars A-C which are pathogens of the eye and serovars D-K, the genital strains (Mabey and Peeling, 2012). The EB is taken into the host cell by endocytosis and remains enclosed in a phagosome within the cytoplasm. Within 8 hours, it transforms into a larger, metabolically active RB, which grows and divides (Murray *et al.*, 2013). Over the course of the following 20–30 hours, binary fission produces >1,000 EBs which mature to the infectious state and are released on lysis of the infected epithelial cell (Mabey and Peeling, 2012).

Chlamydia is one of the most commonly diagnosed sexually transmitted infections worldwide (Mabey and Peeling, 2012). Infection rates are highest in the younger than

25-year-old age group. *C. trachomatis* infection is often sub-clinical and it is difficult to diagnose reliably without robust laboratory tests. This can lead to patients carrying the bacterium undetected for 10 years or more, resulting in chronic damage leading to infertility and pelvic inflammatory disease (PID) in females. The issues with diagnosis have hampered research on its effects during pregnancy (Horne *et al.*, 2011), although there does appear to be clear evidence for at least two possible adverse outcomes, namely pre-term delivery and ectopic pregnancy. Active *C. trachomatis* infection during pregnancy has been associated with pre-term delivery in prospective epidemiological studies (e.g., Rours *et al.* 2011), although the exact interaction between host and microorganism is not clear. There also appears to be a link between long-term undiagnosed *C. trachomatis* infection and an increased risk of ectopic pregnancy (Horne *et al.*, 2011). What is known about the pathogenesis of the latter is largely based on animal studies (Darville and Hiltke, 2010; Horne *et al.*, 2011). In this case, the pathology is considered to be due to the inflammatory response of the infected host epithelial cells in the genital tract mediated by pro-inflammatory cytokines; these include Interleukin (IL)-1, IL-6 and Il-8, tumour necrosis factor (TNF)-α and granulocyte-macrophage colony stimulating factor (GM-CSF). Chemokines, which recruit cells such as neutrophils, natural killer cells and monocytes are also involved (Darville and Hiltke, 2010) and the overall effect is to cause tissue damage. If the infection is untreated or the patient is re-infected, these effects are exacerbated by the specific immune response to the organism – possibly involving a type IV–delayed hypersensitivity response to Chlamydia heat shock protein 60 (Darville and Hiltke, 2010) resulting in scarring (Horne *et al.*, 2011). This damage seems to create the conditions for ectopic implantation of an embryo which is extremely dangerous to the mother and obviously means that the foetus is not able to survive (Horne *et al.*, 2011).

5.3.3.2 Epidemiology

It should be borne in mind that reported data about laboratory confirmed chlamydia infections are likely to be an underestimate – even in countries which have effective screening and awareness programmes and the best diagnostic tests. Nevertheless, it is estimated that global prevalence of *C. trachomatis* infection in females is 4%–6% and slightly lower in males at 3%–4%, with an incidence of over 130 million (Newman *et al.*, 2015). Since about 30% of PID cases are associated with chlamydia (Brunham and Rappuoli, 2013), where these infections are undiagnosed and untreated, this represents a substantial risk to future pregnancies. Infections are subclinical in up to 90% of women and 30%–50% of men (Brunham and Rappuoli, 2013). Due to the long-term sequelae of long-standing untreated infection, a number of countries have implemented systematic screening programmes, usually targeting sexually active young people. Examples of schemes which have been evaluated in detail include the National Chlamydia Screening Programme in England (Dorey *et al.*, 2012) and programmes in British Columbia, Canada (Rekart *et al.* 2013).

Point to consider 5.4: In the United States, it is recommended that chlamydia testing should be part of routine antenatal screening, while in the UK only symptomatic patients or those with a history suggesting a high risk of *C. trachomatis* infection are tested. What is your opinion of these differing policies?

5.3.3.3 Laboratory Diagnosis

Antigen detection is the most useful means of detecting *C. trachomatis* infection and while culture and immunological-based methods (direct fluorescent antibody (DFA) and EIA) are available, a nucleic acid amplification test (NAAT) is recommended (SMI V37, 2014). This is because culture is expensive and specialised and therefore not practical for routine diagnostic laboratories. Fluorescent antigen detection is quite subjective and EIA assays tend to have issues with both sensitivity and, more seriously, specificity (e.g., Bébéar and de Barbeyrac, 2009). In addition, NAAT tests can be set up to detect genetic variants such as new variant *C. trachomatis* (nvCT) which emerged in Sweden in the mid-2000s (Hadad *et al.*, 2009; SMI V37, 2014). Commercially available NAATs (Figure 5.1) are designed to be used in automated systems and there are methods which allow testing for both *C. trachomatis* and *Neisseria gonorrhoea* on a single sample, for optimum convenience and efficiency. Vaginal or endocervical swabs are recommended specimens for females (the former can be collected by the patient themselves if that is more convenient/acceptable). Sensitivity of detection of the bacterium in urine samples is unacceptably low in women (Allstaff and Wilson, 2012) although this sample is suitable and widely used for *C. trachomatis* diagnosis in men (SMI V37, 2014). POCT kits are under evaluation for use in genitourinary medicine (GUM) clinics, with the NAAT-based methods being of particular interest (e.g., Turner *et al.*, 2013). For pregnant patients, testing in the main laboratory is more suitable and allows investigation of strain variation for epidemiological purposes where required (Bébéar and de Barbeyrac, 2009). Antibody detection in serum might be preferable for the patient who would rather undergo collection of peripheral blood than a genital swab. This method also might be considered when investigating chronic, asymptomatic chlamydial infections which may be affecting fertility. While the test parameters are reported to be reasonably reliable, antibody titres tend to fall in the weeks and months after infection, which affects sensitivities (Horner *et al.*, 2013).

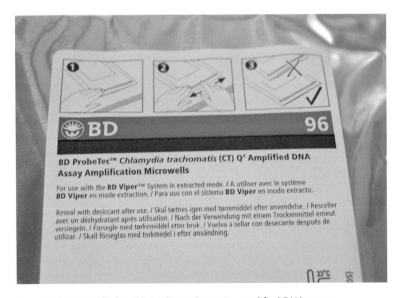

Figure 5.1 Test wells for *Chlamydia trachomatis*–amplified DNA assay.

5.3.3.4 Treatment and Management

Awareness of the risks of *C. trachomatis* infection, leading to diagnosis and treatment before conception through screening and effective GUM services would be ideal. This would increase the chances for successful conception and reduce the risk of ectopic pregnancy. Doxycycline or azithromycin are suitable antibiotic treatments. An active chlamydia infection during pregnancy can result in complications for the baby, such as premature delivery and low birth weight, as well as ophthalmia neonatorum (see Section 5.5.3 below) and pneumonitis (Allstaff and Wilson, 2012). Doxycycline is contraindicated during pregnancy, but erythromycin, amoxicillin and azithromycin are all safe and effective treatments (Allstaff and Wilson, 2012; Mabey and Peeling, 2012). Studies suggest that azithromycin is associated with fewer side effects, which makes compliance better (e.g., Pitsouni *et al.*, 2007). The patient should be followed up to ensure that the infection has been eliminated and further testing during the pregnancy may be considered, if the history suggests continued risk of exposure to *C. trachomatis* (Allstaff and Wilson, 2012).

5.4 Congenital Infections

The list of microorganisms known to be able to cross the placenta and infect the foetus during pregnancy is fairly short. Table 5.2 notes the main ones and shows that they are a variety of different types of organism. The placenta presents physical, biochemical and immunological barriers to any organism which enters from the maternal blood circulation (Robbins and Bakardjiev, 2012) and even if it does reach the growing foetus, infection is not an inevitability. The lack of cell differentiation and organ development may mean that specialised cells with particular receptors are not yet available for invasion. On the other hand, sometimes the effect on the foetus is so devastating as to effectively terminate the pregnancy.

Protocols of investigations of miscarriage, stillbirth or congenital abnormalities where an infectious cause is suspected often use the acronym TORCH. This stands for 'toxoplasma, rubella, cytomegalovirus and herpes' (although the last of these is usually a perinatal infection). Sometimes HIV and also syphilis are added, since these are infections which can be subclinical or overlooked. In addition, other organisms should be considered, such as parvovirus B19 and *Coxiella burnetii*. Another situation where laboratory testing is called upon is where the mother has been in contact with or has symptoms of chickenpox up 10 days before or after giving birth. Where the disease is confirmed or the mother has no evidence of pre-existing immunity to varicella zoster virus, administration of zoster immune globulin (ZIG) is used prophylactically (Breuer, 2009).

In this section, three examples (one bacterium, one virus and one protozoan) are selected for discussion, namely *Treponema pallidum*, rubella virus, and *Toxoplasma gondii*, but the reader is advised to consult other sources of information for accounts of other possible causes of congenital infection. Suggested examples include Manicklal *et al.* (2013) for cytomegalovirus, Marquez *et al.* (2011) for herpes simplex virus and Lamont *et al.* (2011) for varicella zoster virus. Boden *et al.* (2012) provide accounts of Q fever in pregnancy during outbreaks, while Carlier *et al.* (2012) review congenital parasitic infections.

Table 5.2 Organisms associated with congenital infections.

Organism	Possible effect on foetus
Cytomegalovirus	Foetal death, premature labour, live birth with severe abnormalities
Parvovirus B19	Foetal death, stillbirth, premature labour, hydrops foetalis
Rubella virus	Foetal death, live birth with severe abnormalities
Varicella zoster virus	Live birth with severe abnormalities
Lymphocytic choriomeningitis virus	Foetal death, hydrops foetalis, live birth with severe abnormalities
Treponema pallidum	Foetal death, stillbirth, hydrops foetalis, live birth with severe abnormalities
Listeria monocytogenes	Foetal death, stillbirth, premature labour, live birth with severe abnormalities
Coxiella burnetii	Stillbirth, intrauterine growth retardation
Brucella spp.	Foetal death, stillbirth, intrauterine growth retardation, premature labour
Mycobacterium tuberculosis	Premature labour, live birth with severe abnormalities
Toxoplasma gondii	Foetal death, stillbirth premature labour, live birth with severe abnormalities
Plasmodium spp.	Foetal death, stillbirth, intrauterine growth retardation, live birth with severe abnormalities
Trypanosoma spp.	Intrauterine growth retardation, live birth with severe abnormalities

Source: Information taken from Robbins and Bakardjiev (2012)

Point to consider 5.5: Look at the list of organisms in Table 5.2 which are all associated with congenital infection. Do they have any biological features in common which could facilitate transmission from the mother and successful infection of the baby?

5.4.1   *Treponema pallidum*

Domain: Bacteria
Phylum: Spirochaetes
Order: Spirochaetales
Family: *Spirochaetaceae*
Genus: *Treponema*
Species: *Treponema pallidum*

5.4.1.1   Syphilis Infection

Thanks to penicillin and changing social attitudes, the perception of syphilis as a death sentence and something to be ashamed of altered during the twentieth century. Since people were more likely to seek and diagnosis and treatment for the disease, prevalence declined in many countries during the 1950s onwards (Cockayne, 2012). The raised awareness about taking care over sexual health, which arose as a consequence of the spread of HIV in the 1980s, further reduced the incidence of *T. pallidum* towards the end of the last century. A number of factors, including availability of anti-retrovirals and complacency about risk of STIs, appear to have contributed to a re-emergence of syphilis (e.g., Simms *et al.*, 2005).

Between 1997 and 2008, the number of newly diagnosed cases of syphilis reported in England and Wales increased by over 1,000% overall (Jebbari *et al.*, 2011)! The reported incidence has a gender ratio of approximately 10 males: 1 female and the highest rates are recorded in males aged 23 to 44 years old and 25- to 34-year-old females (Jebbari *et al.*, 2011). The incidence of *T. pallidum* infection (new diagnoses, any stage of the disease) amongst women of childbearing age in the United Kingdom is between 200 and 300; for example, in 2013, the number of reported cases was 206 (Simms *et al.*, 2016). While this is a relatively low figure, due to the nature of the clinical disease (see below) as well as the stigma surrounding STIs (which still exists) data collected in the form of laboratory confirmed *T. pallidum* are likely to be an underestimate. Nevertheless the number of cases of congenital syphilis reported in the UK each year remains small (Simms *et al.*, 2016) and this is mainly attributable to the successful antenatal screening programme. Estimates suggest that about 1.4 million women across the world have active *T. pallidum* infection during pregnancy (Newman *et al.*, 2013). Since routine antenatal screening for syphilis is not carried out in every country many infections are unconfirmed. The authors conclude that this number of undiagnosed infected mothers would be expected to result in 215,000 cases where the foetal damage was incompatible with life, 90,000 babies so badly affected that they would die in the neonatal period, 65,000 premature congenitally infected babies and 150,000 children surviving with varying severities of congenital syphilis (Newman *et al.*, 2013).

In adults, the course of syphilis is recognised to have three distinct phases which proceed if the infection is not identified and treated. The acute phase (primary) is associated with the characteristic genital chancre, which appears about three weeks after infection (Peeling and Hook, 2006). It is not painful and it recedes, although the spirochaetes do spread systemically from the initial site of infection (Peeling and Hook, 2006). This is followed, a few months later, by the secondary stage, during which the patient experiences bacteraemia and may have a rash and/or other symptoms, before the bacterium becomes latent (Peeling and Hook, 2006). After a period of time – which may be many years – the patient develops tertiary syphilis; this can be manifest as granulomas infecting internal organs (so-called 'benign' tertiary syphilis), cardiovascular syphilis (involving the aortic values) or neurosyphilis (also known as general paralysis of the insane) due to destruction of organs and tissues by the spirochaete (Cockayne, 2012). Once a person has syphilis, they can be infectious at any stage, even when the bacterium is in the latent phase. If a pregnant woman is bacteraemic, the *T. pallidum* can access and then cross the placenta (Walker and Walker, 2007).

5.4.1.2   Pathogenesis and Clinical Symptoms of Congenital Syphilis

T. pallidum does not appear to be able to cross the placenta during the first trimester of the pregnancy, but from the fourteenth week onwards, it can reach the foetus and cause damage (Walker and Walker, 2007). The risk of vertical transmission when the mother has active syphilis is high (Peeling and Hook, 2006; Herremans *et al.*, 2010) and has been estimated in some studies as up to 100% during the primary stage of the disease (Herremans *et al.*, 2010). The outcome for the baby is generally poor – in about 40% of cases the result will be abortion, stillbirth or death during the immediate perinatal period (Walker and Walker, 2007). Those who survive may be born with a range of symptoms including low birth weight, hepatosplenomegaly, rash and bone abnormalities (Herremans *et al.*, 2010), although the majority are asymptomatic on first examination (Walker and Walker, 2007). Indications of infection which manifest during the first few weeks and months of life ('early congenital syphilis') again include liver and splenic enlargement, jaundice, rash and characteristic bone deformities (Herremans *et al.*, 2010). The latter is particularly apparent in the face, where defects in the nose and hard palate cause breathing difficulties and nasal discharge called, rather euphemistically, "snuffles" (Walker and Walker, 2007). The abnormal anatomical structure makes it hard for the baby to feed. Children whose symptoms appear after the age of two years fall into the 'late congenital syphilis' category (Walker and Walker, 2007; Herremans *et al.*, 2010) and they may have blunted teeth, keratitis and deafness (Walker and Walker, 2007) and sometimes develop serious neurological sequelae in adulthood (Walker and Walker, 2002).

Research into the pathogenic mechanisms involved in congenital syphilis has been hampered by the lack of a suitable *in vitro* culture system for *T. pallidum* and limitations of the rabbit animal model (Peeling and Hook, 2006). Nevertheless, it is thought that the bacterial lipoproteins activate macrophages and set up a strong inflammatory response, coupled with production of inflammatory cytokines such as IL-2, IFN-γ and TNF-α and this causes the damage to the foetus (Peeling and Hook, 2006).

5.4.1.3 Laboratory Diagnosis

Antenatal screening for syphilis usually involves testing blood or serum for the presence of specific anti-*T. pallidum* IgG by immunoassay (SMI V44, 2015). Any reactive samples should be repeated and must be confirmed using a different type of test, such as *Treponema pallidum* haemagglutination (TPHA) or *Treponema pallidum* particle agglutination assay (TPPA). This is very important, since the population of women of childbearing age the UK has a very low prevalence of this infection, which means there is a risk of false-positive results (see Chapter 1). If the initial result is corroborated by further tests, then it is necessary to determine the stage of the syphilis. Immunoassays for *Treponema pallidum* IgM are available and if this class of antibody is detected it would be considered indicative of recent infection. It is worth noting that these assays are subject to high false-positivity rates (SMI V44, 2015) and that IgM has been shown to be detectable for a year or more after successful antimicrobial treatment (SMI V44, 2015). The rapid plasma reagin (RPR) test can be used as a screening test, as it is a simple card agglutination assay (Figure 5.2). It takes about 10 to 15 minutes from receipt of the specimen to produce a result, which is read by eye. It is rather non-specific because it detects serum antibodies raised against cardiolipin which is released by *Treponema* spp. bacteria and other lipins which are produced as a result of cell damage. These antibodies are also found in patients with other conditions such as autoimmune diseases. Thus, alone it is not sufficient for a diagnosis of syphilis. However, the RPR test is quantifiable (through serial dilution of reagents) and it can therefore be helpful in ascertaining the stage of the infection. A titre of 16 or greater is considered a marker of recent or active infection (SMI V44, 2015).

Given that at least 35% of babies with early congenital syphilis who survive the perinatal period are asymptomatic at birth (Walker and Walker, 2007) laboratory investigations

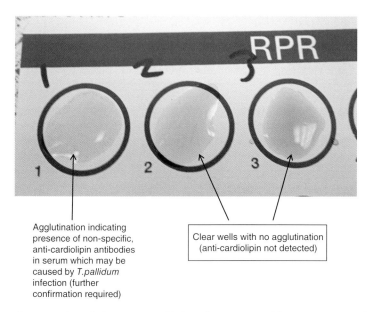

Agglutination indicating presence of non-specific, anti-cardiolipin antibodies in serum which may be caused by *T.pallidum* infection (further confirmation required)

Clear wells with no agglutination (anti-cardiolipin not detected)

Figure 5.2 Rapid plasma reagin (RPR) agglutination card for non-treponemal (cardiolipin) antibodies which may indicate *Treponema pallidum* infection. (*See insert for colour representation of the figure.*)

are a key part of confirming the diagnosis (Herremans *et al.*, 2010). Where the baby is suspected of having congenital infection, detection of the treponemes, in nasal discharge or swabs from skin lesions, by dark ground microscopy is a sensitive and specific method of confirmation (SMI V44, 2015; Herremans *et al.*, 2010), but it requires specialist expertise and is therefore not widely available (SMI V44, 2015). PCR assays have been developed (SMI V44, 2015; Herremans *et al.*, 2010), in some cases as part of a multiplex method (SMI V44, 2015), which can be used on a range of samples including amniotic fluid, the baby's serum, CSF and NPA, as well as the lesion swabs and nasal fluid (SMI V44, 2015; Herremans *et al.*, 2010). Published information about evaluation of PCR protocols is rather limited, but sensitivities of up to 86% and specificities of 100% have been reported (Herremans *et al.*, 2010).

Inconclusive results need to be followed up with serology, but the results need to be interpreted carefully, preferably comparing maternal and neonatal results and titres, to exclude the passive carriage of the mother's antibodies. Where anti-treponemal IgM is detected in the infant's sample, this is a strong indication of congenital infection (Herremans *et al.*, 2010), but it is recommended that this is confirmed with TPPA and RPR (SMI V44, 2015) to ensure that it is not a false-positive. The TPPA and RPA detect IgG antibodies; therefore in a child less than a year old, reaction in either of these assays alone is not conclusive. Nevertheless, if the RPR titre in the baby's serum is determined in parallel with that of the mother and is found to be at least four fold greater and anti-*T. pallidum* antibodies are also detected by the TPPA assay, this is suggestive of congenital syphilis, regardless of the IgM result (SMI V44, 2015). Serological investigations are also advocated for infants who may be at risk of infection but are asymptomatic. Follow-up tests during the first year of life to monitor antibody titres is recommended (SMI V4, 2014; Herremans *et al.*, 2010). For late congenital syphilis, detection of antibody gives a clearer picture (although the unlikely, but not unheard of, possibility of postnatal acquisition through abuse should be taken into account). Current assays to detect IgM or the bacterial DNA in CSF to confirm neurosyphilis are reported to have low specificities (SMI V44, 2015; Herremans *et al.*, 2010).

5.4.1.4 Management and Treatment

Antenatal screening (see above) is a successful and cost effective method of preventing mother to child transmission of syphilis (Walker and Walker, 2007; Newman *et al.*, 2013). Where resources (in terms of both finance and trained laboratory staff) are limited, the RPR can be used as a point-of-care test (Hawkes *et al.*, 2011), since it is relatively inexpensive and is a card latex agglutination assay, so is quick and easy to perform. Where evidence of *T. pallidum* infection is found in a pregnant woman's serum, she can be treated safely and effectively with penicillin (Walker and Walker, 2007). This has been found to reduce rate of foetal and perinatal death attributable to syphilis (Walker and Walker, 2007; Hawkes *et al.*, 2011). In countries such as the UK where blood samples for microbiological screening are usually collected during the initial antenatal visit at 12 weeks gestation, vertical transmission should be largely preventable (since the risk is low before 14 weeks – see above).

In situations where screening is not offered (or not accessed) and a baby is born with the infection, commencing penicillin treatment as soon as possible has shown to be valuable in mitigating against the bacterium causing further damage to the child (Peeling and Hook, 2006; Walker and Walker, 2007). Benzylpenicillin is the only suitable treatment for *T. pallidum* infection. Therefore, clinicians treating patients who are allergic to penicillin are advised to put them through a desensitising programme first and then give the antibiotic (Walker and Walker, 2007).

5.4.1.5 Epidemiology

Reliable data regarding the rate of mother-to-child transmission of syphilis globally is difficult to access; this is at least partly due to the number of pregnancies which do not continue to term as a result of the infection and partly because the of proportion of babies with the infection who are asymptomatic at birth. Epidemiologists therefore tend to concentrate on the actual or estimated maternal infection rates and extrapolate from this (e.g., Newman *et al.*, 2013; WHO [http://www.who.int/repro ductivehealth/topics/rtis/syphilis/en/]). Reported rates of syphilis in pregnant women identified through antenatal screening seem to vary from year to year in some countries and differ widely between countries in the same geographical region (WHO). In general, however, it seems that the WHO Eastern Mediterranean region tends to report the lowest incidence of *T. pallidum* in their antenatal population and the WHO Africa region reports the highest (Newman *et al.*, 2013) and overall the number infected is over a million each year (WHO). It is estimated that in a third of the countries sending data to the WHO, at least 1% of women attending antenatal screening have active syphilis infection (WHO), but screening is far from uniform or universal (WHO). In the UK, congenital syphilis is rare and occurs at a rate of <0.5/1000 live births. Between 2010 and 2015, 17 babies born with the infection were identified (Simms *et al.*, 2016).

Point to consider 5.6: What factors might affect the accuracy of the data collected by the WHO about antenatal screening and *T. pallidum* positivity rates? In terms of control of mother to child transmission of syphilis do the likely inaccuracies matter?

5.4.2   Rubella

Family: *Togaviridae*
Genus: *Rubivirus*
Species: *Rubella virus*

5.4.2.1   Rubella Infection

Rubella (German measles) used to be a common childhood infection and is characterised by a characteristic maculopapular and relatively short-lived rash, which starts on the face and moves first to the trunk and then the limbs (Best *et al.*, 2009). The disease acquired its colloquial name because the first clear descriptions and studies of the clinical condition were reported from Germany in the eighteenth century (Best *et al.*, 2009). There is an incubation period of two to three weeks, but virus shedding begins during the first seven days of this time. In children, rubella is usually less debilitating than measles and apart from the rash (and occasionally lymphadenopathy), which usually appears quite suddenly and lasts for about three days, other symptoms are rarely reported (Hesketh, 2012). It is thought that up to half of childhood German measles infections are actually asymptomatic (Hesketh, 2012). In adults there may be a more distinct prodromal phase, lasting 12 to 21 days, during which the patient has a headache, pyrexia, sore throat, cough and sometimes conjunctivitis (Best *et al.*, 2009, SMI G7, 2014). About 70% of adult females with primary rubella infection also experience arthralgia, which can last for several weeks (Best *et al.*, 2009). Rare complications include encephalitis, which appears within weeks of the primary infection and thrombocytopaenia which is found in approximately 1 in 3,000 cases (Best *et al.*, 2009). German measles is not usually a serious risk to health, except when the mother acquires the primary infection during pregnancy.

5.4.2.2   Pathogenesis and Clinical Symptoms of Congenital Rubella

It was Gregg, who was a paediatric ophthalmologist in Sydney, Australia who deduced the link between primary rubella virus infection in the mother and what is now called congenital rubella syndrome (CRS) in their babies (Hesketh, 2012; Best *et al.*, 2009). He noticed that an unusually high number of children were referred to his clinic with congenital eye defects such as cataracts in early 1941 and on questioning the mothers found them many of them had been affected by German measles during an outbreak in the area during 1940 (Hesketh, 2012). It is now known that rubella virus can infect and subsequently cross the placenta (Best *et al.*, 2009). It is able to enter and replicate in foetal cells causing both tissue destruction and the slowing down of cell division. During the first trimester of pregnancy, when cells are relatively undifferentiated (and therefore vulnerable to infection), organ formation should be occurring rapidly and the immune response is not well developed, the virus can cause widespread damage (Best *et al.*, 2009). A combination of cardiac abnormalities, cataracts, neurological defects, thrombocytopaenia, hepatitis and intrauterine growth retardation (IUGR) occur in up to 90% of babies whose mothers have primary infection during the first 12 weeks of pregnancy (De Santis *et al.*, 2006; Hesketh, 2012) and in many cases, the foetus spontaneously aborts.

After the first trimester, the risk to the baby decreases markedly (De Santis *et al.*, 2006). If the maternal infection occurs between 13 and 16 weeks of pregnancy, the likelihood of the child developing congenital abnormalities is estimated to be 17%, falling to 6% for 17 to 20 weeks and 2% for later gestational ages (Best *et al.*, 2009). Also, after 12 weeks, the effects of CRS are reported to be less extensive, rarely comprising multiple anomalies. The most commonly noted outcomes at birth are deafness and retinopathy, although other eye problems (such as cataracts and glaucoma), heart defects (such as patent ductus arteriosus), and delay in neurological and physiological development may become obvious later in life (Hesketh, 2012). An intriguing complication is type 1 (insulin-dependent) diabetes mellitus (IDDM) which is reportedly seen in up to 20% of CRS children who are followed up into adulthood (Stene and Gale, 2013) and is thought to be due to an autoimmune response initiated in some way by the virus. The exact mechanism remains elusive, although research evidence suggests that CRS predisposes susceptibility (in conjunction with exposure to other risk factors) rather than causes IDDM *per se* (Stene and Gale, 2013). Progressive rubella panencephalitis is clinically similar to the subacute sclerosing panencephalitis associated with measles (Schneider-Schaulies and ter Meulen, 2009); although much rarer, it is similarly invariably fatal (Best *et al.* 2009).

Maternal rubella infection immediately prior to conception is not considered a risk to the baby (De Santis *et al.*, 2006). However, foetal infection has been reported to occur if the virus is acquired just after conception. So it would be technically possible – though incredibly unfortunate – for a woman to be incubating German measles and be in the first few weeks of pregnancy and be unaware of either! Although there is only one type of rubella virus, re-infection has been reported in people who have demonstrable pre-existing anti-rubella IgG. It is not a common occurrence, is often asymptomatic and is usually only investigated when recognised in a pregnant woman (Hesketh, 2012). Reinfection is reported more often in people with low antibody titres, especially when they acquired their immunity through the vaccine rather than natural infection. It has been associated with CRS (e.g., Pitts *et al.*, 2014), although the risk of transplacental infection during a re-infection event is estimated at around 8% in the first 16 weeks and malformations are reported to be rare (Best *et al.*, 2009; Tipples, 2011).

5.4.2.3 Laboratory Diagnosis

Routine antenatal screening comprises testing for serum anti-Rubella IgG levels. It is generally accepted that a titre of 10 mIU/mL provides protective immunity (Best *et al.*, 2009, SMI G7, 2014; SMI V30, 2015). Pregnant women with a low titre or no detectable antibody are advised to very careful about avoiding contact with anyone with a rash and/or possible rubella and would be offered the vaccine postpartum. It is worth noting that some people do not respond to particular vaccines and there are cases of people who have documented evidence of rubella immunisation but no discernible antibody. It is thought likely that such people do nevertheless have some protection against infection (partly due to the stimulation of the immune system by the vaccine) but that the circulating antibody titre is below 10mIU/mL (Best *et al.*, 2009). In countries where the

MMR vaccination programme is deemed to be working well, routine antenatal screening may be discontinued. The rationale is that the resources are better spent on maximising childhood vaccination rates and targeting women more likely to be seronegative for anti-rubella virus antibody (e.g., ECDC, 2016; PHE, 2016)

When a pregnant woman presents with a red rash or a history of contact with someone with a rash and/or arthralgia, it is important to establish the cause. If she has no previous history of rubella infection or vaccination, it is clearly vital to confirm or eliminate rubella virus infection. Similar fleeting rashes and joint pains can be produced during infection with parvovirus B19, which can also result in miscarriage or an abnormality called hydrops foetalis. Other infectious agents causing a red rash include some enteroviruses and of course measles virus (SMI V30, 2015; Tipples, 2011). More unusual viruses such as Dengue should also be considered according to epidemiological risk and individual travel history (SMI G7, 2014; Tipples, 2011). It should be noted that during a documented rubella outbreak, Hukic *et al.* (2012) found that less than half of the clinically suspected cases were subsequently confirmed by laboratory tests.

Serology is the best way to investigate potential rubella virus infection during pregnancy. If it is feasible, the ideal first step would usually be to test a serum sample predating the contact with possible German measles for anti-rubella IgG (or if that is not possible to at least check for previously recorded serological results). If there is credible evidence of a pre-existing antibody titre over 10 IU/ml, then rubella virus is most unlikely to be the cause of any symptoms and the patient would not be considered at particular risk in the case of contact. However, re-infection should be excluded by further tests (see below).

Diagnosis of primary rubella would involve testing an acute blood sample for IgM and there is a range of commercially available EIA assays for this (Tipples, 2011; van Helden *et al.*, 2014). Results must be interpreted with caution, as issues with both sensitivity and specificity have been reported (Best *et al.*, 2009; Tipples, 2011) In spite of the two- or three-week incubation period, IgM does not reach detectable levels until a few days after the appearance of the rash (SMI G7, 2014) and a negative result in the very early stages of infection may be misleading. If not followed up, this can lead to missed diagnoses and under reporting (Hukic *et al.*, 2012). Another factor is the lack of specificity in some assays (van Helden *et al.*, 2014). Antibodies raised against other infections agents and autoimmune antibodies, including rheumatoid factor, are known to give false-positive results in some IgM assays (Tipples, 2011; van Helden *et al.*, 2014), so assay controls must be applied appropriately. Since symptoms of joint pain and swelling are common in primary rubella in postpubertal females and clinically this might be the only presenting feature, it would be important not to confuse German measles with rheumatoid arthritis! Anti-rubella IgG antibody is likely to be undetectable or just detectable (i.e., giving a borderline result) within a week or so of the symptomatic disease (or possible contact), but in a primary infection, titres would be expected to rise within 10 to 14 days (SMI G7, 2014) and this should be quantifiable in an EIA (van Helden *et al.*, 2014). Another test which can be useful, particularly if the timing of the onset of infection is not known or is more than three months previously, is IgG avidity (Best *et al.*, 2009; Tipples, 2011). Recently produced IgG will have low avidity and this will complement the IgM result. Re-infection can be diagnosed by detecting rising titres of IgG in a patient with previous evidence of protective levels of antibody (Tipples, 2011). This antibody would be expected to be of high avidity. In this situation, IgM antibody may or may not be produced to measurable levels (SMI G7, 2014).

Where indicated by maternal serology and facilities allow the possibility, the infection can also be confirmed pre-natally, by detection of the virus in amniotic fluid or foetal cord blood using reverse-transcriptase PCR. Products of conception after a spontaneous or therapeutic abortion would also be tested in this way for the presence of the rubella virus (de Santis *et al.*, 2006; Tipples, 2011). Postnatal laboratory diagnosis of CRS in a child relies on detection of IgM in cord or peripheral blood, since IgM does not cross the placenta (Hesketh, 2012); assays to detect the antibody in oral fluid are also available (Best *et al.*, 2009). The rubella virus can be found in blood, urine and tissue samples collected from a range of sites and this is best achieved by reverse-transcriptase PCR (Best *et al.*, 2009; Hesketh, 2012).

Both German measles and CRS are notifiable diseases in the UK and many other countries. This contributes to monitoring of the effectiveness of both the MMR vaccine and the reduction in congenital infection. When the virus itself is isolated, it is useful for epidemiological purposes to send it to a reference centre for sequencing. There is only one serotype of rubella virus, but there are 13 known genotypes (Abernathy *et al.*, 2011). It is important to know which are circulating as part of surveillance towards elimination of rubella infection and CRS. The most commonly reported genotype in the UK is 2B (Abernethy *et al.*, 2011; Muscat *et al.*, 2014).

5.4.2.4 Management

Due to the high chance of the baby developing severe defects, which may not be compatible with life, when there is laboratory confirmation of primary maternal rubella infection in the first trimester, a therapeutic abortion may be offered to the patient (Best *et al.*, 2009; van Helden *et al.*, 2014). During the later stages of pregnancy, the risk of CRS is considerably lower and guidance to clinicians recommends reassurance and careful monitoring of the foetal development. There have been some trials using large doses of human immunoglobulin after known exposure to rubella virus during pregnancy, but its efficacy in preventing CRS is not proven (De Santis *et al.*, 2006, Hesketh, 2012).

5.4.2.5 Epidemiology

Rubella, and thus CRS, are not commonly seen in most countries of the world now, due to concerted vaccination campaigns and accompanying surveillance (Reef *et al.*, 2011). Prior to the introduction of the vaccine in the late 1960s, German measles was a seasonal (spring/summer) disease, particularly in temperature climates and large outbreaks tended to occur every three or four years (Best *et al.*, 2009). The usual transmission has been disrupted since the licensing of the RA27/3 live attenuated vaccine in 1969. Some countries, such as the United States included the vaccine in their universal childhood immunisation policy straight away. By contrast, in the UK and other European countries, the more targeted approach of vaccinating schoolgirls between the ages of 11 and 14 years and susceptible women of childbearing age was used. This was supported by strategic screening of antenatal patients and of both females and males in occupations likely to bring them into contact with rubella (such as healthcare workers and school teachers) to identify and vaccinate seronegative people (Best *et al.*, 2009). This reduced the incidence of CRS quite dramatically. In the UK there were around 70 CRS live births each year and an estimated 10 times as many therapeutic abortions on the grounds of confirmed maternal rubella in the early 1970s. The rate fell to an annual average of fewer than 10 cases by the 1990s. In line with many other countries, the UK changed to using the trivalent mumps, measles and rubella (MMR) vaccine – which incorporates the RA27/3 – in 1988 as part of the immunisation programme for all pre-school children. The rationale for this is to eliminate circulation of wild-type virus altogether.

By December 2009 it was reported that 130 out of the 193 World Health Organisation (WHO) member countries used a rubella vaccine as part of their childhood immunisation schemes (Reef *et al.*, 2011). The efficacy and practicality of combined vaccines meant that the majority (115) were using the MMR formulation, while 12 utilised the bivalent measles rubella (MR) vaccine. The remaining member states used either measles/mumps/rubella/varicella zoster (two countries) or the individual rubella vaccine (one country). Disease surveillance was a bit more sporadic and the total number of CRS cases reported was 165 – although this data was collected from only 123 countries (Reef *et al.*, 2011). In 2012, the WHO launched a Global Measles and Rubella Strategic Plan (http://www.who.int/mediacentre/factsheets/fs367/en/), which aims to reduce and eventually eliminate CRS (along with measles) through effective vaccination and surveillance. Across the European WHO region, 68 confirmed CRS cases were recorded between 2005 and 2009 (Zimmerman *et al.*, 2011) and 60 in 2012, although this is likely to be an underestimate (Muscat *et al.*, 2014). Reported outbreaks of mumps and measles in Europe during the early twenty-first century associated with less than optimal MMR vaccine coverage, suggest that some younger women of childbearing age may be now susceptible to rubella and that this should be borne in mind during laboratory investigations of symptoms or contact with red rash. It is interesting to note that during a large outbreak of German measles in Poland in 2013, the majority of cases were in 15- to 29-year-old males (Muscat *et al.*, 2014). This was attributed to the Polish vaccination strategy which between 1989 and 2003 was to immunise teenage girls, before changing to universal childhood MMR vaccination in 2004 (Muscat *et al.*, 2014). Thus, in 2013, adult males under the age of 35 would have expected to be more susceptible than other groups due to the partial interruption of transmission during their adolescence afforded by selective vaccination of their sisters and female schoolmates. However, an account of an outbreak in Bosnia and Herzegovena in during the late 2009 and early 2010 (Hukic *et al.*, 2012) similarly reported most cases in 15 to 29 year olds and almost two and a half times more rubella infections in males than females. In this situation, around 80% of the patients were either known to have never been immunised against rubella or their vaccination status was uncertain – that is, both males and females could be considered equally susceptible in terms of lacking protective antibody – suggesting a possible genuine gender difference.

Universal childhood MMR vaccination started in 1980 in Bosnia and Herzegovina, but the programme was interrupted during the Balkans War (1992 to 1995). As a result, it was estimated that nearly 15% of women of childbearing age did not have protective immunity against rubella at the time of this outbreak (Hukic *et al.*, 2012) and some cases of maternal infection during pregnancy were recorded. Rises in reported cases of measles across Europe (Muscat *et al.*, 2014; Pitt, 2013) have highlighted issues with inadequate MMR vaccine cover and this appears to be leaving women susceptible to primary rubella in pregnancy once again. For example, in the West Midlands, UK, susceptibility to rubella infection (rubella IgG titre <10 IU/ml) found amongst antenatal patients through routine screening is reported to have markedly increased from 1.4% to 6.9% between 2004 and 2011 (Skidmore *et al.*, 2014). A survey of 5,413 women attending antenatal care in this region revealed that the non-immune rate amongst those who had been born before 1986 was 2.4%. In contrast almost 18% of the pregnant women screened in this study who had been born in 1987 or later did not have protective levels of rubella IgG (Skidmore *et al*., 2014). People born in 1986 were the first cohort to be given MMR as children in the UK, whereas previously girls were vaccinated as teenagers and would also have been exposed to wild type virus to boost immunity. Another study in West London (UK) reported a change in proportion of susceptible people from 4.1% in 2007 to 6.8% in 2012 (Mortlock, 2015). In this investigation, retrospective analysis of records of 30,756 patients (mostly antenatal clinic attendees and health care staff) who were tested for anti-rubella IgG during this period, showed a significant difference in non-immune rates between people resident in two London boroughs. A greater percentage of those living in the area which had a higher proportion of recent immigrants and a more transient population had titres <10 IU/mL (Mortlock, 2015). This highlights another drawback of universal childhood vaccination in the modern world and shows the value of follow-up testing and/or immunisation. It appears that there is still some work to do before elimination of CRS can be achieved (Skidmore *et al.*, 2014).

Point to consider 5.7: Skidmore *et al*. (2014) question whether the strategy of universal childhood vaccination is the right approach to eliminate CRS. What do you think?

5.4.3   *Toxoplasma gondii*

Kingdom: Protista
Phylum; Apicomplexa
Genus: *Toxoplasma*
Species: *Toxoplasma gondii*

5.4.3.1   Toxoplasmosis Infection

Toxoplasma gondii is usually transmitted to humans zoonotically, through contact with infected cat faeces (directly or via contaminated food or water) or ingestion of poorly cooked contaminated meat (Greenwood, 2012). There are also sporadic cases of human-to-human infection via transfusion or transplant (Robert-Gangneux and Dardé 2012), although this is rare due to screening regimes. The life cycle involves the cat as the definitive host in which sexual reproduction, resulting in oocysts, occurs. Cats excrete oocysts in an underdeveloped (and therefore non-transmissible) form and development to the sporulated (infectious) form usually takes 36 to 48 hours. The intermediate host can be any mammal, including humans (https://www.cdc.gov/dpdx/toxoplasmosis). On ingestion by this secondary host, the sporulated oocysts are released into the small intestine and, via macrophages, enter the blood stream and thus reach cells in many parts of the body (Greenwood, 2012). They are largely indiscriminate in the cell type which they can infect at this point (Dubey, 2005) and they grow and divide rapidly (tachyzoite stage), with progeny parasites invading adjacent cells. When challenged by the host's immune response (which includes necrosis and inflammation around the site of parasite multiplication), they form less active bradyzoites contained within cysts (Dubey, 2005). These are usually lodged in tissues (particularly central nervous system and skeletal and cardiac muscles). If this intermediate host is then preyed upon by a cat, the life cycle can continue (Robert-Gangneux and Dardé, 2012).

In humans, the bradyzoites tend to remain dormant, but can reactivate (and thus cause more damage) if the person is immuncompromised (Pappas *et al.*, 2009). Humans can also ingest *T. gondii* tachyzoites and/or bradyzoites from the muscle tissue of other secondary hosts such as sheep and pigs in the form of undercooked meat. The parasite again reaches the small intestine, where bradyzoites present become activated and, along with the tachyzoites, invade the host tissue as described above (Robert-Gangneux and Dardé, 2012).

People who have regular close contact with cats and other domestic animals and those who regularly consume meat that is still pink, commonly acquire *T. gondii* unwittingly. Human infection is often asymptomatic, but toxoplasmosis can be manifest as lymphadenopathy, along with fairly non-specific symptoms such as lethargy and pyrexia (Pappas *et al.*, 2009). Prevalence studies have shown that in the UK about 22% of the healthy adult population has serological evidence of past *T. gondii* infection, while in the United States this rate is reported to be around 32% (Dubey, 2005). In Central and South America, prevalence can be considerably higher at up to 40%. This has been associated with poorer sanitary conditions leading to more widespread contamination of food and water with parasite oocysts (Dubey, 2005; Robert-Gangneux and Dardé 2012) and consumption of animals at high risk of infection such as rodents and primates (Pappas *et al.*, 2009). In France, where eating rare meat is more common than in the UK, some surveys have found *T. gondii* IgG antibody detection rates at over 80% (Dubey, 2005).

Point to consider 5.8: What advice would you give to a woman in the early stages of pregnancy who is planning to give away her cat in order to prevent congenital toxoplasmosis in her baby?

5.4.3.2 Pathogenesis and Clinical Symptoms of Congenital Toxoplasmosis

When the mother has primary toxoplasmosis during pregnancy, it is likely that she will not experience patent symptoms, but the tachyzoites in her blood may reach the placenta. Here they are able to grow, divide and then move to the trophoblasts in the foetal tissue (Dubey, 2005; Carlier *et al.*, 2012). The chance of foetal infection rises as the pregnancy progresses (De Paschale *et al.*, 2010), since the placenta appears to act as a more effective barrier to transmission in the early gestation period (Robert-Gangneux and Dardé 2012). The risk is estimated to be about 6% to 10% during the first trimester, 22% to 40% in the second and 58% to 72% in the last trimester (Carlier *et al.*, 2012; Robert-Gangneux and Dardé 2012). The extent of foetal damage decreases with gestational age (De Paschale *et al.*, 2010; Carlier *et al.*, 2012). The damage caused by the parasite when the primary maternal infection occurs in early pregnancy can be so extensive as to be incompatible with life, leading to spontaneous abortion or stillbirth (Pappas *et al.*, 2009; Robert-Gangneux and Dardé, 2012). Babies that are born alive will be severely handicapped, with a range of conditions including hydrocephalus, intracerebral calcification, convulsions and intrauterine growth retardation and retinochoroiditis (inflammation of the retina and choroid) and a consequently poor prognosis (Dubey, 2005). When the mother is infected after the first trimester, foetal pathology is often limited to the eye (Carlier *et al.*, 2012). While in some cases the child is blind at birth, the damage is often not apparent until later in life (Dubey, 2005). It has been noted that reactivation of a latent *T. gondii* infection in pregnant women who are immunocompromised (e.g., due to autoimmune disease or HIV infection) can lead to congenital infection (Robert-Gangneux and Dardé, 2012). There are also occasional reports of re-infection causing foetal damage (Robert-Gangneux and Dardé, 2012).

5.4.3.3 Laboratory Diagnosis

Policy on antenatal screening for toxoplasmosis infection varies between countries. Routine serological testing is not carried out in the UK at any stage in the pregnancy. In other European countries, such as Austria, France and Italy, all patients are initially screened for anti-*T. gondii* IgG. Those that are seronegative are encouraged to present for regular (ideally monthly) blood collection throughout the pregnancy; the samples are tested for the presence of anti-*T. gondii* IgM (Cornu *et al.*, 2009; De Paschale *et al.*, 2010; Sagel *et al.*, 2012), thus allowing early detection and treatment of primary infection. This strategy is reported to be effective in allowing early detection (e.g. De Paschale *et al.*, 2010), but compliance appears to be variable in all three countries, particularly in the final trimester when the risk of parasite transmission is greatest (Cornu *et al.*, 2009; de Paschale *et al.*, 2010; Sagel *et al.*, 2012).

Detection of maternal primary infection is based on detection of anti-*T. gondii* antibodies in serum. The recommended strategy is to test for IgG and IgM using an EIA (Robert-Gangneux and Dardé, 2012). IgG results can be confirmed in a reference laboratory using the gold standard Sabin-Feldman dye test, which uses complement to detect the antibody-antigen reaction (Robert-Gangneux and Dardé, 2012). A standard EIA is preferred to an immunosorbent agglutination assay for IgM (IgM-ISAGA), since the latter is highly sensitive and can detect antibody for up to 12 months after the primary infection, which may not be helpful (Robert-Gangneux and Dardé, 2012). Where IgM is detected in maternal serum, then confirmation of this, plus IgG avidity testing are required to ascertain the timing of the infection. This is because most EIAs would be expected to detect IgM up six months post-infection, but when the mother experiences the acute infection prior to conception this is not considered a risk to the baby. Therefore an IgG avidity test – which is capable of discriminating between IgG produced in the preceding six months and that attributable to an infection more than six months previously – is needed (Robert-Gangneux and Dardé, 2012). This would usually be carried out in a reference laboratory due to technical and scientific expertise involved both in performing the assay and in interpreting the results. A follow-up blood sample should also be collected 10 to 14 days later to look for increase in IgG titre (Robert-Gangneux and Dardé, 2012).

Once maternal primary infection is confirmed (or strongly suspected pending confirmation), laboratory investigations can be used to detect foetal infection *in utero*. Cord blood can be analysed for parasite IgG, IgM and IgA; this is not very reliable, as antibody detection rates in cases where the baby is later found to be infected tend to be low. Detection of *T. gondii* DNA in amniotic fluid by PCR is a lot more sensitive and specific, although the performance of commercially available molecular assays is reported to be variable (Robert-Gangneux and Dardé, 2012). Morelle *et al.* (2012) compared a commercial kit which used nested PCR with two 'in house' assays, which were a conventional and a real-time PCR. They used amniotic fluid collected from 33 known or suspected cases of congenital toxoplasmosis and 3 simulated preparations in which *T. gondii* grown *in vitro* was added to amniotic fluid taken from women in whom toxoplasmosis had definitely been excluded (Morelle *et al.*, 2012). The commercial assay did not detect parasite DNA in one of the simulated samples, whereas the 'in house' assays both gave positive results for all three. Amongst the clinical samples, 12 had been taken during pregnancies where the child was later confirmed as having congenital toxoplasmosis. The 'in house' tests detected parasite DNA in 11 of these amniotic fluids, compared with 6 for commercial assay (Morelle *et al.*, 2012). It is for this reason that authors advocate relying on the expertise of a specialised reference laboratory where possible for prenatal diagnosis of congenital toxoplasmosis (Robert-Gangneux and Dardé, 2012).

Investigations in the neonate similarly involve comparative serology of cord blood (or the baby's blood) and a maternal sample. If IgA and IgM are detectable in the baby's blood but not the mother's, then that is indicative of congenital toxoplasmosis (Carlier *et al.*, 2012). Sensitivities for serological tests are reported at 70% at best (even for the IgM-ISAGA) and IgM is more likely to be detectable if the maternal infection occurred during the last trimester of pregnancy (Robert-Gangneux and Dardé, 2012). Thus, where primary toxoplasmosis was diagnosed in the mother and/ or the clinical condition of the baby points to congenital infection with the parasite, regular serological follow-up on the child must be carried out. Where IgG is detectable in both mother and baby, there must be a demonstrably higher titre in the child to confirm the diagnosis. Western blotting can allow qualitative comparison of the IgGs (Carlier *et al.*, 2012; Robert-Gangneux and Dardé, 2012). Also, any maternal IgG which was passively transferred would be expected to decline within 12 months, so persistent antibody would only be found in congenital infection. The parasite can be detected by PCR in amniotic fluid, cord blood or placenta; in some countries, tests involving inoculation of these specimens into mice are used (Carlier *et al.*, 2012; Robert-Gangneux and Dardé, 2012). It should be noted that the presence of *T. gondii* DNA in the placenta is not usually taken as sufficient for a diagnosis of congenital infection on its own, as the parasite may not cross the placenta (see above). However, Robert-Gangneux and Dardé (2012) report some cases where this was indeed the only evidence. This is a clear example of a situation where the laboratory results must be carefully interpreted in the clinical context. Where the child has a less severe infection, which is not recognised until later in life, serology may be of limited use, but parasite detection may be helpful. For example, isolation of *T. gondii* DNA from aqueous humour in patients with retinochoroiditis has been reported (Robert-Gangneux and Dardé, 2012).

5.4.3.4   Management and Treatment

Where primary infection is confirmed during the pregnancy, the family must be carefully counselled with respect to the relative risks of infection and possible outcomes for the child. Ideally, investigations to determine whether the parasite has crossed the placenta would be carried out. Termination may be considered, but safe, effective antibiotic treatments are available. Recommended regimes tend to vary slightly between countries, but if the infection occurs during early pregnancy, spiramycin can be used, as this is non-teratogenic. After 16 to 18 weeks, treatment switches to a combination of sulfadiazine and pyrimethamine, plus folinic acid, daily for the remainder of the pregnancy (Carlier *et al.*, 2012; Hotop *et al.*, 2012). Sulfadiazine and pyrimethamine are considered to be more effective than spiramycin against active tachyzoites (Hotop *et al.*, 2012). Studies have variously shown good evidence of either a better outcome for the baby with maternal treatment or no evidence of any effect at all (Robert-Gangneux and Dardé, 2012)! However study design and treatment programmes vary. In the UK and France, the sulfadiazine, pyrimethamine and folinic acid combination is administered where laboratory confirmation of foetal infection has been obtained (Hotop *et al.*, 2012), which means that treatment does not start until the eighteenth week. Spiramycin is only used where the laboratory diagnosis has not been possible (e.g., due to negative test results or lack of suitable samples), but it is continued until the baby is delivered (Hotop *et al.*, 2012). In Germany, spiramycin is given if maternal infection is diagnosed in the first 16 weeks and this is changed to the combination thereafter, regardless of whether investigations have been conducted to determine whether the parasite has reached the foetus (Hotop *et al.*, 2012). In a trial, this approach was found to be effective in reducing *in utero* transmission of *T. gondii* and the rate of congenital infection (Hotop *et al.*, 2012).

At birth, the baby will be carefully examined for any of the signs and symptoms of congenital toxoplasmosis, particularly ocular damage and brain lesions, which will complement the laboratory investigations. Severely affected children are generally treated with sulfadiazine and pyrimethamine for at least a year (Robert-Gangneux and Dardé, 2012), while if impairment is less obvious, the child would be followed up regularly to ensure that effects which might appear later in life (such as eye disease) are noticed as early as possible.

5.4.3.5 Epidemiology

The global incidence of congenital toxoplasmosis has been estimated at 1.5 per 1000 live births (Torgerson and Mastroiacovo, 2013), with the highest rates reported from the WHO Regions of the Americas and the African and Eastern Mediterranean Regions (Torgerson and Mastroiacovo, 2013); this suggests that it is a considerable public health problem in some countries. Susceptibility to primary toxoplasmosis amongst woman of childbearing age varies across the world (Pappas *et al.*, 2009). Due to the transmission patterns of the disease, where the prevalence of the infection in the overall population is low, infection during pregnancy – and thus congenital infection – is usually similarly uncommon. The European Union average rate of children being diagnosed with congenital toxoplasmosis within the first 12 months of life is about 1.01 per 100,000 children (ECDC, 2013). In the UK, where about 80% of adults do not have serological evidence of past exposure to the parasite, the annual incidence of congenital infection is reported at around 1 per 100,000 population, which is fewer than 10 cases. A study of just over 2,600 serum samples collected from antenatal patients in London (Flatt and Shetty, 2012) found anti-*T. gondii* IgG antibodies in around 17% of participants. Importantly, this figure was under 12% amongst UK-born women, regardless of any other demographic variables – that is, it is solely about exposure to the parasite and in Britain transmission rates are low. This contrasts sharply with France, where the adult susceptibility rate is estimated at about 20%, but there are on average 250 confirmed cases of congenital toxoplasmosis each year (30 per 100,000 population) (ECDC, 2013).

Point to consider 5.9: The screening policies in France and the UK are quite different. Discuss the epidemiological justification for this difference.

5.5   Perinatal Infections

Passage through the birth canal is a chaotic and messy business for both mother and baby. There are many opportunities for the baby to come into contact with maternal blood and bodily fluids, including through swallowing and via small lesions acquired during the frantic movement associated with delivery. The baby is therefore at risk of infection with blood borne microorganisms from its mother during this perinatal period. Examples of these include hepatitis B virus and human immunodeficiency virus. At the time of delivery, the baby is in intimate contact with the maternal genital mucosa and skin surface, thus affording an ideal occasion for transmission of organisms harboured there. The neonate's eye is particularly vulnerable to infection at this time and ophthalmia neonatorum is a fairly common outcome of birth. Examples of organisms associated with this condition include group B *Streptococcus* spp., *Chlamydia trachomatis*, *Neisseria gonorrhoeae* and herpes simplex virus.

5.5.1   Hepatitis B Virus

As discussed in Chapter 4, hepatitis B virus (HBV) is highly infectious, with at least $10^5 \, \text{mL}^{-1}$ copies of viral DNA detectable in serum (Harrison *et al*., 2009). There is considerable risk of perinatal transmission of this virus and it is therefore very important to diagnose HBV infection as early as possible in the pregnancy by screening maternal blood for hepatitis B surface antigen (HBsAg). The likelihood of the baby being infected depends on the viral activity within the mother, as established by her HBV marker profile (see Chapter 4 and below). Specifically, if the mother is hepatitis B e antigen (HBe)– positive, then the risk of transmission is considered to be around 90%; where serum anti-HBe antibodies are detectable, it is less than 10% (Chang, 2007). Acquisition of HBV infection in early childhood is associated with up to a 90% chance of becoming a chronic carrier (Harrison *et al*., 2009), with the potential for developing long-term sequelae such as liver cirrhosis or hepatocellular carcinoma (see Chapter 4). Thus, prevention of transmission early in life is key to the long-term health of the individual and also to reducing incidence of the disease overall.

Protection of the baby against perinatal infection with HBV is based around immunisation, with the additional support of anti-viral chemotherapy where this is suitable. The recommended management of the pregnancy, delivery and neonatal care depends on the serological profile of the mother. Somewhat counter intuitively, research evidence suggests that delivery by caesarean section affords no advantage in terms of reducing transmission risk (e.g., Hu *et al*., 2012) and so this is not explicitly advocated. Where the mother is HBsAg- positive and anti-HBe–positive, the newborn is given the first dose of the HBV vaccine at birth and must complete the course; the baby needs to be followed up to ensure that they did not become infected and have produced adequate levels of anti-hepatitis B surface antigen antibody (anti-HBS). Where the mother is HBsAg and HBe–positive, in addition to the course of vaccine, the baby should be given passive immunisation in the form of immunoglobulin containing a high titre of anti-HBS (HBIG) (Ramsey, 2008). Guidelines in the UK also include using this regime for premature and low-birth-weight babies (≤1500 g) born to mothers with HBV infection (irrespective of the mother's HBe/anti-HBe status), since the immune response in small neonates is less well developed than for full term babies (Ramsey, 2008). Ott *et al*. (2012)

estimated that in 2005, the proportion of women of childbearing age with the HBsAg, HBe−positive serological profile was as high as 50% in some countries. There was wide geographical variation and their data suggest that rates had fallen in most areas since 1990, due to increasingly effective control programmes. Since routine testing for HBeAg is not universally available, these authors advocate vaccination of all newborns where the mother is HBsAg-positive (Ott *et al.*, 2012).

Point to consider 5.10: In situations where laboratory services to test for all the HBV markers are not available, would it be better to direct limited resources into vaccination of all babies born to HBsAg-positive mothers or to strengthen laboratory facilities to allow targeting of patients where the risk of perinatal transmission is greatest?

Evidence suggests that where the mother has a high HBV viral load ($>10^6$ HBV DNA copies mL^{-1}), there is still about a 10% chance of transmission even when these prophylactic measures have been used. Since there are now safe and efficacious anti-HBV chemotherapies available, it is suggested that in these cases, the woman should be treated with a nucleotide analogue, such as lamivudine, in the last trimester where this is feasible (European Association for the Study of the Liver, 2012; ter Borg *et al.*, 2008). This may be beneficial as an extra preventative measure (European Association for the Study of the Liver, 2012). Also, there is some evidence that changes to the immune response during pregnancy can exacerbate liver damage in the mother (European Association for the Study of the Liver, 2012; ter Borg *et al.*, 2008), although this most often manifests immediately postpartum. It is therefore recommended that hepatic markers such as alanine aminotransferase (ALT) should be monitored in women with chronic HBV just after delivery. Anti-viral drug therapy can be administered at this point if necessary (Sinha and Kumar, 2010; ter Borg *et al.*, 2008). Although HBV has been detected in breast milk, numerous studies have concluded that the risk of mother-to-child transmission of the virus via this route is low when the appropriate management and prophylaxis for mother and baby respectively have been put in place (e.g., Sinha and Kumar, 2010; Chen *et al.*, 2013).

It is policy in the UK and many other countries to routinely screen antenatal patients for HBsAg (SMI V4, 2014), using an appropriate serological antigen detection test. In common with anyone who is found to be HBsAg-positive, for the pregnant patient further blood samples should be collected for confirmation and tests for further markers, including HBV viral load (see Chapter 4 and SMI V4, 2014). These results will determine the stage of the infection (i.e., acute or chronic) and whether the risk of transmission is high or low (as indicated by the mother's HBe antigen status), thus guiding the management of mother and baby (see above). It is worth noting that the 'e-antigen-positive, high HBV DNA copy number' profile could be found in someone with acute HBV as well as a 'high-risk' chronic carrier. Therefore in cases where the mother is found to HBsAg-negative at the initial first trimester screening, but is at high risk of infection (and with no history of vaccination), it might be prudent to re-test her later in the pregnancy to prevent unwitting perinatal transmission due to primary infection. The possibility of mutant HBsAg or occult HBV infections should also be considered when analysing the serological results (see Chapter 4). UK guidelines also recommend that infants born to HBV-positive mothers should be screened at 12 months to ensure that the vaccine has elicited a suitable immune response and has afforded protection against HBV infection (SMI V4, 2014).

5.5.2 Human Immunodeficiency Virus

Epidemiological data suggest that the risk of perinatal transmission of human immunodeficiency Virus (HIV) can be up to 50% where the mother is viraemic and there is no intervention (Pillay *et al.*, 2009). The virus can be present in the blood as well as other bodily fluids, including genital tract secretions and breast milk (Pillay *et al.*, 2009). The likelihood of transmission occurring from mother to child appears to be related to viral load (see Chapter 2). Nevertheless, there have been reports of perinatal infection occurring when the maternal serum viral copy number was <500 copies mL^{-1} (Pillay *et al.*, 2009; Tubiana *et al.*, 2010).

Babies who acquire HIV perinatally and are not treated for the virus are subject to similar types and severity of infectious disease as adults with acquired immune deficiency syndrome (AIDS) (see Chapter 2). In particular they tend to experience recurrent bacterial infections and unusual respiratory abnormalities (Taha, 2012). Long-term follow-up of cohorts of children born to HIV-positive mothers since the early 1990s (e.g., Dollfus *et al.*, 2010) show that before availability of antiretroviral therapy, up to 30% of children were likely to die in early childhood. The advent of treatment and specifically highly active antiretroviral therapy (HAART) changed the prognosis greatly. Dollfus *et al.* (2010) followed up a group of 348 children born with HIV before 1994. When they were reassessed at around 15 years old, 210 were alive and in contact with the study organisers. Amongst these, 90% had begun antiretroviral treatment before the age of five years. Those who took their medication as advised and had regular medical check-ups, were generally healthy and their height, weight, BMI and academic achievement were all comparable to the general population for their age. This shows that early diagnosis of perinatal (or neonatal) infection and prompt start of HAART would be crucial in determining a good outcome for the baby. It is also highlights the point that prevention of mother-to-child transmission altogether would be even better.

Antenatal screening for HIV is key, in order to manage the situation in good time before the birth. Evidence suggests that administration of HAART, along with continual monitoring, to ensure optimal HIV viral load and CD4+ counts during the pregnancy, can reduce the risk of mother to child transmission (MTCT) of the virus from over 20% to around 1% (e.g., Tubiana *et al.*, 2010; Townsend *et al.*, 2008). While a planned caesarean section is probably the best choice for delivery, the risk of perinatal infection does not appear to be notably higher with a planned vaginal birth – the key point is that the obstetric and paediatric teams are fully informed and on hand to manage the situation. If the maternal viral load is well controlled, she may have a choice for delivery, but breastfeeding should be avoided. If the mother is not already on HAART at the time of her first antenatal appointment, this should be started as soon as possible (Tubiana *et al.*, 2010; Townsend *et al.*, 2008). Antiretroviral therapies appear to be safe and do not carry a high risk of damaging the foetus (Townsend *et al.*, 2009). Where the infrastructure and facilities do not exist to provide this level of care, the outcomes for children born to HIV-positive mothers are less certain. Car *et al.* (2013) reported a systematic review of data about programmes to prevent perinatal MTCT in middle and low-income countries and found that transmission occurred in up to 18% of cases. This appeared to be due to a complex combination of factors impeding comprehensive antenatal screening schemes, lack of thorough clinical and virological monitoring of pregnant women identified as HIV-positive and inconsistencies in antiretroviral administration (Car *et al.*, 2013). Providing effect anti-retroviral treatment to HIV-positive women during pregnancy is an important part of the strategy to reduce the incidence of the virus worldwide (www.who.int/hiv/en).

5.5.3   Neonatal Conjunctivitis (Ophthalmia Neonatorum)

During delivery, the baby can acquire infections from contact with microorganisms in the mother's genital area. As well as neonatal sepsis (which can be caused by maternal skin bacterial flora) and disseminated herpes simplex virus (HSV) infection, a common site of infection is the newborn's eye(s). It should not be altogether surprising that some sexually transmitted organisms are associated with neonatal conjunctivitis ('ophthalmia neonatorum'), in particular *Chlamydia trachomatis* and *Neisseria gonorrhoeae*, as well as HSV (Palafox *et al.*, 2011). Significant pathogens in this context also include skin flora such as *Staphyloccocus epidermidis*, *Staph. aureus* and some *Streptococcus* species (Palafox *et al.*, 2011). A number of factors need to be in place for ophthalmia neonatorum to occur. The mother must have an active infection, though traumatic birth appears to enhance the risk of transmission (Palafox *et al.*, 2011). Babies which are delivered prematurely or who are 'small for dates' are known to be more vulnerable. This may be because the innate immune mechanisms in the eye and tears which afford some protection to infection are not fully developed in such neonates. Certain organisms are capable of circumventing the lysozyme, complement and antibodies present in the ocular environment and invade the eye, especially *C. trachomatis* and *N. gonorrhoeae* (Palafox *et al.*, 2011). It should be noted that premature or otherwise vulnerable babies who are treated in specialist neonatal units are at risk from hospital-acquired conjunctivitis. However, the profile of causative organisms in this situation is different and includes methicillin-resistant *Staphylococcus aureus* (MRSA) and Enterobacteriaceae (Palafox *et al.*, 2011).

It has been estimated that globally 12% of babies are born with conjunctivitis, but that the infection rate is double that figure in less-developed countries (Palafox *et al.*, 2011). The most commonly reported causes are *C. trachomatis* and *N. gonorrhoeae*; this is likely to be due to the high prevalence of these STIs as well as their invasive ability (Palafox *et al.*, 2011). It is suggested that up to 50% of babies born to mothers with active, untreated chlamydia or gonorrhoea will develop opthlamia neonatorum (Allstaff and Wilson, 2012). Symptoms include inflammation of the eye and an abnormal level of discharge ('sticky eye'). If untreated, the consequences can be keratinal damage – which can result in blindness – as well as dissemination of the infection to other parts of the body (Palafox *et al.*, 2011). Detection and treatment of infections associated with ophthalmia neonatorum in the women during their final trimester of pregnancy therefore appears to be important, but policies on this are variable (Allstaff and Wilson, 2012). A widely advocated approach is to give prophylactic treatment to all babies at birth (Meyer, 2014; Palafox *et al.*, 2011). Use of silver nitrate for this purpose was introduced in the late nineteenth century, but this has been largely replaced by povidone-iodine which is an inexpensive antiseptic. In some situations, antibiotics are used instead, but this sets up conditions favourable to organisms developing resistance (Meyer, 2014).

Where infection has occurred, prompt diagnosis is required. Eye swabs from the neonate should be collected as soon as possible. The recommended laboratory procedure for detection of the main pathogens (*C. trachomatis, N. gonorrhoeae*, HSV) is PCR (SMI S3, 2014). Specific antimicrobial treatment can then be administered either topically or, in a severe infection (which could disseminate), systemically (Palafox *et al.*, 2011)

5.6   Neonatal Infections

The neonatal period is usually considered to be the first 90 days of life. During this time, the newborns are vulnerable to infection, since the immune response is not fully developed at birth. Microorganisms are a particular hazard for those who are born prematurely, are unusually 'small for dates' or who have underlying conditions. It should be remembered that some are a serious danger to healthy babies as well. Neonatal infections usually involve pathogens acquired from the mother, although they can also be acquired from the environment (e.g., healthcare-associated infection amongst babies being treated in neonatal units). The principal presentations of neonatal infection are sepsis and meningitis and the causative agents are usually bacterial. Fungi and viruses should also be taken into account at the diagnostic stage (e.g., disseminated HSV infection).

5.6.1   Neonatal Sepsis

Neonatal sepsis is generally classified as either 'early onset' (occurring in the first four weeks after birth) and or 'late onset' (29–90 days). The most common causative agents of early onset disease are group B streptococci (GBS) and *Escherichia coli*. Acquisition is perinatal – GBS are considered to be normal vaginal flora, while *E. coli* is known to colonise the vaginal canal (Simonsen *et al.*, 2014). Other pathogens include *Listeria monocytogenes* and enteroviruses and parechoviruses, which are all thought to be capable of crossing the placenta – though they could also be transmitted perinatally (Simonsen *et al.*, 2014). Enterobacteriaceae other than *E. coli* (especially *Klebsiella* spp., *Enterobacter* spp. and *Serratia* spp.) are commonly associated with late-onset neonatal sepsis. It can also be a presentation of disseminated neonatal herpes (see below). Nosocomial infections are also an issue, with *Staphylococcus* spp. and *Candida* spp. being the most likely causes (Simonsen *et al.*, 2014). A baby with neonatal sepsis will usually present as seriously ill. They may have respiratory symptoms (apnoea, pneumonia), cardiac problems (bradycardia, cyanosis) and general signs such as poor feeding, lethargy. It should be noted that pyrexia, which is a usual indicator of septicaemia in adults, is extremely unusual, while hypothermia (i.e., the opposite!) is seen fairly often (Simonsen *et al.*, 2014).

5.6.2   Neonatal Meningitis

Neonatal bacterial meningitis is reported to occur at a rate of just over 0.2 in 1000 live births in European countries, with higher rates in the developing world (Okike *et al.*, 2014). Thus, newborns are at significant risk from this type of infection which can be fatal or can result in long term neurological consequences (Okike *et al.*, 2014; Furyk *et al.*, 2011). In situations where cases are not diagnosed or treated in time, mortality rates are at least 50% (Furyk *et al.*, 2011). Causative organisms are usually those associated with perinatal infection or nosocomial acquisition, so the list of the main pathogens is similar to that for neonatal sepsis and for the same reasons. Most cases are associated with either GBS or *E. coli* infection (Okike *et al.*, 2014). A survey of 364 patients from the UK and Republic of Ireland during one year found *L. monocytogenes* meningitis was diagnosed in 4% of neonates who were less than four weeks old; *Streptococcus pneumoniae* and *Neisseria meningitis* were important pathogens in the 29 to 90 days age group – usually in babies who had acquired the infection in the community (Okike *et al.*, 2014). Enteroviruses and parechoviruses are also notable possible pathogens and again they are found more often in neonates becoming ill after going home than those remaining in hospital (Harvala *et al.*, 2011). Clinical presentation of neonatal meningitis does not involve a defined set of symptoms but pyrexia, unusual crying, difficulty in settling to feed and episodes of fitting are common (Furyk *et al.*, 2011). It may be hard to differentiate between neonatal meningitis and sepsis (Furyk *et al.*, 2011), although, as noted above, fever is not a regular feature of the latter.

5.6.2.1 Laboratory Diagnosis

The aetiologies and clinical presentations of sepsis and meningitis in neonates are similar (and an individual baby could be experiencing both types of infection). Therefore, collection of blood cultures and CSF in either situation is recommended (Simonsen *et al.*, 2014; Furyk *et al.*, 2011). Gram stain of CSF and of positive blood cultures may be helpful in the case of bacterial or fungal infection. However, given the diversity of possible pathogens and the requirement to administer effective treatment as soon as possible, employment of techniques leading to rapid, species identification should always be considered. This could include MALDI-TOF and PCR, which can also help to provide detailed information about isolates. Harvala *et al.* (2011), looked at enteroviruses and parechoviruses in specimens collected from patients of all age groups with suspected meningitis, including paediatric patients with possible neonatal sepsis or meningitis. They found that amongst the CSF samples collected from over 1,200 babies younger than three months, almost 3% contained human parechoviruses (HPeV), while enteroviruses (EV) were found in about 8%. The study was conducted over a five-year period (July 2005 to June 2010) and it showed that the incidence of these viruses was seasonal (peak: late spring–early summer). Interestingly HPeVs were only isolated from neonates and only in alternate years; the most common type was HPeV 3 (Harvala *et al.*, 2011).

5.6.3 Group B *Streptococcus* spp. Infection

As noted in the above paragraphs, *Streptococcus agalactiae* strains – also referred to as group B streptococci (GBS) – are a significant cause of neonatal disease. Around half of all cases of neonatal meningitis (Okike *et al.*, 2014) and at least 40% of those of neonatal sepsis (Simonsen *et al.*, 2014) are associated with GBS infection. The categories for GBS infection in newborns are different from those mentioned previously for other pathogens – early onset is defined as the first 7 days of life and late onset is 8 to 90 days (Gilbert, 2004). Early onset GBS disease encompasses all the symptoms and complications of meningitis and sepsis, but also frequently involves pneumonia (Verani *et al.*,

2010). If unrecognised and not treated, case fatality can be up to 50%. In the 1970s the incidence of neonatal GBS disease was reported to be 2 to 3 per 1,000 live births in countries where data was collected (Gilbert, 2004), but with interventions such as testing mothers for GBS colonisation, prompt microbiological diagnosis and antibiotic treatment, rates have fallen to around 0.5 per 1,000 live births in the United States and Europe (Verani *et al.*, 2010; Gilbert, 2004). Issues with diagnosis, laboratory methods and reporting of information mean that true rates in developing countries remain uncertain, but may be up to 3 per 1,000 live births (Dagnew *et al.*, 2012), suggesting that some intervention may be beneficial to reduce this.

Estimates of rates of colonisation with GBS amongst women of childbearing age suggest that about a quarter of U.S. women (26%) carry the bacterium and the proportion in the UK is probably similar (Gilbert, 2004). Other parts of the world report lower figures, for example 22% in sub-Saharan Africa and 12% in India (Johri *et al.*, 2013) but whether this is a real difference or due to problems with data collation is unclear (e.g., Dagnew *et al.*, 2012). This does put a large number of babies at risk of perinatal infection and since administration of prophylactic antibiotics to the mother is a viable option, it seems clear that a policy for obstetricians and paediatricians would be valuable. In the United States, the advice is to routinely collect vaginal-rectal swabs from all mothers in their last trimester (within five weeks of delivery) and treat those with evidence of GBS presence with penicillin or ampicillin intravenously (Verani *et al.*, 2010). Patients who are allergic to penicillin, but with no history of anaphylaxis after taking the drug should be given cefazolin; for those who experience the most serious side effects clindamycin is safer (although the bacterial isolate should be tested for inducible resistance). The guidance advocates use of NAATs tests rather than plate culture for improved sensitivity (Verani *et al.*, 2010). The intervention strategy appears to be well-received and to be reasonably effective (Schrag and Verani, 2013). In contrast, this approach remains controversial in the UK, where it is considered that targeting only high-risk pregnancies (e.g., early rupture of membranes, symptomatic vaginal infection in the mother) is more effective and less likely to result in antimicrobial resistance (Brocklehurst, 2015; Gilbert, 2004). However, some authors disagree with this stance and suggest that screening should be at least offered to all mothers in late third trimester (Steer, 2015).

Point to consider 5.11: What are the issues around routine screening of all pregnant women for GBS at 37 weeks or later from the laboratory's point of view?

5.6.4 Neonatal Herpes Simplex

Global estimates suggest that up to 15% of females between the ages of 15 and 49 years are infected with herpes simplex virus (HSV) type 2 (Looker *et al.*, 2015a) and the prevalence of genital HSV-1 in this demographic group is approximately 3% (Looker *et al.*, 2015b). These rates vary geographically – with highest prevalences in Africa and the Americas – but taken together this represents 360 million women of childbearing age carrying genital HSV infection, 39 million of whom are in Europe (Looker *et al.*, 2015b). Since it is a sexually transmitted infection, pregnant women are at risk of having acquired primary HSV infection. Also, the metabolic changes in pregnancy increase the chances of a reactivation of an existing latent infection. Virus shedding from the genital area often occurs without a patent lesion, so there is a risk of perinatal transmission from undetected maternal HSV around the time of delivery (Pinninti and Kimberlin, 2014). With a primary infection, there is a 50% to 60% chance of passing the virus to the baby during passage through the birth canal; in recurrent infections, this rate is about 3%. Only about 1% of all women infected with HSV will shed virus during the pregnancy, making neonatal herpes infection a relatively uncommon occurrence (Pinninti and Kimberlin, 2014).

If a baby does acquire the virus perinatally or immediately postpartum, the most likely consequences are a skin infection (wide spread vesicular rash), lesions in the mouth and/ or ophthalmia neonatorum (Pinninti and Kimberlin, 2014; Palafox *et al.*, 2011). Disseminated herpes or encephalitis (not always accompanied by skin presentation) can also occur; if untreated, neonates with these two conditions are unlikely to survive for more than a year (Pinninti and Kimberlin, 2014). It is not easy to prevent neonatal herpes infection, although if the mother does have obvious lesions, caesarean section may be beneficial. The main approach to improving outcomes is prompt treatment with aciclovir and this relies on accurate and timely laboratory diagnosis. The virus particles should be detectable by electron microscopy (Figure 5.3) or culture. Unfortunately, there are issues with sensitivity with both methods. Also it is not possible to distinguish between HSV and varicella zoster virus from the microscopic appearance. The best assay is PCR to detect viral DNA in the baby's blood, CSF and/or from lesion swabs (as appropriate to the clinical presentation). The use of antiviral therapy is reported to have reduced the mortality rates considerably (from 85% to 29% for disseminated herpes and 50% to 4% for CNS disease). There is also some evidence that prolonged aciclovir treatment may reduce the number and severity of HSV reactivations in the baby (Pinninti and Kimberlin, 2014). Batra *et al.* (2014) reported their experience from one paediatric centre in the UK where the team cared for 19 cases in eight years. The babies displayed a range of symptoms, which did not necessarily point to

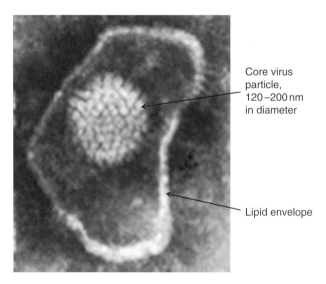

Core virus particle, 120–200 nm in diameter

Lipid envelope

Figure 5.3 Electron micrograph of vesicle fluid showing particles of either herpes simplex virus or varicella zoster virus. Note: It is not possible to identify species from EM appearance. Source: Courtesy of Mr I. Phillips.

HSV infection initially and in some cases, this delayed the start of aciclovir treatment. Nine of the patients died and all of these had non-specific presentations such as sepsis or liver failure and were not tested for HSV until at least 6 days post-delivery (Batra *et al.*, 2014). This shows the support which laboratory scientists can give, by highlighting the possibility of herpes virus infection in such cases and encouraging the collection of appropriate samples to test for HSV DNA.

Point to consider 5.12: It is clear that a high-quality laboratory service contributes to the diagnosis and prevention of perinatal and neonatal infections. This benefits the patients in terms of the medical welfare of the individuals, as well as the wider public health issues. What about the psychological aspects of vertical transmission?

5.7 Exercises

5.1 **A** Construct your own algorithm for the screening of antenatal patients without using any background information to help you.
   B Then read the relevant national policy and/ or your laboratory's SOP and compare these with your algorithm.
   C Do you think that the recommended protocol is adequate to reflect the requirements of the antenatal population in your area? Justify your answer.
   D Are there any alterations to the standard protocol which you think should be considered? Explain your reasoning.

5.2 **A** Find out your laboratory's usual testing procedure for diagnosis of infection in an antenatal patient presenting with a red rash.
   B Consider the protocol in detail and decide whether it is up to date with the current understanding of the pathology of possible causative organisms and available laboratory techniques.

5.3 Imagine that you have been asked to provide expert advice to a multidisciplinary obstetrics and gynaecology team. They have decided that they would like to use point-of-care tests as routine during the 12-week antenatal visit, so that the patient does not have to wait for her results. Write a short report discussing the feasibility of this.

5.8 Case Studies

5.1 Mrs GH is 38 years old and in the fourteenth week of her first pregnancy. She presents with enlarged and tender lymph nodes, pyrexia (38.8 °C), rigours and a history of vomiting in the last 24 hours.
   a) Which infectious causes of these symptoms would you consider testing for and why?
   b) Which samples should be collected and which investigations should be performed?

c) The serological results are as follows:
    CMV IgG detected
    CMV IgM Not detected
    Toxoplasma IgG indeterminate – please send repeat blood sample in two weeks
    Toxoplasma IgM detected
    What do these results indicate?

d) How should Mrs GH be treated and counselled?

5.2 Ms JK is in the thirty-second week of her second pregnancy. She is 27 years old and had her first child when she was 17. She then moved to another part of the England to live with her current partner. She has a recent history (last five years) of intravenous drug abuse. Throughout this pregnancy, the midwife has experienced difficulty in persuading JK to give venous blood samples for the various routine antenatal checks, but a specimen has now been received in the microbiology laboratory. The results are as follows:

Rubella IgG antibody: DETECTED

(Note this is not part of the routine screening, but without the previous notes there was no record of the previous antenatal rubella test or vaccination history)

Human immunodeficiency virus antibody: NOT DETECTED
Treponema pallidum IgG antibody: NOT DETECTED
Hepatitis B surface antigen: DETECTED

a) Provide an interpretation for each of these results
b) Which of these results should be followed up and which further tests should be undertaken?
c) The follow-up tests indicate a pattern of markers indicating a chronic HBV infection and a high risk of transmission of the virus to her baby perinatally. Outline the likely results and what plans should be put in place for the birth.

References

Abernathy ES *et al.* (2011). Status of global virological surveillance for rubella viruses. *Journal of Infectious Diseases*, **204(Suppl 1):** S524–532. doi: 10.1093/infdis/jir099

Allstaff S and Wilson J (2012). The management of sexually transmitted infections in pregnancy. *The Obstetrician and Gynaecologist*, **14:** 25–32. doi: 10.1111/j.1744-4467.2011.00088.x

Anderson TP *et al.* (2013). Comparison of four multiplex PCR assays for the detection of viral pathogens in respiratory specimens. *Journal of virological methods*, **191:** 118–121. doi: 10.1016/j.jviromet.2013.04.005

Batra D *et al.* (2014). The incidence and presentation of neonatal herpes in a single UK tertiary centre, 2006–2013. *Archives of Disease in Childhood*, **99(10):** 916–921. doi: 10.1136/archdischild-2013-305335

Bébéar C and De Barbeyrac B (2009). Genital Chlamydia trachomatis infections. *Clinical Microbiology and Infection*, **15:** 4–10. DOI: 10.1111/j.1469-0691.2008.02647.x

Best JM, Icenogle JP and Brown DWG (2009). Rubella. In: *Principles and Practice of Clinical Virology, 6th edn.* Zuckerman AJ, Banatvala JE, Schoub BD, Griffiths PD and Mortimer P, eds. Chichester: Wiley-Blackwell.

Boden K *et al*. (2012). Maternofetal consequences of *Coxiella burnetii* infection in pregnancy: a case series of two outbreaks. *BMC Infectious Diseases*, **12**: 359. doi: 10.1186/1471-2334-12-359

ter Borg MJ *et al*. (2008). Exacerbation of chronic hepatitis B infection after delivery. *Journal of Viral Hepatitis*, **15**: 37–41. doi: 10.1111/j.1365-2893.2007.00894.x

Brendish NJ, Schiff HF and Clark TW. (2015). Point-of-care testing for respiratory viruses in adults: The current landscape and future potential. *Journal of Infection*, **71**: 501–510. doi: 10.1016/j.jinf.2015.07.008

Breuer J (2009). Varicella zoster. In: *Principles and Practice of Clinical Virology, 6th edn*. Zuckerman AJ, Banatvala JE, Schoub BD, Griffiths PD and Mortimer P, eds. Chichester: Wiley-Blackwell.

Brocklehurst P (2015). Screening for Group B streptococcus should be routine in pregnancy: AGAINST: Current evidence does not support the introduction of microbiological screening for identifying carriers of Group B streptococcus. *BJOG: An International Journal of Obstetrics & Gynaecology*, **122**: 368–368. doi: 10.1111/1471-0528.13085

Brunham RC and Rappuoli R (2013). *Chlamydia trachomatis* control requires a vaccine. *Vaccine*, **31**: 1892–1897. doi: 10.1016/j.vaccine.2013.01.024

Car LT *et al*. (2013). The uptake of integrated perinatal prevention of mother-to-child HIV transmission programs in low-and middle-income countries: a systematic review. *PloS One*, **8**: e56550. doi: 10.1371/journal.pone.0056550

Carlier Y *et al*. (2012). Congenital parasitic infections: A review. *Acta Tropica*, **121**: 55–70. doi: 10.1016/jactatropica.2011.10.018

Chang MH (2007). Hepatitis B virus infection. *Seminars in Fetal and Neonatal Medicine*, **12**: 160–167. doi: 10.1016/j.siny.2007.01.013

Chartrand C *et al*. (2012). Accuracy of rapid influenza diagnostic tests: A meta-analysis. *Annals of internal medicine*, **156**: 500–511. doi: 10.7326/0003-4819-156-7-201204030-00403

Chen X *et al*. (2013). Breastfeeding is not a risk factor for mother-to-child transmission of hepatitis B virus. *PLoS One*, **8**: e55303. doi: 10.1371/journal.pone.0055303

Cockayne A (2012). Treponema and borrelia. In: *Medical Microbiology: A Guide to Microbial Infections: 18th edn*. Greenwood D, Barer M, Slack R and Irving W, eds. London: Churchill Livingston – Elsevier.

Cornu C *et al*. (2009). Factors affecting the adherence to an antenatal screening programme: an experience with toxoplasmosis screening in France. *Eurosurveillance*, **14**: 1–5.

Dagnew AF *et al*. (2012). Variation in reported neonatal group B streptococcal disease incidence in developing countries. *Clinical Infectious Diseases*, **55**: 91–102. doi: 10.1093/cid/cis395

Darville T and Hiltke TJ (2010). Pathogenesis of genital tract disease due to *Chlamydia trachomatis*. *Journal of Infectious Diseases*, **201**: S114–S125. doi: 10.1086/652397

De Paschale M *et al*. (2010). Implementation of screening for *Toxoplasma gondii* infection in pregnancy. *Journal of Clinical Medicine Research*, **2**: 112–116. doi: 10.4021/jcmr2010.05.321w

De Santis M *et al*. (2006). Rubella infection in pregnancy. *Reproductive Toxicology*, **21**: 390–398. doi: 10.1016/j.reprotox.2005.01.014.

Desai M *et al*. (2007). Epidemiology and burden of malaria in pregnancy. *Lancet Infectious Diseases*, **7**: 93–104. doi: 10.1016/S1473-3099(07)70021-x

Dollfus C *et al.* (2010). Long-term outcomes in adolescents perinatally infected with HIV-1 and followed up since birth in the French perinatal cohort (EPF/ANRS CO10). *Clinical Infectious Diseases*, **51(2):** 214–224. doi: 10.1086/653674

Dorey MD *et al.* (2012). Modelling the effect of *Chlamydia trachomatis* testing on the prevalence of infection in England: What impact can we expect from the National Chlamydia Screening Programme? *Sexually Transmitted Infections*, **89**: 272.

Dubey JP (2005). Toxoplasmosis. In: Cox FEG, Wakelin D, Gillespie SH and Despommier DD, eds. *Topley and Wilson's Microbiology and Microbial Infections, 10th edn.* New York: John Wiley & Sons.

ECDC (2016). Antenatal Screening for HIV, Hepatitis B, Syphilis and Rubella Susceptibility in the EU/EEA. Stockholm: European Centre for Disease Prevention and Control.

ECDC (2013). Annual Epidemiological Report: Reporting on 2011 Surveillance Data and 2012 Epidemic Intelligence Data. Stockholm: European Centre for Disease Prevention and Control.

European Association for the Study of the Liver. (2012). EASL Clinical Practice Guidelines: Management of chronic hepatitis B virus infection *Journal of Hepatology*, **57**: 167–185. doi: 10.1016/j.jhep.2012.02.010

Fell DB *et al.* (2013). Influenza vaccination and fetal and neonatal outcomes. *Expert Review of Vaccines*, **12**: 1417–1430. doi: 10.1586/14760584.2013.851607

Flatt A and Shetty N (2012). Seroprevalence and risk factors for toxoplasmosis among antenatal women in London: A re-examination of risk in an ethnically diverse population. *European Journal of Public Health*, **23**: 648–652. doi: 10.1093/eurpub/cks075

Furyk JS, Swann O and Molyneux E. (2011). Systematic review: Neonatal meningitis in the developing world. *Tropical Medicine & International Health*, **16(6):** 672–679. doi: 10.1111/j.1365-3156.2011.02750.x

Gambon M and Potter CW (2009). Influenza. In: *Principles and Practice of Clinical Virology, 6th edn.* Zuckerman AJ, Banatvala JE, Schoub BD, Griffiths PD and Mortimer P, eds. Chichester: Wiley-Blackwell.

Gilbert GL (2004). Vaccines for other neonatal infections: are group B streptococcal infections vaccine-preventable? *Expert Review of Vaccines*, **3**: 371–374. doi: 10.1586/14760584.3.4.371

Gitau GM and Eldred JM (2005). Malaria in pregnancy: Clinical, therapeutic and prophylactic considerations. *The Obstetrician and Gynaecologist*, 7: 5–11. doi: 10.1576/toag.7.1.005.27036

Greenwood D (2012). Protozoa. In: *Medical Microbiology: A Guide to Microbial Infections, 18th edn.* Greenwood D, Barer M, Slack R and Irving W, eds. London: Churchill Livingston – Elsevier.

Gunn A and Pitt SJ (2012). *Parasitology: An Integrated Approach.* Chichester: Wiley-Blackwell.

Hadad R, Fredlund H and Unemo M (2009). Evaluation of the new COBAS TaqMan CT test v2. 0 and impact on the proportion of new variant *Chlamydia trachomatis* by the introduction of diagnostics detecting new variant *C. trachomatis* in Örebro county, Sweden. *Sexually Transmitted Infections*, **85**: 190–193. doi: 10.1136/sti.2008.033142

Harrison TJ, Dusheiko GM and Zuckerman AJ (2009). Hepatitis viruses. In: *Principles and Practice of Clinical Virology, 6th edn.* Zuckerman AJ, Banatvala JE, Schoub BD, Griffiths PD and Mortimer P, eds. Chichester: Wiley-Blackwell.

Harvala H *et al.* (2011). Comparison of human parechovirus and enterovirus detection frequencies in cerebrospinal fluid samples collected over a 5-year period in Edinburgh: HPeV type 3 identified as the most common picornavirus type. *Journal of Medical Virology*, **83**: 889–896. doi: 10.1002/jmv.22023

Hawkes S *et al.* (2011). Effectiveness of interventions to improve screening for syphilis in pregnancy: A systematic review and meta-analysis. Lancet Infectious Diseases, **11**: 684–691. doi: 10.1016/S1473-3099(11)70104-9.

van Helden J (2014). Evaluation of the fully automated assays for the detection of Rubella IgM and IgG antibodies by the Elecsys ® immunoassay system. *Journal of Virological Methods*, **199**: 108–115. doi: 10.1016/j.viromet.2014.01.009.

Herremans T *et al.* (2010). A review of diagnostic tests for congenital syphilis in newborns. *European Journal of Infectious Diseases*, **29**: 495–501. doi: 10.1007/S10096-010-0900-8.

Hesketh LM (2012). Togaviruses. In: *Medical Microbiology: A Guide to Microbial Infections, 18th edn*. Greenwood D, Barer M, Slack R and Irving W, eds. London: Churchill Livingston – Elsevier.

Horne AW *et al.* (2011). Elucidating the link between *Chlamydia trachomatis* and ectopic pregnancy. *Expert Reviews in Obstetrics and Gynaecology*, **6**: 231–233. doi: 10.1586/eog.11.12.

Horner PJ *et al.* (2013). Effect of time since exposure to Chlamydia trachomatis on chlamydia antibody detection in women: a cross-sectional study. *Sexually Transmitted Infections*, **89**(5): 398–403. doi: 10.1136/sextrans-2011-050386.

Hotop A *et al.* (2012). Efficacy of rapid treatment initiation following primary *Toxoplasma gondii* infection during pregnancy. *Clinical Infectious Diseases*, **54**: 1545–1552. doi: 10.1093/cid/cis234.

Hu Y *et al.* (2013). Effect of elective cesarean section on the risk of mother-to-child transmission of hepatitis B virus. *BMC pregnancy and Childbirth*, **13**: 119. doi: 10.1186/1471-2393-13-119

Hukic M *et al.* (2012). An outbreak of rubella in the federation of Bosnia and Herzegovina between December 2009 and May 2010 indicates failure to vaccinate during wartime (1992-1995). *Epidemiology and Infection*, **140**: 447–453. doi: 10.1017/S0950268811000707

Hviid L (2011). The case for PfEMP1-based vaccines to protect pregnant women against *Plasmodium falciparum* malaria. *Expert Review of Vaccines*, **10**: 1405–1414. doi: 10.1586/erv.11.113.

Jaworowski A *et al.* (2009). Relationship between human immunodeficiency virus type 1 coinfection, anemia, and levels and function of antibodies to variant surface antigens in pregnancy-associated malaria. *Clinical and Vaccine Immunology*, **16**: 312–319. doi: 10.1128/CVI.00356-08

Johri AK *et al.* (2013). Epidemiology of group B streptococcus in developing countries. *Vaccine*, **31**: D43–D45. doi: 10.1016/j.vaccine.2013.05.094

Jebbari H *et al.* (2011). Variations in the epidemiology of primary, secondary and early latent syphilis, England and Wales: 1999 to 2008. *Sexually Transmitted Infections*, **87**: 191–198. doi: 10.1136/sti.2009.040139

Kattenburg JH *et al.* (2011). Systematic review and meta-analysis: Rapid diagnostic tests versus placental histology, microscopy and PCR for malaria in pregnant women. *Malaria Journal*, **10**: 321. doi: 10.1186/1475-2875-10-321.

ter Kuile FO and Rogerson SJ (2008). *Plasmodium vivax* infection during pregnancy: An important problem and need of new solutions. *Clinical Infectious Diseases*, **46**: 1382–1384. doi: 10.1086/586744.

Kraus TA, *et al.* (2012). Characterizing the pregnancy immune phenotype: Results of the viral immunity and pregnancy (VIP) study. *Journal of Clinical Immunology*, **32**: 300–311. doi: 10.1007/S10875-011-9627-2

Lalloo DG *et al.* (2016). UK malaria treatment guidelines 2016. *Journal of Infection*, **72**: 635–649. doi: 10.1016/j.jinf.2016.02.001

Lamont RF *et al.* (2011). Varicella zoster virus (chickenpox) infection in pregnancy. *BJOG: An International Journal of Obstetrics & Gynaecology*, **118**: 1155–1162. doi: 10.1111/j.1471-0528.2011.02983.x

Lawn J and Kerber K, eds (2006). *Opportunities for Africa's Newborns: Practical Data, Policy and Pragmatic Support for Newborn Care in Africa.* Cape Town: PMNCH WHO Africa. http://www.who.int/pmnch/media/publications/aonsectionIII_2.pdf

Looker KJ *et al.* (2015a). Global estimates of prevalent and incident herpes simplex virus type 2 infections in 2012. *PLoS One*, **10(1):** e114989. doi: 10.1371/journal.pone.0114989

Looker KJ *et al.* (2015b). Global and regional estimates of prevalent and incident Herpes Simplex Virus Type 1 infections in 2012. *PloS One*, **10(10):** e0140765. doi: 10.1371/journal.pone.0140765

Mabey D and Peeling RW (2012). Chlamydia. **In:** *Medical Microbiology: A Guide to Microbial Infections, 18th edn.* Greenwood D, Barer M, Slack R and Irving W, eds. London: Churchill Livingston – Elsevier.

Manicklal S *et al.* (2013). The "silent" global burden of congenital cytomegalovirus. *Clinical Microbiology Reviews*, **26**: 86–102. doi: 10.1128/CMR.00062-12

Marquez L *et al.* (2011). A report of three cases and review of intrauterine herpes simplex virus infection. *The Pediatric Infectious Disease Journal*, **30**: 153. doi: 10.1097/INF.0b013e3181f5595C

Mayxay M *et al.* (2004). Mixed-species malaria infections in humans. *Trends in Parasitology*, **20**: 233–240. doi: 10.1016/j.pt.2004.03.006.

Memoli MJ *et al.* (2013). Influenza in pregnancy. *Influenza and Other Respiratory Viruses*, 7: 1033–1039. doi: 10.1111/irv.12055

Meyer D (2014). Ophthalmia neonatorum prophylaxis and the 21st century antimicrobial resistance challenge. *Middle East African Journal of Ophthalmology*, **21**: 203–204. doi: 10.4103/0974-9233.134667

Mor G and Cardenas I (2010). The immune system in pregnancy: A unique complexity. *American Journal of Reproductive Immunology*, **63**: 425–433. doi: 10.1111/j.1600-0897.2010.00836.x

Morelle C *et al.* (2012). Comparative assessment of a commercial kit and two laboratory-developed PCR assays for molecular diagnosis of congenital toxoplasmosis. *Journal of Clinical Microbiology*, **50**: 3977–3982. doi: 10.1128/JCM.01959-12

Morley LC and Taylor-Robinson AW (2012). Understanding how *Plasmodium falciparum* binds to the placenta and produces pathology provides a rationale for pregnancy-associated malaria vaccine development. *The Open Vaccine Journal*, **5**: 8–27.

Mortlock S (2015). Rubella: Distribution of susceptible antenatal patients in SW London. *Biomedical Scientist*, **59**: 378–380.

Murray PR, Rosenthal KS and Pfaller MA (2013). *Medical Microbiology, 7th edn.* Philadelphia: Elsevier -Saunders.

Muscat M *et al.* (2014). The state of measles and rubella in the WHO European Region, 2013. *Clinical Microbiology and Infection*, **20(Suppl 5):** 12–18 doi: 10.1111/1469-0691.12584

Newman L *et al.* (2013). Global estimates of syphilis in pregnancy and associated adverse outcomes: Analysis of multinational antenatal surveillance data. *PLOS Medicine,* **10(2):** e10011396. doi: 10.1371/journal.pmed.1001396

Newman L *et al.* (2015). Global estimates of the prevalence and incidence of four curable sexually transmitted infections in 2012 based on systematic review and global reporting. *PloS One,* **10(12):** e0143304. doi: 10.1371/journal.pone.0143304

NICE (2016). NICE Guidelines CG62 Antenatal Care for Uncomplicated Pregnancies. Published March 2008, updated March 2016. https://www.nice.org.uk/guidance/cg62/chapter/1-guidance#screening-for-infections

Okike IO *et al.* (2014). Incidence, etiology, and outcome of bacterial meningitis in infants aged < 90 days in the United Kingdom and Republic of Ireland: Prospective, enhanced, national population-based surveillance. *Clinical Infectious Diseases,* **59:** e150–e157. doi: 10.1093/cid/ciu514

Ott JJ, Stevens GA and Wiersma ST (2012). The risk of perinatal hepatitis B virus transmission: hepatitis B e antigen (HBeAg) prevalence estimates for all world regions. *BMC Infectious Diseases,* **12:** 131. doi: 10.1186/1471-2334-12-131

Palafox SKV, Smith J Tauber AD and Foster SC (2011). Ophthalmia neonatorum. *Journal of Clinical and Experimental Ophthalmology,* **2(1)** 119. doi: 10.4172/2155-9570.1000119

Pappas G *et al.* (2009). Toxoplasmosis snapshots: Global status of *Toxoplasma gondii* seroprevalence and implications for pregnancy and congenital toxoplasmosis. *International Journal for Parasitology,* **39:** 1385–1394. doi: 10.1016/j.ijpara.2009.04.003

Peeling RW and Hook EW (2006). The pathogenesis of syphilis: The great mimicker, revisited. *Journal of Pathology,* **208:** 224–232. doi: 10.1002/path.1903

Pillay D, Geretti AM and Weiss RA (2009). Human immunodeficieny viruses. In: *Principles and Practice of Clinical Virology, 6th edn.* Zuckerman AJ, Banatvala JE, Schoub BD, Griffiths PD and Mortimer P, eds. Chichester: Wiley-Blackwell.

Pinninti SG and Kimberlin DW (2014). Management of neonatal herpes simplex virus infection and exposure. *Archives of Disease in Childhood-Fetal and Neonatal Edition,* **9**(3): F240–F244. doi: 10.1136/archdischild-2013-303762

Pitsouni E *et al.* (2007). Single dose azithromycin versus erythromycin or amoxicillin for *Chlamydia trachomatis* infection during pregnancy: A meta-analysis of randomized controlled trials. *International Journal of Antimicrobial Agents,* **30:** 213–221. doi: 10.10.16/j.ijantimicag.2007.04.015

Pitt SJ (2013). Measles: Potentially fatal but eminently preventable. *Biomedical Scientist,* **57:** 444–448.

Pitts SI *et al.* (2014). Congenital rubella syndrome in child of woman without known risk factors, New Jersey, United States. *Emerging Infectious Diseases,* **20:** 307–309. doi: 10.3201/eid2002.131233

PHE (2016). Rubella susceptibility screening in pregnancy to end in England, January 2016. https://www.gov.uk/government/news/rubella-susceptibility-screening-in-pregnancy-to-end-in-england

Ramsay M (2008). *On behalf of the Hepatitis Programme Board: Policy on the Use of Passive Immunisation With Hepatitis B Immunoglobulin (HBIG) for Infants Born to Hepatitis B Infected Mothers.* London: Health Protection Agency.

Rasmussen SA and Jamieson DJ (2014). 2009 H1N1 influenza and pregnancy—5 years later. *New England Journal of Medicine,* **2014**(371): 1373–1375. doi: 10.1056/NEJMp1403496

Reef SE *et al.* (2011). Progress toward control of rubella and prevention of congenital rubella syndrome: Worldwide 2009. *Journal of Infectious Diseases*, **204**: S24–S27. doi: 10.1093/infdis/jir155

Rekart ML *et al.* (2013). Chlamydia public health programs and the epidemiology of pelvic inflammatory disease and ectopic pregnancy. *Journal of Infectious Diseases*, **207**(1): 30–38. doi: 10.1093/infdis/jis644

Robbins JR and Bakardjiev AI (2012). Pathogens and the placental fortress. *Current Opinion in Microbiology*, **15**: 36–43. doi: 10.1016/j.mib.2011.11.006

Rours GIJG *et al.* (2011). *Chlamydia trachomatis* infection during pregnancy associated with preterm delivery: A population-based prospective cohort study. *European Journal of Epidemiology*, **26**: 493–502. doi: 10.1007/S10654-011-9586-1

Robert-Gangneux F and Dardé M-L (2012). Epidemiology of and diagnostic strategies for toxoplasmosis. *Clinical Microbiology Reviews*, **25**: 264–296. doi: 10.1128/CMR.05013-11

SMI G7: Public Health England. (2014). Investigation of Red Rash. UK Standards for Microbiology Investigations. G 7 Issue 2.1. http://www.hpa.org.uk/SMI/pdf

SMI G8: Public Health England. (2014). Respiratory Viruses. UK Standards for Microbiology Investigations. G8 Issue 1.3. http://www.hpa.org.uk/SMI/pdf

SMI S3: Public Health England. (2014). Conjunctivitis. UK Standards for Microbiology Investigations. S3 Issue 1.2. http://www.hpa.org.uk/SMI/pdf

SMI V4: Public Health England. (2014). Hepatitis B Diagnostic Serology in the Immunocompetent (including Hepatitis B in Pregnancy). UK Standards for Microbiology Investigations. V 4 Issue 5.3. http://www.hpa.org.uk/SMI/pdf

SMI V30: Public Health England. (2015). Investigation of Pregnant Women Exposed to Non-Vesicular Rash. UK Standards for Microbiology Investigations. V30 Issue 5. https://www.gov.uk/uk-standards-for-microbiology-investigations-smi-quality-and-consistency-in-clinical-laboratories

SMI V37: Public Health England. (2014). Chlamydia trachomatis Infection – Testing by Nucleic Acid Amplification Tests (NAATs). UK Standards for Microbiology Investigations. V 37 Issue 3.2. http://www.hpa.org.uk/SMI/pdf

SMI V 44: Public Health England. (2015). Syphilis serology. UK Standards for Microbiology Investigations.V44Issue2.https://www.gov.uk/uk-standards-for-microbiology-investigations-smi-quality-and-consistency-in-clinical-laboratories

Sagel U *et al.* (2012). "Blind periods" in screening for toxoplasmosis in pregnancy in Austria – A debate. *BMC Infectious Diseases* **12**: 118. doi: 10.1186/1471-2334-12-118

Schneider-Schaulies S and ter Meulen V. (2009). Measles virus. In: *Principles and Practice of Clinical Virology*, *6th edn.* Zuckerman AJ, Banatvala JE, Schoub BD, Griffiths PD and Mortimer P, eds. Chichester: Wiley-Blackwell.

Schrag SJ and Verani JR (2013). Intrapartum antibiotic prophylaxis for the prevention of perinatal group B streptococcal disease: Experience in the United States and implications for a potential group B streptococcal vaccine. *Vaccine*, **31**: D20–D26. doi: 10.1016/j.vaccine.2012.11.056

Simms I *et al.* (2005). The re-emergence of syphilis in the United Kingdom: The new epidemic phases. *Sexually Transmitted Diseases*, **32**: 220–226. doi: 10.1097/01.olq

Simms I *et al.* (2016). The incidence of congenital syphilis in the United Kingdom: February 2010 to January 2015. *BJOG: An International Journal of Obstetrics & Gynaecology*, **124**: 72–77. doi: 10.1111/1471-0528.13950

Simonsen KA *et al.* (2014). Early-onset neonatal sepsis. *Clinical Microbiology Reviews*, **27(1)**: 21–47. doi: 10.1128/CMR.00031-13

Sinha S and Kumar M (2010). Pregnancy and chronic hepatitis B virus infection. *Hepatology Research*, **40**: 31–48. doi: 10.1111/j.1872-034X.2009.00597.x

Skidmore S, Boxall E and Lord S (2014). Is the MMR vaccination programme failing to protect women against rubella infection? *Epidemiology and Infection*, **142**: 1114–1117. doi: 10.1017/S0950268813002045

Steer PJ (2015). Screening for Group B streptococcus should be routine in pregnancy: FOR: The case for screening. *BJOG: An International Journal of Obstetrics & Gynaecology*, **122(3)**: 369–369. doi: 10.1111/1471-0528.13086

Stene LC and Gale EAM (2013). The prenatal environment and type 1 diabetes. *Diabetologia*, **56**: 1888–1897. doi: 10.1007/s00125-013-2929-6

Taha YA (2012). Retroviruses. In: *Medical Microbiology: A Guide to Microbial Infections*, *18th edn*. Greenwood D, Barer M, Slack R and Irving W, eds. London: Churchill Livingston – Elsevier.

Tipples GA (2011). Rubella diagnostic issues in Canada. *Journal of Infectious Diseases*, **204**: 659–663. doi: 10.1093/infdis/jir430

Torgerson PR and Mastroiacovo P (2013). The global burden of congenital toxoplasmosis: A systematic review. *Bulletin of the World Health Organization*, **91**: 501–508. doi: 10.2471/BLT.12.111732

Townsend CL *et al.* (2008). Low rates of mother-to-child transmission of HIV following effective pregnancy interventions in the United Kingdom and Ireland, 2000–2006. *Aids*, **22(8)**: 973–981. doi: 10.1097/QAD.0b013e3282f9b67a

Townsend CL *et al.* (2009). Antiretroviral therapy and congenital abnormalities in infants born to HIV-infected women in the UK and Ireland, 1990–2007. *AIDS*, **23(4)**: 519–524. doi: 10.1097/QAD.0b013e328326ca8e

Trotta F *et al.* (2014). Evaluation of safety of A/H1N1 pandemic vaccination during pregnancy: cohort study. *BMJ*, **348**: g3361. doi: 10.1136/bmj.g3361

Tubiana R *et al.* (2010). Factors associated with mother-to-child transmission of HIV-1 despite a maternal viral load < 500 copies/mL at delivery: A case-control study nested in the French perinatal cohort (EPF-ANRS CO1). *Clinical Infectious Diseases*, **50**: 585–596. doi: 10.1086/650005

Turner KM *et al.* (2013). An early evaluation of clinical and economic costs and benefits of implementing point of care NAAT tests for Chlamydia trachomatis and Neisseria gonorrhoea in genitourinary medicine clinics in England. *Sexually Transmitted Infections*, sextrans-2013. doi: 10.101136/sextrans-2013-051147

Van Kerkhove MD (2011). Risk factors for severe outcomes following 2009 influenza A (H1N1) infection: A global pooled analysis. *PLoS Med*, **8(7)**: e1001053. doi: 10.1371/journal.p.med.1001053

Verani JR, McGee L and Schrag SJ (2010). Prevention of perinatal group B streptococcal disease: Revised guidelines from CDC. *Recommendations and Reports, November 19*, **59(RR10)**: 1–32.

Walker DG and Walker GJA (2002). Forgotten but not gone: The continuing scourge of congenital syphilis. *Lancet Infectious Diseases*, **2**: 432–436. doi: 10.1016/S1473-3099(02)00319-5

Walker GJA and Walker DG (2007). Congenital syphilis: A continuing but neglected problem. *Seminars in Fetal and Neonatal Medicine*, **12**: 198–206. doi: 10.1016/j.siny.2007.01.019

Zimmerman LA *et al*. (2011). Status of rubella and congenital rubella syndrome surveillance, 2005-2009, the World Health Organization European Region. *Journal of Infectious Diseases*, **204**: S381–S388. doi: 10.1093/infdis/jir104

6

Sexually Transmitted Infections

6.1   Introduction

Sexually transmitted infections (STIs) primarily affect the genital tract, although since the responsible microorganisms can be passed between people during a variety of forms of intimate sexual contact, it should be remembered that other outcomes, such as rectal and oropharyngeal infection, are also possible. STIs account for significant morbidity – not only because of the physical pain and discomfort, but also through emotional distress (Gottlieb *et al.*, 2014). Most pathogens are carried chronically due to their propensity for asymptomatic infection and (in some cases) recurrent infection, as well as the embarrassment people feel, which makes them reluctant to seek medical help (Goering *et al.*, 2013). This means that many people infected with STIs who could be diagnosed and treated relatively easily are not identified. Global estimates suggest that the annual incidence of trichomoniasis is about 276 million, while for chlamydia and gonorrhoea the figures are around 106 million in each case. There are also thought to be at least 11 million new cases of syphilis (Gottlieb *et al.*, 2014). There are at least 35 million people in the world living with human immunodeficiency virus (HIV), although the rate of new infections does appear to be in decline in most populations (Maartens *et al.*, 2014). For more details about HIV, see Chapter 2. All of these infections would be treatable (Gottlieb *et al.*, 2014), but this requires effective education, diagnostic, screening and treatment programmes. Unfortunately, these are not available for all of those at risk (Gottlieb *et al.*, 2014; Maartens *et al.*, 2014).

   The incidence of STIs in the UK has shown some interesting trends. For example, the number of reported cases of both syphilis and gonorrhoea doubled between 1918 and 1919. Sharp increases in incidence were also seen in the mid-1940s (Hughes and Lowndes, 2014). The introduction of antibiotics to treat these infections in the late 1940s, along with more comprehensive diagnostic and screening programmes, then reduced the rates of these STIs during the 1950s (Hughes and Lowndes, 2014). Regular fluctuations have been seen since, attributable to changes in peoples' behaviour. Incidence of most STIs increased the 1960s due to various societal changes and decreased during the 1980s, as people took more precautions during sexual activity because of the concerns about acquired immune deficiency syndrome (AIDS) (see Chapter 2). After the development of anti-retroviral drugs, an AIDS diagnosis was no longer a death sentence. This appears to have relaxed peoples' attitudes towards STIs in general, resulting in increased incidences of most of them during the 1990s. Similar trends have been seen

Clinical Microbiology for Diagnostic Laboratory Scientists, First Edition. Sarah J. Pitt.
© 2018 John Wiley & Sons Ltd. Published 2018 by John Wiley & Sons Ltd.

across the rest of Europe (Fenton and Lowndes, 2004). Antibiotic resistance has also contributed to higher transmission rates and morbidity in individuals. Higher reported incidences also reflect improvements in diagnostic testing to some extent. For instance, early results from the National Chlamydia Screening Programme (NCSP) in England unexpectedly indicated that around 10% of sexually active 16 to 24 years olds were asymptomatically infected with the bacterium (LaMontagne *et al.*, 2004). Clearly the scheme, which offered more opportunities for testing and raised awareness of the infection in target groups, provided more accurate data about prevalence. The programme has continued, so it must be borne in mind that the increased reported rates compared to the early 2000s do not necessarily mean that transmission of *Chlamydia trachomatis* has drastically changed. Equally, since the data collected before the NCSP scheme started in 2002 is not accurate, this cannot be proved either way.

Point to consider 6.1: Why were increases in incidences of STIs seen in the UK in 1919 and 1946? Could this information help public health planners in future and in other countries?

The microbiology service plays a vital role in preventing long-term sequelae in individuals with STIs and in reducing transmission, through a combination of diagnosis and screening. An important tool in the public health aspect of STI management is 'contact tracing' or 'partner notification' – that is, identifying and providing information to people who have recently had sexual contact with a patient who has been newly diagnosed with such an infection (Bell and Potterat, 2011). Laboratory support is valuable in this context in terms of testing people potentially at risk from an STI, but also in linking cases epidemiologically through detailed analysis of isolates, which can elucidate the course of an outbreak. Sequence data can provide evidence in criminal prosecutions, including as support for cases of knowing and malicious infection of others with the human immunodeficiency virus (e.g., Brooks and Sandstrom, 2013) and sexual assault of a child, leading to gonorrhoea in the victim (e.g., Whaitiri and Kelly, 2011). Laboratory-derived data can only be used as suitable evidence in court cases if the highest possible standards of quality management have been used throughout and all procedures have been appropriately documented. This situation emphasises the value of good laboratory practice and accreditation in diagnostic microbiology services (see Chapter 1). The goals of timely diagnosis and effective interventions for STIs would lend themselves well to the use of point-of-care tests (POCT). There are POCT kits available for detection of *Treponema pallidum* and human immunodeficiency virus antibodies in blood, including some duplex assays (Tucker *et al.*, 2013). In the main laboratory, nucleic acid amplification tests (NAATs) for simultaneous detection of *C. trachomatis* and *Neisseria gonorrhoeae* in genital and urine samples are widely used in screening and diagnosis (Parra-Sánchez *et al.*, 2012). The turnaround times for this can be fairly rapid and the technology has been adapted to develop bench-top NAAT POCT systems (Tucker *et al.*, 2013; Gaydos *et al.*, 2013).

A list of some common STIs and their clinical presentations is given in Table 6.1. It illustrates how anatomical differences between the male and female genital tracts can mean that pathology caused by a particular organism can be quite different according to gender. For example while most adults infected with *C. trachomatis* are asymptomatic, males with overt disease commonly experience urinary tract dysfunction (pain on micturition, urethritis). Females may have urinary symptoms, but abdominal pain and bleeding after intercourse are more common (Mabey and Peeling, 2012). In males,

Table 6.1 Selected common sexually transmitted infections and usual clinical presentations.

Organism	Site of infection in females (F) and males (M)	Most common clinical presentations in females (F) and males (M)
Chlamydia trachomatis	Cervix, urethra, uterus, Fallopian tubes (F) Ureter, epididymis (M)	Usually asymptomatic (F&M) Cervical discharge, post coital bleeding, urethritis, pelvic inflammatory disease (F) Non-specific urethritis, epididymitis (M)
Neisseria gonorrhoeae	Cervix, vagina (F) Penis, ureter (M)	Usually asymptomatic (F&M) Vaginal discharge, pelvic pain, pelvic inflammatory disease (F) Urethral discharge, dysuria (M)
Treponema pallidum	Cervix and systemic (F) Penis and systemic (M)	Chancre (primary) Rash – palms of hand, soles of feet (secondary) Tissue damage (neurological, cardiac), tumour-like lesions (gummas)
Herpes simplex virus (types 1 and 2)	Vulva (F) Penis (M) Oral cavity (F&M)	Genital lesions Oral lesions
Human Papillomavirus	Cervix (F) Penis (M) Throat and neck (F&M)	Genital warts, Cervical, penile, oropharyngeal cancer
Human immunodeficiency virus	CD4+ white blood cells (F&M)	Acquired immune deficiency syndrome
Hepatitis B virus	Liver (F&M)	Acute or chronic hepatitis
Haemophilus ducreyi	Vulva, cervix (F) Penis (M)	Genital ulcers (chancroid)
Candida albicans	Vagina, vulva (F) Penis (M)	Rash, itching, whitish discharge
Trichomonas vaginalis	Vagina (F) Urethra (M)	Dysuria, frequency of micturition, genital soreness Itching, frothy, greenish vaginal discharge (F) Penile soreness, thin discharge (M)
Phthirus pubis	Genital area via pubic hairs	Itching, rash
Sarcoptes scabei	Genital area	Itching, encrusted lesions

Information taken from Goering *et al.* (2013).

particularly young men, the bacterium can cause epididymitis, but suggested links with infertility have not been proven. By contrast, up to 30% of women with chronic, undiagnosed and untreated *C. trachomatis* infection develop pelvic inflammatory disease (PID), which is associated with difficulty in conceiving and ectopic pregnancy (Mabey and Peeling, 2012). This is considered in detail in Chapter 5. Rectal *C. trachomatis* infection in both male and female patients is also a possibility that is often overlooked (Sethupathi *et al.*, 2010; Annan *et al.*, 2009). Continued sexually activity leads to minor, even imperceptible abrasions, which can provide an entry point for a sexually transmitted organism. There is also evidence that the damage to the genital tissues caused by one infection in that part of the body can pre-dispose the person to acquiring another sexually transmitted organism – in particular human immunodeficiency virus – more readily (Goering *et al.*, 2013; Gottlieb *et al.*, 2014).

It should also be noted that all organisms which are transmitted via blood and body fluids could in theory be passed from person to person via sexual intercourse. This was illustrated by the cases of recovering patients monitored during the 2014–2015 Ebola outbreaks in West Africa. Follow-up tests for viral RNA found that it took many months for some people to clear the virus completely and in some cases it was found persisting in the semen (e.g., MacKay and Arden, 2015). Also, some non-genital pathogens can be transmitted sexually if the opportunity arises. For example, there is some evidence for sexual transmission of the gastrointestinal protozoan parasite *Entamoeba histolytica*, particularly via oral-anal intercourse (e.g., Hung *et al.*, 2012), but also through direct genital contact (e.g., Asano *et al.*, 2014). This type of infection is often asymptomatic, but laboratory support in making the diagnosis is important because the invasive sequelae of the gut infection including ulceration and amoebic liver disease (Gunn and Pitt, 2012) have been reported in patients infected via this unusual route (Hung *et al.*, 2012). Another important link to remember is the risk posed by an STI in a pregnant woman to her baby. Most STIs are associated with perinatal infections, through exchange of blood and body fluids (human immunodeficiency virus, hepatitis B virus, hepatitis C virus) or through direct contact with active lesions during birth (*Chlamydia trachomatis*, herpes simplex virus, *Neisseria gonorrhoea*), although *Treponema pallidum* can cross the placenta.

Despite that fact that transmission requires the sort of intimate contact that most individuals only have with a few other people during their life time, STIs are very successful pathogens. This chapter will consider genital herpes, gonorrhoea, genital warts and trichomoniasis in detail. However, there are a number of pathogens which can be sexually transmitted and are of public health importance, which are discussed in other chapters. These include HIV (see Chapter 2 and www.who.int/hiv/en and www.unaids.org), *T. pallidum* and *C. trachomatis* (see Chapter 5 and www.chlamydiascreening.nhs.uk) and HBV and HCV (see Chapter 4 and www.who.int/topics/hepatitis/en/).

6.2   Herpes Simplex Virus

Order: *Herpesvirales*
Family: *Herpesviridae*
Sub-family: *Alphaherpesvirinae*
Genus: *Simplexvirus*
Species: *Human herpesvirus 1* (Herpes simplex type 1), *Human herpesvirus 2* (Herpes simplex type 2).

6.2.1   Introduction

Herpes viruses are large (120–200 nm), enveloped, double-stranded DNA viruses and the *Herpesvirales* order contains a number of clinically important pathogens. They have relatively complex replication cycles and exhibit both lytic and latent stages (Murray *et al.*, 2013; Forsgren and Klapper, 2009). Herpes simplex virus (HSV) infects epithelial cells during primary infection and remains latent in neurones (Forsgren and Klapper, 2009). Although the 'received wisdom' is that HSV type 1 is associated with oral infections ('cold sores') and HSV type 2 causes genital herpes, epidemiological studies suggest that this distinction is no longer valid (Bernstein *et al.*, 2013; Forsgren and Klapper, 2009).

Point to consider 6.2: Why might the possible ways of transmission of HSV-1 now include the genital route?

6.2.2   Pathogenesis and Clinical Symptoms

HSV particles are known to express at least 11 surface glycoproteins, labelled in an incomplete sequence of gB to gE and gG to gM (Forsgren and Klapper, 2009). The mechanisms of interactions with host cells have not been elucidated for all of them, but they are each considered to be involved in pathogenesis in some way. For example, gB and gC are involved attachment and entry into receptive host cells. Variation in some of these glycoproteins (e.g., gG) accounts for the biochemical and antigenic differences between HSV types 1 and 2 (Johannessen and Ogilvie, 2012; Forsgren and Klapper, 2009).

On primary infection, the virus infects host epithelial cells at the site of infection; in genital infection these are the glans and shaft of the penis in males and the labia, vagina and/or cervix in females (Johnanessen and Ogilivie, 2012). Replication is complex, with the viral genome adopting a closed, rolling circle formation. It involves three distinct stages – immediate early, early and late – and results in the destruction of the cell and release of virus particles (Forsgren and Klapper, 2009). This is manifested as vesicular lesions containing viable virus and, in some patients, a noticeable discharge. The immune system responds by producing cytotoxic CD4+ and CD8+ T-cells and an inflammatory response mediated by cytokines (Shin and Iwasaki, 2013). This can cause pyrexia, lethargy and lymphadenopathy at the site of infection (Forsgren and Klapper, 2009). Symptoms can last for several weeks (Johannessen and Ogilvie, 2102), but in some cases, they are not troubling enough to seek medical intervention – so the condition is not diagnosed until a later reactivation event. A B-cell response, resulting in detectable HSV-type-specific antibody does also occur, but this does not appear to have a protective function (Shin and Iwasaki, 2013; Forsgren and Klapper, 2009).

During the lytic process, virus particles come into close proximity to local sensory neurones, which in genital herpes would be the lumbro-sacral nerves (Shin and Iwasaki, 2013; Forsgren and Klapper, 2009). Herpes virus can enter and replicate inside these neuronal cells, establishing latency within a few days. The virus replication cycle inside nerve cells appears to bypass the immediate early stage and produces a piece of viral DNA (the 'LAT transcript') which does not form into viable particles (Murray *et al.*, 2013). The non-specific immune response employing various cells and interferon (IFN) is not particularly effective at preventing the virus establishing latency, but the specific response – involving cytokines and CD8+ T cells – is thought to keep virus activity in check (Shin and Iwasaki, 2013; Forsgren and Klapper, 2009). The precise events leading to reactivation are not well understood, but considering that stress, being run down and menstruation are known to precipitate genital herpes episodes, it is thought that the cause is alterations in the equilibrium established between virus activity and host immunity (Johannessen and Ogilvie, 2012; Murray *et al.*, 2013). As a general rule, reactivation events are generally less severe and more short-lived than the primary infection (though this is not always true). It is also important to note that virus shedding may occur without clinically manifest symptoms (Forsgren and Klapper, 2009), meaning transmission can be unwitting. Johnston *et al.* (2014a,b) studied 28 subjects known to have HSV-2 infection and noted that reactivation occurred regularly and evidence of virus activity was often at more than one particular site in the genital region. They confirmed the involvement of CD4+ and CD8+ T cells in the immune response, but found raised T-cell counts during asymptomatic shedding of the virus (Johnston *et al.*, 2014a,b). They suggested that this would lead to inflammation and tissue damage in the genital area.

It has been observed that co-infection with HSV-2 and HIV-1 is common (Freeman *et al.*, 2006). The inflammatory damage caused by recurrent genital herpes episodes is thought to create an environment conducive to HIV infection. It has also been shown that HIV-1 viral replication and viral load in plasma and genital discharge are higher in the presence of HSV-2 (Corey, 2007). In addition, the suppression of the immune response caused by the HIV virus leads to the conditions associated with reactivation of the HSV – meaning that the patient has more frequent genital herpes episodes (Awashthi and Friedman, 2014).

A consequence of the regular reactivations of genital herpes is the risk of transmission of HSV from mother to baby during birth (see Chapter 5). Infection most commonly results in eye and skin infections in the neonate, but more widespread disease involving multiple organs and also encephalitis can occur (Johannessen and Ogilvie, 2012).

6.2.3 Epidemiology

HSV is endemic in human populations. Transmission of oral herpes usually occurs early in life and in many parts of the world, seroprevalence to HSV-1 in young adults is thought to be over 90% (Murray, 2013; Forsgren and Klapper, 2009). Acquisition of genital herpes appears to be less common, partly due to the requirement for intimate sexual contact (as opposed to only kissing or touching) and also because the immune response mounted against one HSV type appears to afford protection against infection with the other – thus a person infected with HSV-1 as a baby will be less likely to contract HSV-2 as an adult (Forsgren and Klapper, 2009). Although most HSV-2 infections are associated with genital herpes, it does sometimes cause oral infections and the opposite is also true for HSV-1.

It is difficult to collect accurate data about the prevalence of genital herpes for both logistical and virological reasons. It is reported that many people are unaware of their primary episode (Forsgren and Klapper, 2009) and therefore studies about prevalence have tended to rely on serology. This may be inadequate for species differentiation (see below) and also it cannot be assumed that a person with anti-HSV-2 antibodies has a genital infection. Looker *et al.* (2008), using a systematic review, estimated that over 500 million people worldwide are infected with HSV-2, with an annual incidence of 23 million (Looker *et al.*, 2008). However, HSV-1 is now a significant cause of genital infections and in some populations it may account for between one half and two thirds of cases (Hughes and Lowndes, 2014; Bernstein *et al.*, 2013). Bernstein *et al.* (2013) followed the control group in a herpes vaccine trial for evidence of primary HSV infection. These were sexually active women between 18 and 30 in the United States and Canada. Over the course of almost two years, amongst over 3,000 subjects, just over 5% seroconverted to HSV. The authors reported that type 1 infections accounted for more than twice the number of type 2 infections (Bernstein *et al.* 2013), although the absolute numbers were quite small (183 HSV cases, 127 HSV-1 infections, 56 HSV-2 infections). In the UK, it is thought that around 3% of males and 5% of females are infected with HSV-2 (Hughes and Lowndes, 2014). There are around 30,000 newly diagnosed cases of genital herpes annually (Hughes and Lowndes, 2014) and the majority of patients are between 20 and 24 years of age. It is worth bearing in mind that people may have been infected for some time (even years) before presenting, so this first clinically significant episode may not always be a primary infection. The rate of recorded reactivations is over 24,000 each year, but two points must be noted about this information. The first is that many episodes may be mild or subclinical and therefore not brought to the attention of a medical practitioner. The second is that – conversely – one individual could have several severe, clinically apparent reactivations in a particular 12-month period. Despite the change in epidemiology of the two HSV types, the majority of debilitating genital herpes cases are attributable to HSV-2 (Forsgren and Klapper, 2009).

Accurate data about incidence of neonatal herpes is not always collected or collated, but it is known that it can vary greatly. In the UK it is estimated at 1.65 per 100,000 live births (Johannessen and Ogilvie, 2012), whereas in some parts of the United States it is recorded at 1 in 2,500 births (Forsgren and Klapper, 2009). Screening and treating of pregnant mothers can reduce the rate of neonatal acquisition of HSV (see Chapter 5), but the case fatality rate in babies where the infection is not treated can be 60%. This highlights the importance of sensitive and specific laboratory methods for detecting virus in asymptomatic carriers.

Point to consider 6.3: Do you think pregnant women should be screened routinely for genital HSV? If so, how would you design such a programme?

6.2.4 Laboratory Diagnosis

Although a diagnosis of genital herpes can be made clinically, laboratory confirmation is preferable because lesions attributable to other causes can appear similar (Le Goff *et al.*, 2014). Diagnosis is based on detection and characterisation of the virus from lesions. Genital swabs are the appropriate specimens, but they should scrape the base of the lesion for maximum efficacy (SMI V17, 2013). For a rapid diagnosis, cellular material (e.g., deposit after centrifuging the sample in viral transport medium or scrapings taken directly from a lesion) can be tested with fluorescent-labelled anti-HSV antibodies (SMI V17, 2013); commercial kits generally contain monoclonal antibodies to both HSV-1 and HSV-2. Virus isolation in fibroblasts in monolayer cell culture usually produces cytopathic effect in less than 48 hours (often overnight). The observed CPE is characteristic of the HSV type, but needs to be confirmed using immunofluorescence (SMI V17, 2013). This approach has now largely been replaced by the use of real time PCR for HSV DNA (Le Goff *et al.*, 2014) which is a more sensitive method. Most available assays distinguish between the HSV types, some provide a quantified result and it is also possible to buy kits which will do virus typing, which is an important aspect of the diagnosis of genital herpes (Le Goff *et al.*, 2014). Although drug resistant strains of HSV rarely arise in otherwise healthy adults, sequencing of isolates would help in the case of treatment failure. Most mutations occur in the UL23 gene coding for the viral thymidine kinase (Le Goff *et al.*, 2014).

Serological assays for detection of anti-HSV antibodies do exist, but they have limited value in a diagnostic situation. Detection of IgG could, in theory, be used to confirm the diagnosis of genital herpes if collection of genital swabs was not feasible (e.g., if the person was not symptomatic at presentation), but kits tend not to be sufficiently accurate in distinguishing between types 1 and 2 (Le Goff *et al.*, 2014). While the use of an IgM assay to ascertain recent primary infection appears useful, it is known that IgM can be raised during reactivation so it will not provide a definitive answer (Le Goff *et al.*, 2014).

6.2.5 Treatment and Management

Patients should be advised to refrain from intercourse during symptomatic episodes and can be treated with aciclovir ('Zovirax'), which was the first really successful antiviral, due to its specificity for HSV thymidine kinase (see Chapter 1). A course of oral aciclovir is usually prescribed on initial diagnosis, regardless of the disease history (Forsgren and Klapper, 2009). Subsequent treatment depends on the frequency of reactivations (which in turn may be influenced by the HSV type). When the patient experiences genital herpes fewer than six times per year, then treatment of each episode, as it occurs, is recommended. The drug should be taken just before the lesions appear (there is usually a tingling sensation in the area where the sore is about to develop). If the occurrences are more frequent, then a daily dose of the drug taken for 6 to 12 months is used as a suppressive treatment (Forsgren and Klapper, 2009). This is intended to reduce the activity of the virus, thus relieving symptoms and reducing the risk of transmission to others, but it does not eliminate the virus (Awashthi and Friedman, 2014). Fortunately, this approach to patient management does not appear to have precipitated the emergence of drug resistant strains of HSV. The knowledge that HSV can never be eliminated can be psychologically devastating and therefore counselling may be required to help the person come to terms with this and to manage their condition. A study of women with HSV-2 and HIV-1 co-infection demonstrated that anti-retroviral therapy, which reduced HIV viral load, unfortunately had no effect on HSV-2 shedding (Péré *et al.*, 2015). Although the number of subjects was small (22 patients), the trends were clear and this implies that although the pathogenesis is synergistic, treatment needs to be planned separately! The risk of neonatal herpes and the association between HSV-2 and HIV-1 are driving work to develop a vaccine against herpes simplex virus – drawing on the success of the vaccine for the closely related varicella zoster virus (e.g., Awashthi and Friedman, 2014; Johnston *et al.*, 2014a,b).

6.3   *Neisseria gonorrhoeae*

Classification
Kingdom: Bacteria
Phylum: Proteobacteria
Class: Beta Proteobacteria
Order: Neisseriales
Family: Neisseriaceae
Genus: *Neisseria*
Species: *Neisseria gonorrhoeae, N. meningitidis*

6.3.1   Introduction

The two important pathogens in the genus *Neisseria* are *N. meningitidis* and *N. gonorrhoeae*. The bacteria are Gram positive and they appear as kidney-shaped diplococci (Oldfield and Ala'Aldeen, 2012). Commensal carriage of *N. gonorrhoeae* does not occur (in contrast to *N. meningitidis*), although asymptomatic infection is fairly common. A range of virulence factors have been identified and strains which are completely resistant to penicillin are now found worldwide (Oldfield and Ala'Aldeen, 2012). This resistance is attributable to acquisition by the bacteria of plasmids coding for a β-lactamase, but was driven by overuse and misuse of this antibiotic to treat gonorrhoea during the twentieth century (Oldfield and Ala'Aldeen, 2012). In the UK and many other countries, gonorrhoea is a notifiable disease.

6.3.2   Clinical Symptoms and Pathogenesis

In women, the endocervical region is the main site of infection and symptoms include vaginal discharge and dysuria (Murray *et al.*, 2013). In 10% to 20% of patients, the bacterium travels up the genital tract and affects the ovaries and uterus, causing salpingitis, abscesses in the ovaries and pelvic inflammatory disease (Murray *et al.*, 2013). Men usually experience acute urethritis with dysuria and an unpleasant penile discharge. Penile infection is nearly always symptomatic in men, although rectal and oropharyngeal infections are less likely to be clinically obvious (Oldfield and Ala'Aldeen, 2012). In a very small proportion of cases, the infection disseminates, in the form of bacteraemia, which can spread the gonococci to the joints and into skin lesions (Murray *et al.*, 2013). It is thought that some strains of *N. gonorrheae* are more likely to cause these (Oldfield and Ala'Aldeen, 2012). Other rare complications include peritonitis and a condition called Fitz-Hugh-Curtis syndrome which is an inflammation around the liver (Oldfield and Ala'Aldeen, 2012).

N. gonorrhoeae has several known virulence factors. They have been characterised biochemically and have been found to be prone to considerable antigenic variation. The exact mechanisms of some of their functions are not fully understood. The bacteria have pili which are important in attachment and entry into host epithelial cells (Murray *et al.*, 2013; Jerse *et al.*, 2014). They are also involved in exchange of genetic material between bacteria and appear to afford resistance to the activity of host neutrophils (Murray *et al.*, 2013). The bacteria cell outer membrane contains lipo-oligosaccharide

(LOS), the porin B proteins (Por B1A and Por B1B) and the opacity proteins (Opa). These enhance invasion of host cells since the LOS and porin Bs form complexes with host complement C3, while Opas have been found to bind very specifically to human carcinoembryonic antigen-related cell adhesion molecules (CEACAMs) on genital epithelial cells (Jerse *et al.*, 2014). Gonococci secure the iron molecules they need for metabolism by having receptors to capture human transferrin, haemoglobin and human lactoferrin (Murray *et al.*, 2013; Jerse *et al.*, 2014).

The non-specific host immunity to gonococcal infection includes a pro-inflammatory response, induced by tumour necrosis factor (TNF)-α. Also LOS appears to induce Th17 cells via stimulation of toll-like receptor (TLR)-4 (Jerse *et al.*, 2014). This causes the production of another inflammatory cytokine, interleukin (IL)-17. The results in the observed exudate, which contains a mixture of sloughed off epithelial cells, polymorphonuclear leukocytes (PMNs) and bacteria (Murray *et al.*, 2013; Jerse *et al.*, 2014). It is thought that gonococci can enter PMNs, thus evading the immune response. Although the pilin and por B proteins are highly immunogenic, they are also antigenically very variable. In addition, experimental data (in animals and *in vitro*) suggest that the bacterial activity can suppress the host's Th1 and Th2 response. This means that the adaptive immune response is not very effective – as evidenced by individuals having multiple gonorrhoeal infections in spite of suitable diagnosis and treatment each time (Jerse *et al.*, 2014). The effects of any antibodies which are produced are also resisted by the bacterium through mechanisms such as the IgA1 protease (Murray *et al.*, 2013; Jerse *et al.*, 2014).

Point to consider 6.4: How might the apparent deficiencies in the host immune response to *N. gonorrhoeae* affect transmission of the bacteria between individuals?

Strains of *N. gonorrhoeae* are able to respond relatively quickly to hostile changes in their environment. This makes them particularly susceptible to developing resistance to antimicrobials. They can exchange genetic material through transformation; mutations in the bacterial genome appear to develop rapidly (Unemo and Shafer, 2014). Although many of the resulting strains will be less biologically 'fit', if the genetic alteration affords resistance to an antimicrobial under selection pressure from that drug, then a new viable, transmissable strain can arise. The ease with which genetic changes can occur seems to have made it possible for examples of most types of antimicrobial resistance mechanism to be found in *N. gonorrhoeae* strains. Enzymes which alter the antibiotic structure, bacterial protein changes (which mean the antibiotic can no longer easily bind to its target) and alterations to the bacterial cell to interfere with influx or efflux of metabolites (including antibiotics) have all been noted in this species (Unemo and Shafer, 2014). Mechanisms of antimicrobial resistance are both plasmid-borne and chromosomal and indeed for some drugs both types are observed. For instance, resistance to penicillin can be conferred by a β-lactamase associated with the plasmid-mediated bla_{TEM-1} gene, as well as through mutations in the chromosomal *penA* gene (Unemo and Shafer, 2014). An altered form of the *penA* gene – which is characterised as mosaic – is linked with resistance to cephalosporins. It is thought that the changes have come about through plasmid-mediated exchange of partial genetic sequences with commensal *Neisseria* spp. and a number of different patterns have been observed. For a detailed review of this topic the reader is directed to Unemo and Shafer (2014).

6.3.3 Epidemiology

Since a significant proportion of cases of gonorrhoea are not laboratory confirmed (due to difficulties with encouraging people to access medical care and the asymptomatic nature of many infections), global epidemiological data can only be estimated. Calculations indicate that in 2012, there were 78 million new cases of *N. gonorrhoeae* in people aged between 15 to 49 years and that the prevalence was over 26 million (Newman *et al.*, 2015). This is lower than the estimated cases of chlamydia (incidence 131 million and prevalence around 127 million), but higher than syphilis (incidence 5.6 million and prevalence just over 17 million). In the UK, the annual incidence of *N. gonorrhoeae* infection is about 29,000 (Hughes and Lowdnes, 2014). Amongst females, the highest rates of infection are seen in those aged 15 to 19, while for males it is those in the 20- to 24-year-old age group. Cases tend to be concentrated in particular geographical areas or socially connected groups and the overall prevalence of gonorrhoea in the UK is less than 0.1%. Identified risk factors include younger age, reluctance to use protection, multiple sexual partnerships at any one time and frequent turnover of partners (Hughes *et al.*, 2013). Around 60% of confirmed cases in males are amongst men who have sex with men (Hughes and Lowdnes, 2014). As noted above, some individuals remain vulnerable to re-infection despite seeking timely clinical advice, laboratory confirmation of the infection and prescription of suitable treatment. A study of 1,650 genitourinary clinical attendees in one UK city over four years found that approximately 8% of patients returned with a new *N. gonorrhoeae* infection within a year of being diagnosed and treated for gonorrhoea (Hughes *et al.*, 2013). This was attributed to being in particular demographic groups and repeatedly undertaking the behaviour which exposed them to the bacterium. Although 8% seems a relatively low proportion, this was a single GUM clinic in one country; if this was extrapolated to all new cases of gonorrhoea worldwide, the global figure for rapid re-infection could be over 6 million!

Biological factors also contribute to individual *N. gonorrhoeae* infections and transmission through populations. The apparently poor (and poorly understood) specific immune response is one of these, while antibiotic resistance is another. The widespread use of penicillin for this infection, once the drug became available in the 1940s, led to the development of resistant strains fairly quickly (Ison *et al.*, 2013). The policy to prevent this happening for subsequent antibiotic treatments is to monitor circulating isolates. When 5% or more are identified as resistant to the recommended first-line treatment, it is changed. In response to growing resistance to ciprofloxacin by the early 2000s, the UK guidelines for gonorrhoea treatment were amended in 2004, to advocate the cephalosporins cefixime and ceftriaxone instead (Ison *et al.*, 2013). Trends amongst *N. gonorrhoeae* isolates are routinely analysed, through programmes such as the Gonococcal Resistance to Antimicrobials Surveillance Programme (GRASP) in England and Wales (Ison *et al.*, 2013) and the European Gonococcal Antimicrobial Surveillance Programme (Euro-GASP) across the EU/EEA area (Cole *et al.*, 2014). Data from these programmes suggest that cephalosporin-resistant strains have been spreading and that certain clones are found in epidemiologically linked populations (Ison *et al.*, 2013; Cole *et al.*, 2014). This illustrates how surveillance is invaluable in informing public health interventions as well as treatment policy.

6.3.4 Laboratory Diagnosis

Bacteria can be readily detected in urethral swabs; in females, endocervical swabs are also appropriate, while in men with symptomatic urethritis, the exudate should be collected (SMI B28, 2014). If anorectal gonorrhoea is a possible diagnosis, then rectal swabs can be examined. Direct smears can be prepared from the swabs and stained with Gram stain; *N. gonorrhoeae* will appear as Gram positive, kidney-shaped diplococci (SMI ID6, 2015). While the bacteria does usually grow on blood agar, it is rather fastidious and therefore the specific gonococcal (CG) medium is recommended for culture of specimens. This is a chocolate blood agar, to which antifungal agents have been added (for improved selectivity, given that genital swabs may contain *Candida* spp). New York City (NYC) medium is another specially designed formula for the isolation of *Neisseria* spp. which could be used. The plates should be incubated at 35 to 37 °C, in CO_2 for two days (SMI B28, 2014). The bacteria grow as greyish/brown colonies, which are oxidase and catalase positive (SMI ID6, 2015). Further identification can be done through detection of specific enzymes, particularly proline iminopeptidase (Alexander *et al.*, 2006) for which commercial kits are available. However, caution must be exercised since mutations in some circulating strains have been associated with false negative results (SMI ID6, 2015; Alexander *et al.*, 2006).

Use of NAATs is advocated for *N. gonorrhoeae* detection, due to the improved sensitivity and specificity and turnaround times compared to culture (Bignell and Unemo, 2013). Urine is a suitable test specimen, which allows flexibility and may improve acceptability to patients, particularly in screening programmes. For diagnosis in someone with clinically apparent symptoms, a targeted sample (i.e., genital, pharyngeal or rectal swab) is usually more appropriate. More reliable results are obtained from endocervical swabs than urine for females with patent gonorrhoea (Bignell and Unemo, 2013). There are a range of commercially available assays, including those which are set up as duplex system designed to test for *N. gonorrhoeae* and *C. trachomatis* simultaneously. Automated systems (Figure 6.1) facilitate testing of large numbers of samples to support both diagnosis and screening. The details of the assays vary, but the probe targets are all conserved regions of the bacterial genome. Evaluation studies suggest that while sensitivities are good in these assays, specificity, in terms of distinguishing between species of *Neisseria*, may be less so. Tabrizi *et al.* (2011) assessed the outcomes of tests for 450 laboratory characterised bacterial isolates in six different NAATs kits. Two hundred and sixteen isolates were *N. gonorrhoeae*, while 234 were other *Neisseria* spp. or *Moraxella* spp. (including 75 *N. meningitdis* and 14 *M. catarrhalis*). All the assays provided false positive results by detecting some of the non-gonoccocal *Neisseria*, although for two of the kits this error was made in respect of more than 10% of isolates (Tabrizi *et al.*, 2011). While noting that microbiological factors could have contributed to the findings (through exchange of genetic material between commensal and pathogenic *Neisseria* spp.),

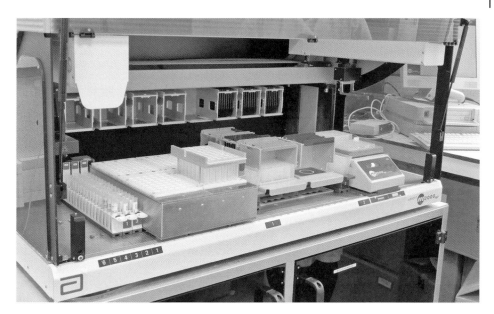

Figure 6.1 Automated laboratory system used to prepare samples for nucleic acid amplification analyses, including polymerase chain reaction. (*See insert for colour representation of the figure.*)

the authors also found technical issues with the assays. The suggestion is that positive results may need further tests to confirm species. Another drawback of NAATs is that without bacterial cultures from the specimens, antimicrobial sensitivities are not determined (Low *et al.*, 2014), since the genetic sequences relating to particular antibiotic resistance patterns have yet to be defined.

Point to consider 6.5: What research and development needs to be undertaken to improve the specificity of the NAATs test for *N. gonorrhoeae*? Think of both species determination and antimicrobial sensitivity.

6.3.5 Treatment

Antibiotic treatment is often prescribed without reference to the susceptibility information (since it may not be determined at all – see paragraph above). Oral cefixime or intramuscular ceftriaxone are recommended, in combination with azithromycin in each case (Bignell and Unemo, 2013). The azithromycin is included to minimise the chances of resistance arising in the *N. gonorrhoeae* strain, but it has the added advantage of being effective against any asymptomatic *C. trachomatis* infection the patient may also have (Bignell and Unemo, 2013; Ison *et al.*, 2013). Collection and testing of laboratory samples after the course of treatment has been finished, to ascertain whether the treatment has been effective is an important part of the management of the individual patient. Contact tracing is also vital for the sake of that person's sexual partners themselves and for public health in that population (Bignell and Unemo, 2013).

6.4 Human Papilloma Virus

Classification
Family: *Papillomaviridae*
Genus: *Alphapapillomavirus*
Species: Human papillomavirus 6, Human papillomavirus 11, Human papillomavirus 16, Human papillomavirus 18.

6.4.1 Introduction

Papillomaviruses are double-stranded DNA viruses, around 55 nm in diameter. The genome is described in sections, according to the functions of the various gene products: E1 to E7 and L1 and L2 (Cubie *et al.*, 2012). The classification is quite complex, with viruses grouped into at least 49 genera (from *Alphapapillomavirus* to *Zetapapillomavirus*!) on the basis of the sequence of the gene for the L1 capsid protein (McCance, 2009). Human Papillomavirus (HPV) species are found in five of the genera, namely *Alphapapillomavirus Betapapillomavirus, Gammapapillomavirus, Mupapillomavirus* and *Nupapillomavirus*. They are usually grouped according to the site of infection. There are cutaneous HPV types (examples of which are found in all the genera) and mucocutaneous types, which are all alpha papillomaviruses (Cubie, 2013). Patients with the rare, genetically inherited disorder epidermodyplasia verruciformis (EV) have an usually high susceptibility to HPV (McCance, 2009). They develop many, polymorphous lesions and these are similarly associated with distinct HPV types (Cubie, 2013). Cutaneous HPV infections are apparent as 'warts' often on hands and feet. Some alpha papillomaviruses cause external genital warts, which is a common STI, the reported incidence of which is steadily rising (Cubie, 2013). Since the lesions are visible and palpable, their presence is likely to cause distress and they may be painful, but they are usually benign. They are linked with specific HPV types – including HPV 6 and HPV 11 – and these are categorised as 'low risk'. In contrast, anogenital and oral infections are associated with different, 'high-risk' alpha papillomaviruses (such as HPV 16 and HPV 18). These may be carried asymptomatically, but have a strong chance of becoming carcinogenic (McCance, 2009; Cubie, 2013).

6.4.2 Clinical Symptoms and Pathogenesis

External genital warts are usually very small skin lesions caused by low-risk alpha HPV types. They are found at sites where contact is made during sexual intercourse, such as the penis in males and the vulva and vagina in females (Cubie, 2013). They tend to range in appearance from pale, through to dark and pigmented. They may be itchy, with a propensity to bleed and can cause pain during sexual activity. Entry of the virus particles into host epithelial cells is enhanced by tiny cuts of the skin in the genital area (often caused by the friction of intercourse). The exact mechanism of cell entry has yet to be elucidated. Virus replication generates several early (E) and late (L) proteins, although precise functions have not been assigned to all of them. L1 and L2 are viral capsid proteins, so would be expected to have a role in binding to the host cell, but the host receptors involved have not been determined (Doorbar *et al.*, 2012). One of the reasons for this is the difficulty of growing HPV *in vitro* (McCance, 2009). It is thought that virus replication within the cell is fairly rapid at first and regulated by the E1 and E2 proteins. Once the viral genome copy number within the cell has reached a certain level (possibly between 50 and 100 copies) then a lower, maintenance rate of genome replication becomes established (Doorbar *et al.*, 2012). Proteins E5, 6 and 7 affect the host cell metabolism to favour virus persistence and replication inside the cell. For example, E5 is thought to prevent apoptosis, while E6 and E7 appear to alter the host cell cycle (McCance, 2009; Doorbar *et al.*, 2012). Inflammation occurs at the site of infection and the skin epithelium thickens, resulting in the visible lesion (Cubie, 2013). This contains viable virus particles, which can be transmitted to another human host on contact. The lesion may resolve on its own, but sometimes it grows large enough to cause discomfort and then treatment is required. Individuals usually acquire genital warts through sexual contact, but sharing contaminated towels or clothing has also been implicated in transmission and perinatal infection is possible (Cubie, 2013). Virus activity can lead to genital warts developing into pre-cancerous lesions, but it is rare.

In contrast, anogenital infection with high-risk alpha HPV types is often asymptomatic, but viral metabolism leads to pre-cancerous and cancerous pathology. The main site of malignancy is the cervix; HPV DNA is found in 99% of cervical cancers. The connection between HPV infection and the development of cervical cancer was proven by zur Hausen in the early 1980s (Chelimo *et al.*, 2013). High-risk types of the virus are also implicated in anal (in both males and females) and head and neck cancers. Cancers of the vaginal, vulval and penile regions are unusual, although about 50% of penile cancers are associated with HPV infection (Cubie, 2013). The key symptom which would lead to investigations for cervical cancer is unexpected vaginal bleeding (i.e., outside of the normal menstrual period or post menopause). This is due to the tissue damage, which in this situation has been caused by the viral activity. It is important to note that there are a range of reasons for unusual menstrual bleeding patterns, so this observation alone does not necessarily mean that the patient has a malignancy. A physical examination of the cervical area (colposcopy) and laboratory analysis of biopsy samples would be required to confirm the suspicion. Some patients also experience pelvic pain, discomfort after intercourse and abnormal vaginal discharge (Cubie, 2013). Rectal bleeding and irritation around the anal area are seen in anal cancer, again due to the effects of the virus on local tissue. However, the patient could simply have haemorrhoids. Penile, vulval and vaginal cancers are all characterised by lesions, soreness and atypical discharges (Cubie, 2013). As noted above, these types of cancer are rare and require thorough investigation before a diagnosis can be made.

The viral replication cycle for high-risk HPV types starts similarly to that outlined above for the low risk types, except that the 'maintenance' genome copy number is thought to be around 200 copies per cell (Doorbar *et al.*, 2012). Another difference is that the E6 and 7 proteins of the high-risk types have been implicated in cell proliferation, thus potentially leading to neoplasia (Doorbar *et al.*, 2012). Through mechanisms that are not entirely understood, the virus transforms basal epithelial cells. It is thought that E6 and E7 genes may become integrated into the host cell genome, thus subverting and deregulating replication. This leads to inappropriate metabolic processes and genetic errors (Doorbar *et al.*, 2012). Oncogenesis is also enhanced by cofactors such as number of sexual partners and smoking (Chelimo *et al.*, 2013; Cubie, 2013). For oral cancer, risk is further increased with regular high alcohol consumption and chewing tobacco or betel nuts (Cubie, 2013). Women who use the contraceptive pill for prolonged periods are more likely to develop cervical cancer (Chelimo *et al.*, 2013), although the cause for this observation has not yet been proven. The progression of HPV infection to cervical and anal cancers appears to be assisted by the presence of HIV. Since they are both STIs, it is not surprising that a person might have a co-infection. The immunosuppression caused by the HIV activity is thought to favour persistent HPV infection and the development of the early stages of tissue abnormalities (Cubie, 2013).

6.4.3   Epidemiology

It is thought that 11% to 12% of the global population are infected with HPV (Forman *et al.*, 2012).It should be noted that this figure is an estimate made by extrapolating from available data, which is mostly obtained from screening of females, but it is considered that the rates of HPV carriage in men within a given population reflect those in women (Gottlieb *et al.*, 2014). Prevalence varies enormously, from over 30% in the Caribbean and some parts of Africa, to less than 10% in Europe (Forman *et al.*, 2012).

The global annual incidence of genital warts is estimated to be around 200 cases per 100,000 population, with the rate being slightly higher in males than females (Patel *et al.*, 2013) and greatest in young adults, since it is linked to the start of sexual activity. The worldwide prevalence is reported as between 1% and 5% (Patel *et al.*, 2013). It appears that the virus persists in some people, while in other individuals, the infection can resolve (Wiley *et al.*, 2002). Laboratory investigations suggest that approximately 90% of genital warts are associated with HPV types 6 and 11, while HPV types 42 and 81 are frequently found in the remaining 10% of cases. Infection with more than one HPV type is also possible (Cubie, 2013). There are at least 80,000 new cases of genital warts in the UK each year.

It has been calculated that HPV infection results in 610,000 cases of cancer each year worldwide (Forman *et al.*, 2012). Of these, 87% are malignancies of the cervix uteri (cervical cancer), just under 4% are anal and around 3.5% are oropharyngeal (head and neck cancer). The remaining cases of HPV-associated cancer are vulval, vaginal and penile (Forman *et al.*, 2012). Cervical cancer is the second most common cancer in women (Cubie, 2013; Murray et al., 2013) – there are a greater number of cases of breast cancer. Across Europe there are at least 58,000 newly diagnosed cases of cervical cancer each year and 24, 000 deaths (Ferlay *et al.*, 2013). In the United States this figure is around 12,000 cases and 4,000 deaths (Murray *et al.*, 2013). The implementation of screening programmes (see below) has considerably reduced the incidence and mortality (Peto *et al.*, 2004). The incidence of head and neck cancer varies across the world, but is decreasing overall. However, the fraction of these which are oropharyngeal is actually rising. HPV infection is implicated in about a quarter of cases (Forman *et al.*, 2012), usually HPV type 16 (Cubie, 2013). This implies that acquisition is likely to occur during sexual activity (Marur *et al.*, 2010).

Point to consider 6.6: Suggest some experiments that could be done to elucidate the pathogenic mechanisms of HPV further.

6.4.4   Laboratory Diagnosis

Patients with genital warts have visible, characteristic lesions. Since treatment targets the growth itself and is not specifically antiviral, laboratory diagnosis for individuals is not usually necessary (Wiley *et al.*, 2002). In patients with suspected HPV-related cancer, virus detection may be a useful adjunct to the histological examination of tissue from the relevant site (Cubie *et al.*, 2012). Nucleic acid detection tests are available to detect HPV DNA and type the isolate on the basis of the L1 gene sequence. The possible oncogenicity can be ascertained through assays for RNA from the E6 and E7 gene products (Cubie *et al.*, 2012). Monitoring for the presence of viral DNA after a course of treatment can also be helpful.

The bulk of the contribution of the laboratory service to HPV detection comes in the form of cervical cancer screening programmes. These involve inviting women between the ages of 20 or 25 (depending on the country) and 64 to be tested regularly for the early signs of cervical cancer. Evidence suggests that the optimal recall interval is three to five years (Sasieni *et al.*, 2003). The screening test is intended to detect indicators showing a risk of developing cancer in cervical specimens from healthy women. In the mid-twentieth century, the Papanicolaou ('Pap') histological method was introduced. This requires a well-collected cervical swab and the specific Papanicolaou stain (Cubie, 2013). Slides are examined for the presence of abnormal cells, with a view to identifying pre-cancerous lesions and thus preventing individuals from progressing to cervical cancer, through careful monitoring and, where necessary, early intervention. Observations are categorised according to the degree of cervical intraepithelial neoplasia (CIN) present on a scale of 1 to 3, where CIN-3 is the severest form of dysplasia. Figure 6.2a shows the histological appearance of normal cervical squamous cell epithelium and Figure 6.2b shows tissue at CIN-2. Screening programmes are considered to be successful and cost-effective in the prevention of cervical cancer. For example, the UK scheme has been calculated to have prevented at least 12,000 cases of cervical cancer and 6,000 deaths amongst women born since 1950 (Peto *et al.*, 2004).

Issues with quality of the specimens and sensitivity of the slide reading were partially resolved with the change to liquid based cytology (LBC) in the 2000s (e.g., Monsonego *et al.*, 2001). Histological examination of samples collected by LBC may not be more accurate than Pap smear tests in all situations (e.g., Sigurdsson, 2013). There is still some

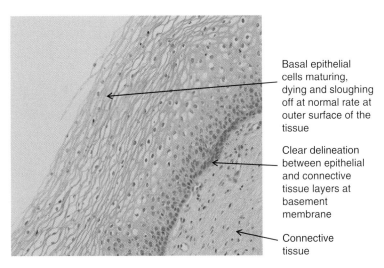

Basal epithelial cells maturing, dying and sloughing off at normal rate at outer surface of the tissue

Clear delineation between epithelial and connective tissue layers at basement membrane

Connective tissue

Figure 6.2a Section of normal cervical squamous epithelium stained with Papanicolaou stain. Source: Courtesy of Ms L. Dixon. (*See insert for colour representation of the figure.*)

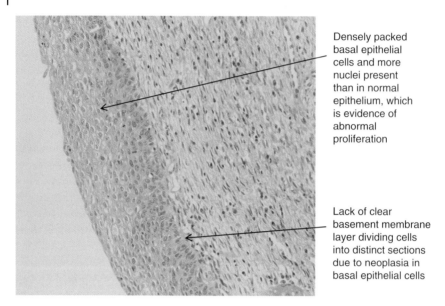

Densely packed basal epithelial cells and more nuclei present than in normal epithelium, which is evidence of abnormal proliferation

Lack of clear basement membrane layer dividing cells into distinct sections due to neoplasia in basal epithelial cells

Figure 6.2b Section of cervical squamous epithelium showing CIN-2 stained with Papanicolaou stain. Source: Courtesy of Ms L. Dixon. (*See insert for colour representation of the figure.*)

subjectivity in the cytological evaluation (Cuzick *et al.*, 2006) but LBC affords a significant advantage, because the specimens can be used in HPV detection assays. Evaluations have shown that the virological test is more sensitive than cytology for screening (Ronco *et al.*, 2014; Cuzick *et al.*, 2006). This extra sensitivity means that it safe to test women less frequently (Ronco *et al.*, 2014). Along with the improvement in sample test around times that is afforded by automation of molecular methods, use of HPV assays can make the overall screening programme more cost-effective. Some commercially available HPV detection assays are designed to detect viral DNA in terms of the L1 gene from HPV types 16 and 18 and a selection of at least 10 other types which have been associated with carcinogenesis (Cubie *et al.*, 2012; Arbyn *et al.*, 2013). Studies have shown that this approach gives better results than cytology or histology for the identification of patients with CIN-2 or -3 (Arbyn *et al.*, 2013). There are obvious difference between the histological appearances of the cervical tissue sections in Figure 6.2a and 6.2b. In particular, it is clear that the malignant changes in Figure 6.2b are characterised by the abnormal proliferation of cells and growth of transformed epithelium into the connective tissue layer. Figure 6.2 shows CIN-2; in CIN-1, the proliferation is less marked and seen at the edge of the basal layer, while in CIN-3 the differentiation between cells types is lost, as the neoplasia takes over the tissue. However, examining slides appropriately takes time and expertise. Screening involves cytology and in the earlier stages towards the possible development of cancer, cytological changes tend to be less clearly defined. CIN-1, also recorded as 'atypical squamous cells of undetermined significance' (ASC-US) or 'low-grade squamous intraepithelial lesions' (LSIL). Patients with these cytological pictures are more likely to develop cancer at some time in the future than those with completely normal cytology, but the risk is still low (Arbyn *et al.*, 2013). The policy in cytological screening would be to recall the woman for a further examination of the cellular appearance after a short time period (e.g., six months). The changes can also be caused by a transient genital infection or a 'low-risk' HPV type such as 6 and 11 (Cubie *et al.*, 2012) and can therefore be short-lived. In this situation, repeat cytology appears to be at least as good as HPV DNA detection for supporting clinical management of patients (Cubie *et al.*, 2012; Arbyn *et al.*, 2013). Therefore assays to detect mRNA transcribed from the E6 and E7 genes have been designed and evaluated. These proteins are particularly associated with invasive, carcinogenic virus activity and thus would be expected to be more useful indicators of the risk of progression to CIN- 2 or -3 (Cubie *et al.*, 2012). The idea would be to avoid waiting for the recall smear, by more accurately identifying those patients whose slightly abnormal cytology (CIN-1) represented a high chance of becoming malignant and investigating them further. Arbyn *et al.* (2013) conducted a systematic review of data published data on two kits – one detecting HPV DNA and the other mRNA. They concluded that the mRNA assay had similar sensitivity to the DNA method for detection of CIN-2 and -3 changes in patients' samples, so would be acceptable in routine screening regimens. They also suggested that it offers higher specificity for identification of patients with CIN-1 (Arbyn *et al.*, 2013). Immunocytochemical methods for detection of specific cell surface markers associated with oncogenic change (e.g., p16INK4a) may also provide extra information to help clinical decision making (Cubie *et al.*, 2013; Arbyn *et al.*, 2012).

6.4.5 Treatment

Genital warts may not be particularly noticeable and do resolve without intervention in some cases, so treatment is not necessarily recommended unless they are causing distress to the patient. Methods for physical removal of the lesions are recommended and they include surgery, freezing with liquid nitrogen and laser therapy (Wiley *et al.*, 2002). Topical treatments are also available. Anti-mitotic agents are available and these are intended to stimulate necrosis in the area surrounding the wart. These are quite toxic and some guidelines suggest they should be avoided (e.g., Lacey *et al.*, 2013). Immunomodulating agents are designed to enhance the local immune response to the virally infected cells, by inducing monocytes and macrophages to produce Il-2 and IFN-α. All of these treatments ought to destroy the wart itself, but may not eliminate the virus and recurrence is a strong likelihood. Therefore, patients need regular follow up to ensure that the infection has been cleared.

Patients with HPV-associated cancer are carefully assessed to determine the nature of the malignancy (squamous cell or adenocarcinoma) and the stage of the cancer. Treatment will compromise a combination of chemotherapy, radiotherapy and/or surgery, as appropriate. Chances of survival are affected by how far advanced the cancer is at the time the diagnosis is made. For a detailed consideration of clinical issues in cancer treatment see, for example, Hanna *et al.* (2015).

6.4.6 Vaccination

In the late 2000s, HPV vaccines became licenced and many countries have implemented programmes to vaccinate pre-pubescent girls (Cubie *et al.*, 2012). The theory is to provide protection before they start sexual activity, thus reducing the incidence of cervical cancer. In the long term, this would be expected to decrease transmission of HPV in the whole population. There are two formulations – the bivalent (which contains antigens intended to raise a protective immune response to HPV types 16 and 18) and the quadrivalent (which includes HPV types 6 and 11 as well). The vaccines have proved safe and uptake is generally good. An interesting outcome of the vaccination programmes has been reducing in the incidence of external genital warts in young adults. In Australia, vaccination with the quadrivalent vaccine was first offered to 12- and 13-year-old girls in 2007 (Ali *et al.*, 2013) and this national programme has continued. Older schoolgirls (13 to 18 year olds) and young women between 18 and 26 were also vaccinated in 2007 and 2009. Analysis of data obtained from over 85,000 patients attending sexual health clinics for the first time at eight sites across the country allowed Ali *et al.* (2013) to study trends in HPV-related disease. They found that the incidence of genital warts in women under 21 years old decreased by 93% between 2007 and 2011. There was also a notable decline in the number of cases of genital warts in heterosexual males (82%). This trend was not observed in the other demographic groups (females and males older than 21 years of age, men who have sex with men) and therefore appears to be attributable to the protection provided by the vaccine (Ali *et al.*, 2013). The reduction in genital warts was not surprising since coverage for HPV 6 and 11 were included in the vaccine. Since several years usually elapse between acquisition of HPV infection and the clinical appearance of cancer, the effects of the vaccine in prevention of cancer remain to be seen, but these results suggest a robust protective effect. The UK began vaccinating against HPV in 2008, taking a more conservative approach. The ongoing programme is in schoolgirls aged 12 and 13, with girls aged 13 to 18 also offered the vaccine in the first years (Howell-Jones *et al.*, 2013). The bivalent vaccine was chosen, because prevention of HPV-related cancer was the specific aim. Rather unexpectedly, data showed that by 2011 there was a reduction in diagnoses of genital warts amongst young adults in England. Similarly to the findings reported from Australia, the decline in cases was seen in 16- to 19-year-old females (i.e., those who had been offered the vaccine at 12 or 13, from 2008 onwards) and their male peers (Howell-Jones *et al.*, 2013). Rates were modest (around 13% decrease overall), but this is nevertheless an intriguing finding. The main target of anti-HPV vaccines is cervical cancer, although it could protect against other cancers such as head and neck, which are more common in males. Since the preparations have also been shown to reduce transmission of genital warts, some countries are now offering the vaccine to boys as well.

Point to consider 6.7: If you were planning a national HPV prevention programme which of the available tools (screening, vaccination) would you choose and why? Which demographic groups would you target?

6.5   *Trichomomas vaginalis*

Classification
Kingdom: Protista
Phylum: Metamonada
Order: Trichomonadida
Genus: *Trichomonas*
Species: *Trichomonas vaginalis, T. tenax, T. hominis*

6.5.1 Introduction

Trichomonas vaginalis is a flagellate protozoan parasite (https://www.cdc.gov/dpdx/trichomoniasis) and is a very common sexual pathogen in both males and females (the species name is misleading!). The majority of infections are asymptomatic, but patients can experience vaginitis, with a noticeable discharge, or urethritis – according to gender. The parasite does not appear to form cysts and is only found in the form of a trophozoite, which is oval-shaped and has five flagellae (Greenwood, 2012). *T. vaginalis* isolates in clinical specimens are reported to vary in size from 7–32 μm in length and 5–12 μm in width (Gunn and Pitt, 2012). Since it grows and divides on the epithelial cells, it is likely that the different-sized organisms represent different stages of maturity rather than any strain variation.

6.5.2 Pathogenesis and Clinical Symptoms

Where genital swabs are taken for investigations of genital infection (or screening) *T. vaginalis* is found in up to 50% of asymptomatic women (Lewis, 2014). Females who have clinically apparent infection experience pain and itching in the vagina, along with a burning sensation on micturition (Murray *et al.*, 2013). This is attributable to a local inflammatory response (Gunn and Pitt, 2012). The protozoan feeds on sloughed off genital epithelial cells and bacteria. A frothy, white or greenish, smelly discharge may also be produced (Lewis, 2014). The symptoms may be more severe around the time of menstruation (Lewis, 2014). While trichomoniasis may not trouble people as much as some other STIs, unrecognised and untreated infection has been associated with infertility (Mielczarek and Blaszkowska, 2015) and the tissue damage may pre-dispose the patient to cervical cancer (Lewis, 2014). There is also evidence that the infection may adversely affect the outcome of pregnancy, resulting in premature labour and intrauterine growth retardation (Mielczarek and Blaszkowska, 2015). Perinatal transmission may occur (Lewis, 2014). *T. vaginalis* infection is also associated with increased risk of acquiring and transmitting HIV, particularly in females (Kissinger and Adamski, 2013). In males, the parasite appears to affect the urinary tract; while the majority of cases are asymptomatic, patients can experience causing urethritis or prostatitis (Murray *et al.*, 2013). Infection has also been associated with male infertility because it might affect sperm function (Lewis, 2014). While the overwhelming majority of patients who present with symptoms are female (Lewis, 2014), males can experience severe consequences. Although it is extremely rare, damage due to chronic *T. vaginalis* infection has been reported to be the cause of genital ulcers and tissue damage in males (e.g., Gosnell *et al.*, 2015).

6.5.3 Epidemiology

Most cases are asymptomatic and screening for prevalence is not routinely carried out, so available data about the number of cases trichomoniasis are very much estimates. Another problem is the relatively low sensitivity of the wet preparation, which is the most widely used diagnostic test (Mielczarek and Blaszkowska, 2015). Nevertheless, it is thought that global prevalence may be up to 20% in females and 10% in males (Mielczarek and Blaszkowska, 2015), with an incidence of over 270 million each year

(Lewis, 2014). Rates of infection in the UK are considered to be low, but, for example, over 6,000 cases are diagnosed in England each year (Lewis, 2014; Hathorn *et al.*, 2015), which is probably an underestimate for the reasons outlined above.

Point to consider 6.8: What are the public health considerations for control and monitoring of an infection where most of the cases are asymptomatic?

6.5.4 Laboratory Diagnosis

Microscopic examination of a wet preparation from the genital swab is the most commonly used diagnostic method. It is a simple and rapid test, but its sensitivity can be as low as 50% (SMI B28, 2014). This is likely to be due to a combination of factors, including time between specimen collection and examination (searching for a trophozoite amongst epithelial cells is always easier when the parasite is alive and active) and – where there is a delay – mode of specimen transport to the laboratory, as well as low parasite shedding rates. Performance can be improved by using a stain such as acridine orange (SMI B28, 2014; SMI TP39, 2015) or specific fluorescent-labelled monoclonal antibodies (Murray *et al.*, 2013). Culture in Diamond's broth or Trichomonas medium is currently regarded as the gold standard method (SMI B28, 2014) and is advocated where the parasite has not been detected by microscopy but circumstances warrant further investigations – for example, clinically indicative symptoms or pregnancy (SMI B28, 2014).

Culture requires specialised reagents and takes several days, but these limitations could be overcome by using molecular methods. A NAAT, the transcription-mediated amplification (TMA) for detection of *T. vaginalis* has been developed. It can use a wide range of appropriate specimens including urine and vaginal swabs collected by the patient themselves (Hathorn *et al.*, 2015), which are more acceptable to participants, improving uptake. Evaluations suggest acceptable performance with sensitivity of 97% and specificity at 98% (Hathorn *et al.*, 2015) and that they would be especially useful in diagnosis of trichomoniasis in males (Lewis, 2014). Since NAATs are used to screen for chlamydia and gonorrhoea in the UK and the data about the epidemiology of *T. vaginalis* is deficient, it seems logical to try this TMA in a screening programme. Hathorn *et al.* (2015) offered screening for *T. vaginalis* to all patients presenting to a GUM clinic with a newly recognised STI (regardless of the presumptive microbiological cause). They collected 3503 samples for testing during a six-week period; over half of the samples (58%) were from female patients. Comparison of the TMA with culture indicated a slightly better detection rate in the former. *T. vaginalis* was detected in 93 specimens; this was a prevalence rate of 2.7% (Hathorn *et al.*, 2015) which confirms that it is low in the UK. Taking males and females separately, the positivity rates were 1.4% and 3.6% respectively (Hathorn *et al.*, 2015), which indicates that women are more likely to carry the parasite – irrespective of whether they are symptomatic. The authors of this study suggest that the prevalence in the UK does not warrant the use of the relatively expensive TMA to screen groups of asymptomatic people but that it could be more cost-efficient to specifically test people in high-risk groups (Hathorn *et al.*, 2015).

Point to consider 6.9: Would you advocate screening of asymptomatic people for *T. vaginalis* infections using TMA?

6.5.5   Treatment and Management

A course of metronidazole is the recommended treatment in the UK and this is usually successful (Lewis, 2014). Another nitroimidazole drug such as tinidazole would also be useful. There is very little evidence for drug resistant strains of the protozoan arising and it is considered safe to administer during pregnancy (Lewis, 2014). Nevertheless, the patient should be advised to take precautions to avoid re-infection due to the observation that despite a detectable serum antibody response, this does not provide protective immunity (Lewis, 2014).

6.6   Exercises

6.1   A Without looking at any sources of information, devise a schedule for routine testing of patients presenting at a genitourinary medicine clinic for all possible sexually transmitted infections.

   B Consult your laboratory's SOPs, the PHE SMIs and other relevant sources of information to refine your plan. Select any available test methods, but organise the schedule so that it allows you to use the minimum number of specimens which could be collected and provide results in the shortest turnaround times.

   C Discuss how realistic it would be to implement your proposal. What are the issues and challenges which would need to be addressed in order to make it work?

6.2   A Investigate and compare the current policies for HPV vaccination from different countries around the world.

   B Discuss whether boys should be vaccinated as well as girls.

   C Is the bivalent vaccine (protection against HPV 16 and 18) adequate or should or the quadrivalent vaccine (protection against HPV 8, 11, 16 and 18) be used? Would you advocate another vaccine formulation (i.e., protecting against other HPV types) if it was available?

6.3   A Investigate the literature about how co-infection with human immunodeficiency virus (HIV) can affect the pathology and prognosis for other sexually transmitted infections (STIs) in an individual.

   B Then discuss whether testing for HIV automatically when a patient presentation provides grounds for testing for another STI (i.e., when patient shows symptoms of genital discharge and or there is a history of possible exposure) simply from a clinical point of view to enhance the care of this individual.

   C Now discuss the ethical issues of such a policy and decide whether it would be feasible to apply it.

6.7   Case Studies

6.1   Ms EP is a 38-year-old teacher who has been living with her partner FT a 45-year-old freelance writer for six months. She has been experiencing vaginal discomfort and an unusual slight discharge for a few days. She is examined by her general practitioner

(GP) who collects an endocervical swab and sends it to the laboratory. The clinical information given on the request form is "?? gonorrhoea infection".

a) Which laboratory investigations would you recommend and why?
   The result is: *Neisseria gonorrhoeae* detected by NAAT.

b) What further laboratory tests might be required in order to treat the individual patient effectively and from a public health point of view?

c) What are the implications for EP and FT of this result?

6.2 Mr GT is 28-year-old single man who regularly changes sexual partner and therefore presents at the genitourinary medicine (GUM) clinic every two to three months for a health check. On this occasion, he has no symptoms and appears to be well. Blood samples and genital swabs are collected to test for a range of STIs.
The results are:

Serological tests:
HIV antigen/antibody: Not detected
HBsAg: Not detected
Anti-HCV antibody: Not detected
Treponema pallidum: Not detected

NAAT result:
Chlamydia trachomatis: Not detected
Neisseria gonorrhoeae: Not detected

Smear examination and culture
Candida spp.: Not detected
Trichomonas vaginalis: DETECTED

This is a shock to GT who is unsure of the implications of the diagnosis.

a) What treatment and management would be recommended for the patient?

b) Staff at the GUM clinic decide to try and contact as many of GT's recent sexual contacts as possible and test them for infection with the parasite. Which test method/assay would you recommend?

References

Alexander S, Martin IMC, Fenton K and Ison CA (2006). The prevalence of proline iminopeptidase negative *Neisseria gonorrhoeae* throughout England and Wales. *Sexually Transmitted Infections*, **82**: 280–282. doi: 10.1136/sti.2005.018424

Ali H *et al.* (2013). Genital warts in young Australians five years into national human papillomavirus vaccination programme: National surveillance data. *BMJ*, **346**: f2032. doi: 10.1136/bmj.f2032

Annan NT *et al.* (2009). Rectal chlamydia—A reservoir of undiagnosed infection in men who have sex with men. *Sexually Transmitted Infections*, **85**: 176–179. doi: 10.1136/sti.2008.031773

Arbyn M *et al.* (2012). EUROGIN 2011 roadmap on prevention and treatment of HPV-related disease. *International Journal of Cancer*, **131**: 1969–1982. doi: 10.1002/ijc.27650

Arbyn M *et al.* (2013) The APTIMA HPV assay versus the hybrid capture 2 test in triage of women with ASC-US or LSIL cervical cytology: A meta-analysis of the diagnostic accuracy. *International Journal of Cancer*, **132**: 101–108. doi: 10.1002/ijc.27636

Asano H *et al.* (2014). Genital infection caused by *Entamoeba histolytica* confirmed by polymerase chain reaction analyses. *Journal of Obstetrics and Gynaecology Research*, **40**: 1441–1444. doi: 10.1111/jog.12351

Awashthi S and Friedman HM (2014). Status of prophylactic and therapeutic genital herpes vaccines. *Current Opinion in Virology*, **6**: 6–12. doi: 10.1016/j.coviro.2014.02.006

Bell G and Potterat J (2011). Partner notification for sexually transmitted infections in the modern world: a practitioner perspective on challenges and opportunities. *Sexually Transmitted Infections*, **87(S2)**: ii34–ii36. doi: 10.1136/sextrans-2011-050229

Bernstein DI *et al.* (2013). Epidemiology, clinical presentation, and antibody response to primary infection with herpes simplex virus type 1 and type 2 in young women. *Clinical Infectious Diseases*, **56**: 344–351. doi: 10.1093/cid/cis89.1

Bignell C and Unemo, M. (2013). 2012 European guideline on the diagnosis and treatment of gonorrhoea in adults. *International Journal of STD & AIDS*, **24**: 85–92. doi: 10.1177/0956462412472837

Brooks JI and Sandstrom PA. (2013). The power and pitfalls of HIV phylogenetics in public health. *Canadian Journal of Public Health*, **104**: e348–e350.

Chelimo C *et al.* (2013). Risk factors for and prevention of human papillomaviruses (HPV), genital warts and cervical cancer. *Journal of Infection*, **66**: 207–217. doi: 10.1016/j.jinf.2012.10.024

Cole MJ *et al.* (2014). Risk factors for antimicrobial-resistant Neisseria gonorrhoeae in Europe. *Sexually Transmitted Diseases*, **41**: 723–729. doi: 10.1097/OLQ.0000000000000185

Corey L (2007). Synergistic copathogens—HIV-1 and HSV-2. *New England Journal of Medicine*, **356**: 854–856. doi: 10.1056/NEJMe068302

Cubie HA (2013). Diseases associated with human papillomavirus infection. *Virology*, **445**: 21–34. doi: 10.1016.jvirol.2013.06.007

Cubie HA, Cuschieri KS and Tong CYW (2012). Papillomaviruses and polyomaviruses. In: *Medical Microbiology: A Guide to Microbial Infections, 18th edn*. Greenwood D, Barer M, Slack R and Irving W, eds. London: Churchill Livingston – Elsevier.

Cuzick J *et al.* (2006). Overview of the European and North American studies on HPV testing in primary cervical cancer screening. *International Journal of Cancer*, **119**: 1095–1101. doi: 10.1002/ijc.21955

Doorbar J *et al.* (2012). The biology and life-cycle of human papillomaviruses. *Vaccine*, **30**: F55–F70. doi: 10.1016/j.vaccine.2012.06.083

Fenton KA and Lowndes CM (2004). Recent trends in the epidemiology of sexually transmitted infections in the European Union. *Sexually Transmitted Infections*, **80**: 255–263. doi: 10.1136/sti.2004.009415

Ferlay J *et al.* (2013). Cancer incidence and mortality patterns in Europe: Estimates for 40 countries in 2012. *European Journal of Cancer*, **49**: 1374–1403. doi: 10.1016/j.ejca.2012.12.027

Forman D *et al.* (2012). Global burden of human papillomavirus and related diseases. *Vaccine*, **30**: F12–F23. doi: 10.1016/j.vaccine.2012.07.055

Freeman EE *et al.* (2006). Herpes simplex virus 2 infection increases HIV acquisition in men and women: systematic review and meta-analysis of longitudinal studies. *AIDS*, **20**: 73–83.

Forsgren M and Klapper PE (2009). Herpes simplex virus type 1 and type 2. In: *Principles and Practice of Clinical Virology, 6th edn.* Zuckerman AJ, Banatvala JE, Schoub BD, Griffiths PD and Mortimer P, eds. Chichester: Wiley-Blackwell.

Gaydos CA *et al.* (2013). Performance of the cepheid CT/NG Xpert rapid PCR test for the detection of Chlamydia trachomatis and Neisseria gonorrhoeae. *Journal of Clinical Microbiology,* JCM-03461. doi: 10.1128/JCM.03461-12

Goering RV *et al.* (2013). *Mims' Medical Microbiology, 5th edn.* Philadelphia: Elsevier Saunders.

Gosnell BI *et al.* (2015) Case report: Trichomonas vaginalis associated with chronic penile ulcers and multiple urethral fistulas. *American Journal of Tropical Medicine and Hygiene,* **92**: 943–944. doi: 10.4269/ajtmh.14-0597

Gottlieb SL *et al.* (2014). Toward global prevention of sexually transmitted infections (STIs): the need for STI vaccines. *Vaccine,* **32**: 1527–1535. doi: 10.1016/j.vaccine.2013.07.087

Greenwood D (2012). Protozoa. In: *Medical Microbiology: A Guide to Microbial Infections, 18th edn.* Greenwood D, Barer M, Slack R and Irving W, eds. London: Churchill Livingston – Elsevier.

Gunn A and Pitt SJ (2012). *Parasitology: An Integrated Approach.* Chichester: Wiley-Blackwell.

Hanna L, Crosby T and Macbeth F, eds (2015). *Practical Clinical Oncology, 2nd edn.* Cambridge: Cambridge University Press.

Hathorn E *et al.* (2015). A service evaluation of the Gen-Probe APTIMA nucleic acid amplification test for Trichomonas vaginalis: Should it change whom we screen for infection. *Sexually Transmitted Infections,* **91**: 81–86. doi: 10.1136/sextrans-2014-051514

Howell-Jones R *et al.* (2013). Declining genital warts in young women in England associated with HPV 16/18 vaccination: An ecological study. *Journal of Infectious Diseases,* **208**: 1397–1403. doi: 10.1093/infdis/jit361

Hughes G *et al.* (2013). Repeat infection with gonorrhoea in Sheffield, UK: predictable and preventable? *Sexually Transmitted Infections,* **89**: 38–44.

Hughes G and Lowndes CM (2014). Epidemiology of sexually transmitted infections: UK. *Medicine,* **42**: 281–286. doi: 10.1016/j.mpmed.2014.03.002

Hung C-C, Chang S-Y and Ji D-D (2012). *Entamoeba histolytica* infection in men who have sex with men. *Lancet Infectious Diseases,* **12**: 729–736. doi: 10.1016/S1473-3099(12)70147-0

Ison CA *et al.* and GRASP Collaborative Group. (2013). Decreased susceptibility to cephalosporins among gonococci: Data from the Gonococcal Resistance to Antimicrobials Surveillance Programme (GRASP) in England and Wales, 2007–2011. *Lancet Infectious Diseases,* **13**: 762–768. doi: 10.1016/S1473-3099(13)70143-9

Jerse AE, Bash MC and Russell MW (2014). Vaccines against gonorrhea: Current status and future challenges. *Vaccine,* **32**: 1579–1587. doi: 10.1016/j.vaccine.2013.08.067

Johannessen I and Ogilvie MM (2012). Herpesviruses. In: *Medical Microbiology: A Guide to Microbial Infections, 18th edn.* Greenwood D, Barer M, Slack R and Irving W, eds. London: Churchill Livingston – Elsevier.

Johnston C, Koelle DM and Wald A (2014a). Current status and prospects for development of an HSV vaccine. *Vaccine,* **32**: 1553–1560. doi: 10.1016/j.vaccine.2013.08.066

Johnston C *et al.* (2014b). Virologic and immunologic evidence of multifocal genital herpes simplex virus 2 infection. *Journal of Virology,* **88**: 4921–4931. doi: 10.1128/JVI.03285-13

Kissinger P and Adamski A (2013). Trichomoniasis and HIV interactions: A review. *Sexually Transmitted Infections*, **89**: 426–433. doi: 10.1136/sextrans-2012-051005

Lacey CJN *et al.* (2013). 2012 European guideline for the management of anogenital warts. *Journal of the European Academy of Dermatology and Venereology*, **27**: e263–e270. doi: 10.1111/j.1468-3083.2012.04493.x

LaMontagne DS *et al.* (2004). Establishing the National Chlamydia Screening Programme in England: Results from the first full year of screening. *Sexually Transmitted Infections*, **80**: 335–341. doi: 10.1136/sti.2004.012856

Le Goff J, Péré H and Bélec L (2014). Diagnosis of genital herpes simplex virus infection in the clinical laboratory. *Virology Journal*, **11**: 83. doi: virologyj.com/content/11/1/83

Lewis D (2014). Trichomoniasis. *Medicine*, **42**: 369–371. doi: 10.1016/j.mpmed.2014.04.004

Low N *et al.* (2014). Molecular diagnostics for gonorrhoea: Implications for antimicrobial resistance and the threat of untreatable gonorrhoea. *PLoS Med*, **11**: e1001598. doi: 10.1371/journal.pmed.1001598

Looker KJ, Garnett GP and Schmid GP (2008). An estimate of the global prevalence and incidence of herpes simplex virus type 2 infection. *Bulletin of the World Health Organization*, **86**: 805–813. doi: 10.2471/BLT.07.046128

McCance D (2009). Papillomaviruses. In: *Principles and Practice of Clinical Virology, 6th edn.* Zuckerman AJ, Banatvala JE, Schoub BD, Griffiths PD and Mortimer P, eds. Chichester: Wiley-Blackwell.

Maartens G, Celum C and Lewin SR (2014). HIV infection: Epidemiology, pathogenesis, treatment, and prevention. *Lancet*, **384(9939):** 258–271. doi: 10.1016/S0140-6736(14)60164-1

Mabey D and Peeling RW (2012). Chlamydia. In: *Medical Microbiology: A Guide to Microbial Infections, 18th edn.* Greenwood D, Barer M, Slack R and Irving W, eds. London: Churchill Livingston – Elsevier.

MacKay IM and Arden KE (2015). Ebola virus in the semen of convalescent men. *Lancet Infectious Diseases*, **15**: 149–150. doi: 10.1016/S1473-3099(14)71033-3

Marur S, D'Souza G, Westra WH and Forastiere AA (2010). HPV-associated head and neck cancer: A virus-related cancer epidemic. *Lancet Oncology*, **11**: 781–789. doi: 10.1016/S1470-2-45910)70017-6

Mielczarek E and Blaszkowska J (2015). *Trichomonas vaginalis:* pathogenicity and potential role in human reproductive failure. *Infection*, **44**: 447–458. doi: 10.1007/s15010-015-0860-0

Monsonego J *et al.* (2001). Liquid-based cytology for primary cervical cancer screening: A multi-centre study. *British Journal of Cancer*, **84**: 360–366. doi: 10.1054/bjoc.2000.1588

Murray PR, Rosenthal KS and Pfaller MA (2013). *Medical Microbiology, 7th edn.* Philadelphia: Elsevier Saunders.

Newman L *et al.* (2015). Global estimates of the prevalence and incidence of four curable sexually transmitted infections in 2012 based on systematic review and global reporting. *PloS One*, **10**: e0143304. doi: 10.1371/journal.pone.0143304

Oldfield NJ and Ala'Aldeen DAA (2012). Neisseria and Moraxella. In: *Medical Microbiology: A Guide to Microbial Infections, 18th edn.* Greenwood D, Barer M, Slack R and Irving W, eds. London: Churchill Livingston – Elsevier.

Parra-Sánchez M *et al.* (2012). Evaluation of the cobas 4800 CT/NG Test for detecting Chlamydia trachomatis and Neisseria gonorrhoeae DNA in urogenital swabs and urine specimens. *Diagnostic Microbiology and Infectious Disease*, **74**: 338–342. doi: 10.1016/j. diagmicrobio.2012.08.004

Patel H, Wagner M, Singhal P and Kothari S. (2013). Systematic review of the incidence and prevalence of genital warts. *BMC Infectious Diseases*, **13**: 1. doi: 10.1186/1471-2334-13-39

Péré H *et al*. (2015) Herpes simplex virus type 2 (HSV-2) genital shedding in HSV-2-/ HIV-1-co-infected women receiving effective combined antiretroviral therapy. *International Journal of STD & AIDS*, **27**: 178–185: doi: 10.1177/0956462415577727

Peto J, Gilham C, Fletcher O and Matthews FE. (2004). The cervical cancer epidemic that screening has prevented in the UK. *Lancet*, **364(9430):** 249–256. doi: 10.1016/ S0140-6736(04)16674-9

Ronco G *et al*. (2014). Efficacy of HPV-based screening for prevention of invasive cervical cancer: Follow-up of four European randomised controlled trials. *Lancet*, **383(9916):** 524–532. doi: 10.1016/S0140-6736(13)62218-7

Sasieni P, Adams J and Cuzick J. (2003). Benefit of cervical screening at different ages: Evidence from the UK audit of screening histories. *British Journal of Cancer*, **89**: 88–93. doi: 10.1038/sj.bjc.6600974

Sethupathi M, Blackwell A and Davies H. (2010). Rectal *Chlamydia trachomatis* infection in women. Is it overlooked? *International Journal of STD & AIDS*, **21**: 93–95. doi: 10.1258/ijsa.2008.008406

Shin H and Iwasaki A (2013). Generating protective immunity against genital herpes. *Trends in Immunology*, **34**: 487–494. doi: 10.1016/j.it.2013.08.001

Sigurdsson K (2013). Is a liquid-based cytology more sensitive than a conventional Pap smear? *Cytopathology*, **24**: 254–263. doi: 10.1111/cyt.12037

SMI ID6: Public Health England (2015). Identification of *Neisseria* species. UK Standards for Microbiology Investigations. ID6 Issue 3. https://www.gov.uk/ uk-standards-for-microbiology-investigations-smi-quality-and-consistency-in-clinical- laboratories

SMI B28: Public Health England. (2014). Investigation of Genital Tract and Associated Specimens. UK Standards for Microbiology Investigations. B28 Issue 4.5. https://www. gov.uk/uk-standards-for-microbiology-investigations-smi-quality-and-consistency-in- clinical-laboratories

SMI TP39: Public Health England. (2015). Staining Procedures. UK Standards for Microbiology Investigations. TP39 Issue 2.1. https://www.gov.uk/uk-standards- for-microbiology-investigations-smi-quality-and-consistency-in-clinical-laboratories

SMI V17: Public Health England. (2013). Isolation of Herpes Simplex Virus Associated with Herpes genitalis. UK Standards for Microbiology Investigations. V17 Issue 4.3. http://www.hpa.org.uk/SMI/pdf

Tabrizi SN *et al*. (2011). Evaluation of six commercial nucleic acid amplification tests for the detection of Neisseria gonorrhoeae and other Neisseria species. *Journal of Clinical Microbiology*, **49(10):** 3610–3615. doi: 10.1128/JCM.01217-11

Tucker JD, Bien CH and Peeling RW (2013). Point-of-care testing for sexually transmitted infections: Recent advances and implications for disease control. *Current Opinion in Infectious Diseases*, **26**: 73. doi: 10.1097/QCO.0b013e32835c21b0

Unemo M and Shafer WM (2014). Antimicrobial resistance in Neisseria gonorrhoeae in the 21st century: past, evolution, and future. *Clinical Microbiology Reviews*, **27**: 587–613. doi: 10.1128/CMR.00010-14

Whaitiri S and Kelly P. (2011). Genital gonorrhoea in children: Determining the source and mode of infection. *Archives of Disease in Childhood*, **96**: 247–251. doi: 10.1136/adc.2009.162909

Wiley DJ *et al.* (2002). External genital warts: Diagnosis, treatment, and prevention. *Clinical Infectious Diseases*, **35(S2):** S210–S224. doi: 10.1086/342109

7

Infections in Immunocompromised Patients

7.1   Introduction

For all animals and plants, interaction with microorganisms is an unavoidable consequence of living on Earth – and it seems, possibly other planets as well! All life forms are vulnerable to pathogens (even bacteria can be attacked by bacteriophages) and therefore must have strategies to deal with them. In humans, both the innate and the specific immune systems are important in defence against microorganisms (Delves *et al.*, 2011) and they combine to ensure that most healthy adults only experience infectious diseases occasionally. However, microorganisms have developed a number of 'virulence factors', which are designed to overcome, subvert or avoid the immune response (Delves *et al.*, 2011). Readers are advised to refer to a standard immunology textbook, such as Delves *et al.* (2011) or Murphy *et al.* (2011), for a reminder of the usual host-pathogen interactions. Changes in the effectiveness of the host immune system – either transiently or chronically – can be exploited by microorganisms. A familiar example of the result of this is the common cold. There are several hundred different types of virus which can cause the common cold – including at least 100 strains of rhinovirus (McIntyre *et al.*, 2013). Humans are therefore thought to be continually exposed to common cold viruses and yet rarely experience symptomatic illness. People frequently succumb to a common cold when overworked, run down, or while on holiday after a stressful period at work; this appears to be linked to a (usually) temporary disruption to the balance of the immune system due to factors such as stress (e.g., Cohen *et al.*, 1991).

The activity of infectious agents within the body can also alter the immune response. The idea is to suppress or divert T- or B-cell responses to favour the growth of that particular pathogen, but as a side effect it can also favour other organisms as well. For example, a rhinovirus infection can disrupt the equilibrium between the host immune response and a latent oral herpes simplex virus (HSV) – hence the name for the manifestation of the latter as the 'cold sore'. Colds and cold sores are usually a passing inconvenience, but there are some infectious conditions which cause serious and long-lasting compromise to the immune system. Human immunodeficiency virus (HIV) is the most prevalent of these (Pillay, *et al.*, 2009 and Chapter 2), while another retrovirus, human T-cell lymphotropic virus type 1 (HTLV-1) is associated with adult T-cell leukaemia/lymphoma (Taylor, 2009). Parvovirus B19 has been implicated in a number of blood cell deficiencies, including neutropaenia, leucopaenia, aplastic anaemia and myelodisplastic

Clinical Microbiology for Diagnostic Laboratory Scientists, First Edition. Sarah J. Pitt.
© 2018 John Wiley & Sons Ltd. Published 2018 by John Wiley & Sons Ltd.

disorders (Kerr, 2015). Interplay between competing effects of microorganisms can also lead to lasting disorders such as cancers (de Martel *et al.*, 2012). Examples of this include Burkitt's lymphoma, a B-cell malignancy associated with infection with Epstein Barr virus – a virus known to have the capacity to be oncogenic. The cancerous cells are able to proliferate due to immunosuppression and in the majority of (but not all) cases of Burkitt's lymphoma this is facilitated by *Plasmodium falciparum*. The protozoan appears to stimulate B cells and suppress T cells, although the mechanism is still not fully elucidated (Thorley-Lawson *et al.*, 2016).

There are also significant non-infectious causes of immunodeficiency. These leave the person open to rarely seen and opportunistic infections, along with unusual presentations of some relatively common diseases. It is important for the laboratory scientist to be aware of these conditions and the kinds of infections which can be experienced by patients. An understanding of the underlying immune disorder can help with appreciating the range of possible manifestations of infections with particular organisms and could be useful in troubleshooting problems with laboratory diagnosis. The main causes of persistent immunodeficiency are congenital genetic diseases ('primary immunodeficiency'), bone marrow disorders, T- and B-cell malignancies, deliberate immunosuppression (e.g., chemotherapy to treat cancer or autoimmune diseases; post-transplant) and infection with HIV. It is worth noting that a compromised immune system can also be a consequence of malnourishment. This can make a patient vulnerable to specific presentations of some infections. For example, giant cell pneumonia is a complication of measles virus infection seen in malnourished children (Schneider-Schaulies and ter Meulen, 2009). This chapter will consider the most important infections seen in patients in each category of immunodeficiency and then consider selected organisms – namely *Staphylococcus aureus*, cytomegalovirus, *Aspergillus* spp. *Cryptococcus* spp. – in detail.

Point to consider 7.1: Make a list of the key components of the human immune system. Then think of the physiological and immunological consequences for the host of a deficiency in each one.

7.2 Congenital Genetic Immunological Disorders

There are a number of congenital genetic abnormalities associated with immunodeficiency. They occur relatively rarely, but can have serious consequences for the patient's quality of life (Blann and Ahmed, 2014). Selected diseases are shown in Table 7.1, along with the immunological problem and common infecting organisms. This is not an exhaustive list and readers requiring further information are directed to Aguilar *et al.* (2014) or Blann and Ahmed (2014). Recurrent, disseminated infections are likely to be presenting features in such conditions and so good communication between the microbiology laboratory and the paediatric ward is essential. The component of the immune response which is affected can have a significant bearing on the types and presentations of infections. For example, in chronic granulomatous disease (CGD) the phagocytic cells are affected, making the patient vulnerable to deep seated infections such as abscesses caused by bacteria and fungal lung infections (Aguilar *et al.*, 2014). Children with X-linked agammaglobulinaemia cannot make antibodies and are particularly

Table 7.1 Selected congenital immunological disorders and commonly-associated infections.

Genetic disorder	Immunodeficiency	Common infections
Severe congenital neutropaenias (e.g., mutation in neutrophil elastase gene, ELA2)	Promyelocytes in bone marrow fail to develop into neutrophils	Bacterial ulcers, abscesses, cellulitis (*Staphylococcus* spp., Enterobacteriaceae, *Streptococcus pneumoniae*); fungal infections –respiratory (*Aspergillus* spp.) and disseminated (*Candida* spp.)
DiGeorge syndrome (22q11.2 deletion syndrome)	Small or absent thymus therefore lack of T- lymphocyte production	Disseminated viral infections (varicella-zoster virus); bacterial pneumonia (*Streptococcus pneumoniae*); oral and disseminated infections with *Candida* spp.
Chronic granulomatous disease (CGD) (mutation in subunit of NADPH oxidase enzyme)	Phagocytic cells (neutrophils and macrophages) fail to produce reactive oxygen species	Bacterial skin infections and abscesses (*Staphylococcus aureus*, *Norcardia* spp. *Actinomyces* spp.); fungal infections (*Aspergillus* spp.)
X-linked agammaglobulinaemia (mutation in tyrosine kinase enzyme)	Lack of maturation of B cells	Bacterial meningitis (*Streptococcus pneumoniae*, *Neisseria menigitidis*); viral meningitis (Enteroviruses including polio vaccine strains); lung infections (*Pseudomonas aeruginosa*); severe gastrointestinal infections (*Campylobacter* spp., *Giardia duodenalis*)
Hyper–IgM syndrome (mutation in CD40 or CD154 T cells)	Inability of antibody response to change from IgM to other classes	Lung infections (*Pneuomocystis jirovecii, Streptococcus pneumoniae, Pseudomonas aeruginosa*); severe gastrointestinal infections (*Campylobacter* spp., *Giardia duodenalis, Cryptosporidium* spp.)
Common variable immunodeficiency (CVID)	Low concentrations of antibodies	Bacterial meningitis (*Streptococcus pneumoniae*, *Neisseria menigitidis*); lung infections (*Pseudomonas aeruginosa*); severe gastrointestinal infections (*Campylobacter* spp., *Giardia duodenalis*)
Severe combined immunodeficiency (SCID) (mutation in cytokine receptors (x-linked) or adenosine deaminase enzyme)	Lack of functional T and B cells and interactions between them	Lung infections (*Pneuomocystis jirovecii, Streptococcus pneumoniae, Pseudomonas aeruginosa, Mycobacterium* spp.); respiratory viruses, *Aspergillus* spp.; disseminated infections (herpes simplex virus, *Candida* spp.)

Information taken from Blann and Ahmed (2014) and Aguilar *et al.* (2014).

susceptible to viral infections, especially CNS complications involving enteroviruses (Halliday *et al.*, 2003). Where the genetic condition is known about, it is important to bear the immune deficiency in mind when choosing diagnostic strategies. For instance, if B-cell maturation and activation are affected, then serological tests for antibodies will always give negative results.

7.3 Blood Cell and Bone Marrow Disorders

Since disorders of the bone marrow and blood cell function adversely affect the patient's immune response, increased vulnerability to infection and/or unusual presentations of infectious disease can be a presenting feature of leukaemias and lymphomas. These 'blood cancers' are characterised by abnormal proliferation of one type of cell, which in most cases has undergone a genetic transformation leading to malignant malfunction (Blann and Ahmed, 2014). In leukaemias, the source of the anomalous cell type is the bone marrow, while in lymphomas, the problem is in the lymph nodes (Hoffbrand and Moss, 2016). Myeloid leukaemias are diseases of adults. The average age for developing symptoms of acute myeloid leukaemia (AML) is 65, while chronic myeloid leukaemia (CML) is most commonly seen in 40 to 60 year olds (Hoffbrand and Moss, 2016). As a consequence of these conditions, the production of neutrophils, eosinophils and basophils is disrupted (Blann and Ahmed, 2014). Patients with acute lymphocytic leukaemia (ALL) typically present between the ages of three and seven, with the majority being under six years old (Hoffbrand and Moss, 2016). In contrast, chronic lymphocytic leukaemia (CLL) is again a disease of older adults (it is rare in people under 40 years). Blood films show lymphocytosis, caused by the overproduction of a transformed lymphocyte cell line; in most conditions it is a B-cell line and this leads to over production of (non-specific) antibody (Blann and Ahmed, 2014). Lymphomas similarly involve the uncontrolled manufacture of an abnormal type of cell, to the detriment of the normal immune function and are generally grouped into Hodgkin's (HL) or non-Hodgkin's (NHL) lymphoma. In HL, the abnormal cells are a B-lymphocyte line with specific characteristics called Reed-Sternberg cells. This is a condition of young adults, with males are twice as likely to develop HL as females (Hoffbrand and Moss, 2016). Although B cells are more commonly implicated in NHL, some forms do involve T cells; this category includes adult T-cell lymphoma, caused by HTLV-1 (see Chapter 2). Patients with myeloma make excessive amounts of aberrant immunoglobulins (paraproteins), while in people with myelodisplasia, the lack of control over blood cell manufacture leads to decreased levels of red blood cells and usually an increased white cell count – although this condition can also be associated with neutropaenia (Blann and Ahmed, 2014). For more details about the diagnosis, classification, treatment and prognosis of leukaemias and lymphomas the reader is referred to haematology textbooks such as Blann and Ahmed (2014) or Hoffbrand and Moss (2016).

Treatment for these conditions may involve immunosuppressive chemotherapy to reduce the activity of the abnormal cells. Cytotoxic drugs (alkylating agents, antimetabolites, cytotoxic antibiotics) are intended to affect the malignant cells more readily than normal cells, because the former are growing and metabolising more rapidly (Hoffbrand and Moss, 2016). They are generally used in combination, with the pattern and dose of administration varying with the particular underlying condition and the patient's age. However, they do also kill normal cells – including appropriately functioning parts of the immune system! More recently, specifically targeted therapies have been developed. These are designed to block proteins being expressed by the cancerous cells. A number of preparations comprising monoclonal antibodies raised against CD proteins are available (Hoffbrand and Moss, 2016). An example is Rituximab, which targets B cells expressing CD20 that are known to proliferate in NHL and some forms of CLL, and is therefore recommended in the treatment of these conditions (https://www.nice.org.uk/guidance/conditions-and-diseases/cancer/blood-and-bone-marrow-cancers). The drug binds to both normal and abnormal B cells and therefore both types are removed during the therapy. Patients' regimes can also include general treatments against T- and B-cell over-activity, such as corticosteroids and interferons (Hoffbrand and Moss, 2016). All of this inevitably creates a situation where the immune system is compromised and neutropenic patients tend to be vulnerable to a wide range of infections (e.g., Cannas *et al.*, 2012). Peripheral and central lines used to deliver drugs often become contaminated with skin flora resulting in localised infection, which can develop into bacteraemia. It should be noted that possible pathogens here include coagulase-negative staphylococci and viridans streptococci (Gudiol *et al.*, 2013). There is also a high risk of Enterobacteriaceae and *Pseudomonas aeruginosa* moving from the gut into the peripheral blood stream to cause septicaemia (Gudiol *et al.*, 2013). Herpes viruses can reactivate and produce disseminated infections, or lesions in unusual sites. There is also a danger from invasive fungal infections, particularly involving *Aspergillus* spp. and *Candida* spp. (Hoffbrand and Moss, 2016; Cannas *et al.* 2012).

Another course of action to treat leukaemias and lymphomas is a bone marrow or peripheral stem blood cell transplant (http://pathways.nice.org.uk/pathways/blood-and-bone-marrow-cancers). This is performed after a course of a very high dose of chemotherapy and/or radiotherapy treatment, designed to eliminate the malignant cell lines entirely. The transplanted cells are then used to replace them. The bone marrow usually comes from a donor, matched as far as possible for major histocompatibility complex (MHC) human leucocyte antigen (HLA) type (Chinen and Buckley, 2010). This is known as allogenic haematopoietic stem cell transplantation (HSCT) and it can also be used to treat some congenital immunodeficiency disorders and autoimmune conditions (Jenq and van den Brink, 2010). Autogenic grafts are possible – depending on the nature of the patient's malignancy or deficiency – and these tend to involve peripheral blood stem cells (Jenq and van den Brink, 2010). While the patient is recovering from the procedure and their body is adjusting to the transplanted cell populations, they are vulnerable to opportunistic infections. There is a risk that the patient's underlying immune mechanisms will react against the transplanted cell populations (graft versus host disease, GVHD) and management of this involves immunosuppressive chemotherapy (Chinen and Buckley, 2010), which also increases the possibility of infection. Opportunistic and unusual infections, such as *Pneuomocystis jirovecii* pneumonia (see Chapter 3) and disseminated varicella zoster virus infection are seen (Hoffbrand and Moss, 2016) and invasive fungal infections occur in up to 20% of HSCT (Jenq and van den Brink, 2010).

Solid organ transplantation (SOT) is now a relatively safe and routine method of managing organ failure. Worldwide, at least 100,000 procedures are carried out each year, with kidney transplants being the most common (Chinen and Buckley, 2010). In the UK, there are between 2000 and 3000 kidney transplants each year, around 900 liver transplants and 150 to 200 heart transplants (NHS Blood and Transplant Organ Donation and Transplantation [www.odt.nhs.uk]). Although tissue matching is done, profound and prolonged immunosuppression (using drugs such as cyclosporine and prednisolone) is required after SOT. Failure of the transplant most often occurs through rejection, but opportunistic infections can also adversely affect the outcome, particularly during the first year post-transplant (Gavaldà *et al.*, 2014). Commonly reported infections include deep-seated mycoses, multidrug resistant bacterial infections (such as methicillin-resistant *Staphylococcus aureus* and vancomycin-resistant enterococci), mycobacterial infections and respiratory virus infections (Gavaldà *et al.*, 2014). Polyoma viruses (including BK and JC) are a significant risk in renal transplant patients, while reactivation of Epstein-Barr virus to cause malignancy is more often associated with heart and lung transplants (Gavaldà *et al.*, 2014). In both HSCT and SOT patients, primary, re-infection or reactivation of cytomegalovirus infection is a significant risk, which needs to be managed (Griffiths *et al.*, 2015 and see below). Before this was accounted for, rates of HSCT failure and patient mortality due to CMV activity were reported to be around 25% (Jenq and van den Brink, 2010).

Point to consider 7.2: What advice is generally given to adults being treated for immunological disorders or post-transplant procedures to minimise their risk of infection? Do you think the information provided is suitable for patients?

7.4 HIV/AIDS

The human immunodeficiency virus (HIV) infects CD4+ white blood cells, and, if uncontrolled, causes severe suppression of T-cell–mediated immunity, leading to acquired immune deficiency syndrome (AIDS) (see Chapter 2). This leaves the patient susceptible to rare infections, unusual presentations of more common ones and reactivation of latent infections. Table 7.2 shows some examples of these.

Table 7.2 Examples of infections associated with HIV/AIDS, indicating nature of the presentation.

Microorganism	Infection	Rare infection	Unusual presentation	Reactivation of existing infection
Candida spp.	Disseminated infection		X	
Cryptococcus spp.	Meningitis	X		
Cryptosporidium spp.	Prolonged gastroenteritis		X	
Isospora belli	Prolonged gastroenteritis	X		
Microsporidia	Prolonged gastroenteritis	X		
Cytomegalovirus	Retinitis, encephalopathy, pneumonitis, hepatitis and/or polyradiculopathy		X	X
Herpes simplex virus	Bronchitis, pneumonitis, oesophagitis and/or disseminated lesions		X	X
Human herpes virus 8	Kaposi's sarcoma	X		
Mycobacterium spp.	Disseminated infection		X	X
Pneumocystis jirovecii	Atypical (pneuomocystis) pneumonia ('PCP')	X		
Toxoplasma gondii	Brain lesions			X

Information taken from Pillay *et al.* (2009).

The remainder of this chapter will discuss some of the key organisms to consider in immunocompromised patients in detail. Please note that more information on several other important pathogens is available in other chapters within this book: HIV (Chapter 2), *Toxoplasma gondii* (Chapter 5), *Pneumocystis jirovecii* and *Mycobacterium tuberculosis* (Chapter 3).

7.5   Cytomegalovirus

Order: *Herpesvirales*
Family: *Herpesviridae*
Sub-family: *Betaherpesvirinae*
Genus*: Cytomegalovirus*
Species: *Human herpes virus 5*

7.5.1   Introduction

Cytomegalovirus (CMV) is a virus in the *Herpesviridae* family. It is taxonomically most closely related to human herpes virus 6, the cause of slapped cheek syndrome, but it is more commonly linked with Epstein-Barr virus, since clinically apparent CMV disease in healthy adults is usually a lymphadenopathy syndrome similar to glandular fever (Johannessen and Ogilvie, 2012). Most primary infections are asymptomatic or mild, but since it is a herpes virus, the pathogen persists; reactivation, leading to shedding of virus, is thought to occur regularly throughout life (Griffiths, 2009). Again this is often unwitting, thus aiding transmission. Primary and secondary CMV infection is important in two main groups of patients, namely pregnant women and the immunocompromised. Infection with the virus has also been linked with the development of cancer and inflammatory bowel disease (Atkinson and Emery, 2011). For more information about the pathogenesis, epidemiology and management of congenital CMV, see Griffiths (2009), Manicklal *et al.* (2013) and Griffiths *et al.* (2015).

7.5.2   Pathogenesis and Clinical Symptoms

It is considered that CMV spreads widely throughout the body during primary infection. A variety of cells are susceptible to CMV infection including epithelial, endothelial, fibroblastic and smooth muscle cells, as well as white blood cells (Griffiths *et al.*, 2015). At least 11 viral glycoproteins have been identified and implicated in functions such as entry into or replication within the host cell. Glycoprotein B (gB) is considered to be important in the attachment to host cell surface markers and in the fusion of viral envelope with the host cell membrane (Griffiths, 2009; Griffiths *et al.*, 2015). It is also highly immunogenic and thus forms the basis of the vaccines which are in development (Griffiths *et al.*, 2013). A complex formed between gH and gL is also known to be a key requirement for viral entry (Griffiths *et al.*, 2015). Inside the host cell, viral replication occurs in a lytic cycle, producing an array of proteins. Some of these are involved in virus production, others in establishment of a latent phase and still others in evasion of the host immune system (Griffiths, 2009). Mechanisms to avoid the immune response include blocking of antigen presentation (by down-regulation of expression of MHC molecules) and production of viral proteins which mimic cytokines (reducing T-cell activation). Healthy adults are nevertheless able to mount an effective response to the initial infection (Griffiths *et al.*, 2015). However, this does not clear the virus completely from the body and it enters into a state of latency; although the site for this has not been definitely established, it is thought to be myeloid cells. The normal immune response is presumed to keep the virus 'in check' through cellular immunity. This is borne out by the observation that patients with CD4+ and/or CD8+ depletion are at high risk for CMV disease (Griffiths *et al.*, 2015), while hypogammaglobulinaemia does not appear to be an important factor (Kotton, 2010). Viral load is an extremely important determinant of CMV disease; there seems to be a 'threshold' above which the infection becomes clinically apparent (Griffiths *et al.*, 2015). Thus, the dynamics between host immune system and virus activity are crucial – where the cellular response is not adequate to control viral replication, then the patient will experience symptoms.

Primary infection often goes unnoticed in healthy children and adults, although it can sometimes cause overt disease, such as infectious mononucleosis (Johannessen and Ogilvie, 2012). While this is debilitating, the patient would be expected to recover uneventfully – albeit with the awareness that they are now carrying CMV if a laboratory diagnosis has been made. The exception to this is when primary infection occurs during pregnancy, which poses a risk of congenital transmission (Griffiths, 2009). The immune system is constantly responding to the virus, which means that a proportion of circulating T lymphocytes are primed to react to CMV antigens. After primary infection, the latent virus does become active intermittently and seropositive people appear to shed virus at intervals throughout life, while remaining asymptomatic. When a seropositive person becomes immunocompromised (e.g., through medical immunosuppression, AIDS or old age), the equilibrium between virus and host is disrupted and CMV viral load can increase sufficiently to cause symptoms. Serious acute disease requiring intensive care treatment is also associated with CMV (Griffiths *et al.*, 2015; Griffiths *et al.*, 2013). Although the virus can infect a range of cell types, the clinical outcome can differ depending on the underlying cause of the immunosuppression. For example, in AIDS, retinitis is the most common manifestation (Griffiths, 2009), while patients in intensive care are prone to CMV-related pneumonia (Griffiths *et al.*, 2015).

It is therefore not desirable to introduce CMV to an immunologically vulnerable patient. Nevertheless this is sometimes unavoidable in transplants because the majority of people in most populations carry the virus. Both patient and donor are screened (for anti-CMV antibody) before transplant so that the recipient can be managed accordingly. There are four possible combinations, often written in terms of Donor (D) and Recipient (R):

1) D-/ R-
2) D-/R+
3) D+/R-
4) D+/R+

Clearly scenario 1 should not lead to disease, although care must be taken to ensure that any blood products used also come from CMV seronegative donors. Situation 3 is not ideal, but given the scarcity of stem cell donors/available organs and the importance of tissue matching in the success of transplants, it is occasionally necessary. The (high) risk of primary infection in the recipient can be minimised using prophylactic anti-viral therapy (Griffiths *et al.*, 2015). In combination 2, the immunosuppression required post-transplant can result in reactivation of the patient's own latent virus. This can also happen in scenario 4, although since there are different strains of CMV, re-infection (i.e., with the donor's strain) is also a possibility. In addition, activated CMV has been implicated in graft rejection, through the adverse effect on the patient's immune response (Griffiths *et al.*, 2013). Virological laboratory support is therefore extremely important throughout the whole process of transplantation (see below).

In solid organ transplants, D+/R- is considered to pose the greatest risk to the recipient (Kotton, 2010; Griffiths *et al.*, 2015), since it lays the recipient open to primary CMV infection. This is usually clinically apparent in the form of generalised symptoms such as pyrexia, joint and muscle inflammation and bone marrow suppression, but may also manifest as gastrointestinal or respiratory illness, retinitis or encephalitis (Beam and Razonable, 2012). The transplanted organ may also be affected with corresponding symptoms – for example, CMV-related nephritis in kidney transplant patients. In contrast, for HSCT recipients, CMV disease is more likely to arise in patients who are seropositive beforehand (through reactivation of their existing virus) – that is, in D-/R+ and D+/R+ cases (George *et al.*, 2010; Griffiths *et al.*, 2015). The extra immunosuppression in the drug regime required to manage graft versus host disease in these patients also increases the risk of clinical CMV (George *et al.*, 2010). Survival rate has been shown to be decreased for the D+/R- combination in stem cell recipient patients as well (Ljungman *et al.*, 2014). Most common symptoms include gastroenteritis, pneumonitis and hepatitis; retinitis, encephalitis and myelosuppression are also sometimes seen (George *et al.*, 2010).

Understanding of reactivation and reinfection of CMV has been enhanced by tracing strain variation (Sijmons *et al.*, 2014). This was initially done by serology – looking for variations in gB. After the publishing of the CMV genome sequence in the early 1990s, methods involving viral DNA sequencing followed by restriction fragment length polymorphism analysis were developed. These have now been replaced by PCR-based protocols, but the focus remains on looking for differences in gB amongst isolates (Pignatelli *et al.*, 2004; Manuel *et al.*, 2009). In a study of 239 SOT recipients with CMV disease, real-time PCR which allowed distinction between gB types (gB1–gB4) demonstrated that D+/R+ patients were more likely to have a mixed CMV infection (Manuel *et al.*, 2009). While not surprising, this level of detail does provide confirmation that both infection with the donor's strain and reactivation of the recipient's own strain contribute to disease. Despite the ability to distinguish between CMV genotypes, it is not at all clear whether there are any differences in virulence afforded by variation in gB (Pignatelli *et al.*, 2004; Manuel *et al.*, 2009) or indeed other viral genes/ gene products implicated in pathogenicity, such as glycoprotein N and immediate early protein 1 (Lisboa *et al.*, 2011). This is an active area of research, which whole genome and next generation sequencing may be able to enhance (Sijmons *et al.*, 2014).

7.5.3   Epidemiology

Cytomegalovirus (CMV) appears to be a ubiquitous human pathogen. It can be transmitted *in utero*, perinatally (through contact with maternal blood and secretions during birth and from breast milk) and at any time subsequently, via saliva or iatrogenically (Griffiths *et al.*, 2013; Griffiths, 2009). It is assumed to be a common childhood infection; although it is rarely symptomatic, prevalence of anti-CMV IgG antibody in teenagers and young adults is consistently shown to be between 60% and 100% (Griffiths, 2009; Griffiths *et al.*, 2015). It is highest in developing countries and poorer areas in Western countries (where people live in very close proximity). Data from two longitudinal studies involving people born in the first quarter of the twentieth century (one of pairs of siblings in the Netherlands and the other of twins in Denmark) has indicated that living environment in early childhood is a more important determinant of acquisition of CMV than genetics or exposure in adult life (Mortensen *et al.*, 2012). Amongst those populations where not everyone acquires the virus in childhood, incidence amongst adults is about 1% per year (Griffiths, 2009; Mortensen *et al.*, 2012). Due to the latency, CMV can reactivate and re-infection is also possible (Griffiths, 2009; Griffiths *et al.*, 2013). As mentioned above, disease caused by CMV is therefore a considerable risk in patients with a compromised immune system, particularly post-transplant and in AIDS (Griffiths *et al.*, 2015; Kotton, 2010; George *et al.*, 2010). Studies involving solid organ transplant patients suggest that for the D+/R- scenario, over 40% of recipients will develop primary CMV infection if they do not receive preventative prophylactic anti-CMV treatment (Kotton, 2010). The rate is higher in heart transplants (up to 50%), lung transplants (as high as 90%) and is also increased in patients who receive more than one organ (Kotton, 2010). Similarly, about half of recipients of HSCT who are CMV-seropositive before the procedure are found to reactivate their virus (George *et al.*, 2010).

7.5.4   Laboratory Diagnosis

Pre-transplant screening for CMV status is carried out by EIA to detect IgG antibody (SMI V28, 2015). Well-established assays, with good sensitivity and specificity are available (Griffiths, 2009) and automated serological systems are widely used in laboratories where routine pre-treatment screening is a significant part of the workload (e.g., Delforge *et al.*, 2015). These analysers often also incorporate tests for anti-CMV IgM, for the diagnosis of primary infection and for the measurement of anti-CMV IgG avidity, which can help to determine whether an infection has occurred in the previous six months (SMI V28, 2015; Delforge *et al.*, 2015). The latter can be useful during investigation of congenital infections, but is rarely applied to other diagnostic situations.

In immunocompromised patients, both investigation of symptomatic disease and screening for CMV activity are carried out by virus detection (Griffiths and Lumley, 2014). During primary infection, reactivation or re-infection, the virus can be found in a wide range of body fluids, but whole blood, urine and respiratory samples are collected for testing most often (Griffiths, 2009). Methods for virus detection have included virus isolation in monolayer cell culture, the shell vial assay, the detection of early antigen fluorescent foci (DEAFF) test and the pp65 antigenaemia assay (Atkinson and Emery, 2011). With the advent of PCR to detect CMV viral DNA came improved turnaround times and better sensitivity – meaning any viral activation could be detected sooner (Atkinson and Emery, 2011). Viral load measurement, by real-time PCR, is useful for assessing the severity of a newly diagnosed infection, as well as monitoring of treatment (Atkinson and Emery, 2011; Razonable and Hayden, 2013). The threshold for starting anti-viral therapy may be determined by local experience, but a value of 3000 genomes/mL (2520 IU/mL) of whole blood has been recommended (Griffiths and

Lumley, 2014; Atkinson and Emery, 2011). This allows clinicians to make judgements about patient management. Setting such a quantitative 'cut-off' requires accuracy, consistency and standardisation of result outputs, which is why there is now a WHO International Standard for CMV Viral load PCR (Fryer *et al.* 2010) for laboratories to use alongside the usual controls.

Point to consider 7.3: How often should blood samples be monitored for CMV viraemia after transplant (daily, weekly)? Should other samples be routinely tested as well?

7.5.5 Treatment and Control

There are a limited number of licensed, effective anti-CMV drugs. The most commonly used are the nucleoside analogues ganciclovir and valganciclovir, with foscarnet (a DNA polymerase inhibitor) as an alternative treatment if required (Griffiths *et al.*, 2015) and the treatment regimes are well established. There are two main approaches to management of the risk of CMV infection in transplant patients, which are either prophylactic or pre-emptive treatment. In prophylaxis, the drug is administered for a period of time (usually 100 or 200 days), with a view to preventing CMV activity (primary or reactivation) as a result of the transplant procedure or immunosuppression (Griffiths *et al.*, 2015; Kotton, 2010). This approach is generally used for solid organ transplant patients and it is reported to be effective (Griffiths *et al.*, 2015) and to possibly reduce the activity of other *Herpesviridae* (Kotton, 2010). The drawback is that the drugs are quite toxic, thus they cannot be taken indefinitely. Withdrawal of the treatment can result in CMV disease, mainly in the D+/R- situation (Griffiths *et al.*, 2015). Pre-emptive therapy involves regular monitoring of patient blood samples for viraemia and treating when the CMV viral load reaches the 3000 genomes/mL threshold. This is the recommended approach in bone marrow/stem cell transplants because ganciclovir and valganciclovir are most toxic to white blood cells (Griffiths *et al.*, 2015). In AIDS, the best approach is to enhance the host immune response to the CMV through use of anti-retrovirals to reduce the HIV viral load and enhance the CD4+ count (Griffiths *et al.*, 2015). Drug resistance is an issue with anti-CMV therapies. Mutations are most commonly detected in in the *UL97* gene, which codes for the protein kinase involved in activation of ganciclovoir and valganciclovir inside the host cell. Changes in the *UL54* gene – the product of which is the CMV DNA polymerase – are also reported and this affords resistance to foscarnet as well as ganciclovoir/valganciclovir (Kotton, 2010). Some new drugs are in clinical trials including maribavir (which targets the *UL97* gene product kinase) and cidofovir/brincidofovir, (which are CMV DNA polymerase inhibitors). Another treatment with potential is letermovir. This targets the terminase enzyme complex, which is involved in a different part of the CMV replication cycle to the DNA polymerase and reports suggest that it is not toxic to myeloproliferative cells (Griffiths and Lumley, 2014).

There are several candidate anti-CMV vaccines (Griffiths *et al.*, 2013). The most widely trialled of them use either an attenuated version of the established Towne control strain of CMV or well-characterised viral subunits including gB and pp65. The main aim of any vaccination programme would be to reduce congenital CMV, but if a suitable vaccine was available it could have a role in preventing CMV disease in immunocompromised patients (Griffiths *et al.* 2013). Given time, if the rates of primary infection in early life were lowered and lasting protection afforded by a vaccine, then prevalence of CMV in the transplant donor and recipient populations would be lessened, thus reducing the risk of CMV disease.

Point to consider 7.4: Since CMV enters into latency within the host after primary infection, are there any special considerations to be made in respect of vaccine design?

7.6   *Staphylococcus* spp.

Classification
Domain: Bacteria
Kingdom: Eubacteria
Phylum: Firmicutes
Class: Coccus
Order: Bacillales
Family: Staphylococcaeceae
Genus: *Staphylococcus*
Species: *Staph. aureus, Staph. epidermidis, Staph. saprophyticus, Staph. lugdunensis*

7.6.1   Introduction

While *Staphylococcus* spp. are found widely in the environment, they are also part of the normal flora in humans. They are Gram positive, catalase positive cocci; they are facultative anaerobes and resistant to dry conditions and situations where the salt concentration is high (Humphreys, 2012). *Staph. epidermidis* is considered to be a usual component of a normal skin flora, while it is estimated that 30% of the normal healthy population are also *Staph. aureus* carriers (Humphreys, 2012). Members of the genus are also clinically important opportunistic pathogens, which are of particular concern in immunocompromised patients. The pathogenicity of *Staph. aureus* is well established, although there is considerable strain variation within this species. This is manifest in the production of different virulence factors, such as cell surface factors and toxins and also in antibiotic resistance profiles. Methicillin- (or meticillin)-resistant *Staphylococcus aureus* (MRSA) is a strain which emerged during the 1980s. It is resistant to a whole suite of antibiotics and can therefore be very difficult to treat (Murray *et al.*, 2013). Other species ('coagulase-negative staphylococci' [CoNS]) are being increasingly recognised as having pathogenic potential (Humphreys, 2012; Murray *et al.*, 2013).

Many of the virulence factors and proteins involved in antibiotic resistance are coded for by genes contained within staphylococcal cassette chromosomes (SCC), which are mobile genetic elements (Shore and Coleman, 2013). The genes identified include arginine catabolic mobile element (ACME-*arc*) which codes for arginine deaminase enzymes that attack host factors involved in preventing the bacteria to colonise the skin (Tong *et al.*, 2015; Shore and Coleman, 2013) and SCC*mec* the gene products of which are involved in antibiotic resistance, specifically to methicillin (Shore and Coleman, 2013). These mobile genetic elements show considerable variety and therefore their sequences are used to classify strains, which is particularly important for epidemiological purposes. This is a complex and changing area but there are some key groupings. The multilocus sequence typing method yields a sequence type (ST), denoted by an Arabic numeral (e.g., ST 8). Methicillin resistance was previously considered to be afforded by the *mecA* gene, but it is now known that there are three *mec* gene complexes (labelled A–C). The gene product is penicillin binding protein 2a (PBP2a). The SSC*mec* cassette contains genes for other pathogenic factors including the cassette chromosome recombinase *ccr*, of which there are also a number of different forms. The combination of *mec* and *ccr* gene sequences is used in categorisation (Shore and Coleman, 2013), giving a type indicated by a Roman numeral (e.g., SSC*mec* IV). There are at least 11 SSC*mec* types (SMI B29, 2014; Shore and Coleman, 2013), although subtypes within those are also recognised. Strains (or clones) involved in significant epidemics are also given an overall identification code (such as 'USA 300').

7.6.2   Pathogenesis and Clinical Symptoms

Staphylococcus spp. are implicated in various different clinical conditions in both immunocompetent and immunocompromised people. For *Staph. aureus* these include septicaemia, meningitis, lung infections, bone disease, endocarditis, food poisoning and skin infections. The pathogenic mechanisms involve different selections from the range of toxins, cell surface proteins and enzymes (Murray *et al.*, 2013). Food poisoning and toxic shock syndrome are mediated by enterotoxins which are 'superantigens' – that is, strong activators of the T-cell response. The activity of epidermolytic toxins can result in 'staphylococcal scaled skin syndrome', a rare condition which mainly affects young children (Humphreys, 2012). A full discussion of the clinical syndromes associated with *Staph.aureus* infection and their causative virulence factors is beyond the scope of this book, so readers are directed to review articles such as Tong *et al.* (2015).

Since *Staph. aureus* is a skin organism, it can invade at sites of local skin vulnerability, resulting in boils, abscesses and wound infections. The production of a polysaccharide capsule allows the bacteria to evade opsonin-binding and phagocytosis, while protein A on the cell surface binds to the Fc receptor of IgGs (Tong *et al.* 2015; Murray *et al.*, 2013). Immunocompromised patients are at high risk of developing *Staph. aureus* skin infections for two reasons. The first is their regular contact with healthcare workers' hands during the administration of treatment and personal care for their underlying illness. This potentially exposes them to MRSA. The second is that they will be less able to respond immunologically to the bacteria at the site of entry, which can lead to a chronic infection (Murray *et al.*, 2013). The increasing prevalence of strains which are capable of producing the neutrophil-destroying Panton-Valentine leucocidin (PVL) within skin lesions is a concern (Murray *et al.* 2013). While the pattern differs around the world (see below), it is clear that both methicillin-sensitive *Staphylococcus aureus* (MSSA) and MRSA strains can be PVL toxigenic (Tong *et al.*, 2015; Shallcross *et al.*, 2013). Manufacture of PVL is linked to a particular genes, *lukS-PV* and *lukF-PV*. The gene products, Luk-S PV and Luk-F PV (where S = slow-eluting and F = fast-eluting) work in tandem to lyse membranes of neutrophils (Murray *et al.*, 2013). The mechanism of action has not been fully elucidated. In the laboratory, bacterial isolates are found to secrete varying amounts of PVL. An intriguing aspect of the research to date is that it has been consistently shown that the concentration of toxin produced by the bacterium does not correlate with the severity of the infection (Tong *et al.*, 2015; Shallcross *et al.*, 2013).

In contrast, CoNSs tend to be introduced through the insertion of medical devices such as central venous lines (catheters) and fluid drainage shunts, which are used on hospitalised, immunologically compromised patients. The bacteria are thought to attach to the device inside the host using a combination of physical forces and cell surface proteins (Becker *et al.*, 2014). Healthy adults can usually prevent CoNS from adhering to internal surfaces, but neutropenic patients do not have this capacity. Once established, the bacteria form biofilms through a series of steps involving proteins, polysaccharides and extra-cellular matrices. The process is described in detail by Becker *et al.* (2014). This colonisation is often a precursor to a systemic infection. Iatrogenic introduction of staphylococci can also result in ventilation-associated or aspiration pneumonia (Tong *et al.*, 2015).

Staphylococcus spp. are an important cause of septicaemia and infective endocarditis (both naturally occurring and prosthesis-related). The pathogenic mechanisms of these conditions are discussed in Chapter 2. It is notable that the incidence of bacteraemia attributable to staphylococcal infection has been steadily rising since the middle of the twentieth century (Tong *et al.*, 2015), with a major contribution from MRSA. In immunocompromised patients, CoNs are a frequent cause of bacteraemic episodes (Becker *et al.*, 2014). Although both *Staph. aureus* and CoNSs can be involved in infective endocarditis, more detailed species differentiation has highlighted *Staph. lugdunensis* as an emerging causative agent (Becker *et al.*, 2014).

Point to consider 7.5: 'Coagulase-negative staphylococci' are regarded as normal skin flora, which can occasionally cause disease in severely immunocompromised and are often grouped together using this collective term. How important do you think it is to distinguish between the individual species?

7.6.3   Epidemiology

Members of the *Staphylococcus* genus are important causes of nosocomial infection. Strains with particular virulence characteristics and antimicrobial susceptibilities can spread readily, particularly in healthcare settings, amongst immunologically vulnerable people. Published epidemiological information usually groups isolates as: methicillin sensitive *Staphylococcus aureus* (MSSA), healthcare- associated methicillin-resistant *Staphlyococcus aureus* (HA-MRSA), community-acquired MRSA (CA-MRSA) and coagulase-negative staphylococci (CoNS). Distinguishing between members of the last group is not routinely done, but evidence suggests that individual species have different pathogenicities (Becker *et al.*, 2014), and therefore full identification can be valuable in some situations.

Staphylococcus spp. are responsible for significant number of cases of septicaemia around the world. Where data is collected systematically, the incidence of *Staph. aureus* bacteraemia in the whole population (i.e., not only immunocompromised patients) is reported to be 10 to 30 per 100,000 (Tong *et al.*, 2015). It is recorded as higher in specific risk groups, such as in people living with HIV – especially if they are intravenous drug users (Tong *et al.*, 2015). While MRSA is a significant cause of staphylococcal septicaemia, rates appear to have peaked in the mid-2000s (Tong *et al.*, 2015). In the

UK, the incidence of *Staph. aureus* septicaemia is about 20 per 100,000 population, with a ratio of MSSA to MRSA of around 5:1 (e.g., PHE, 2015). Data is not collected as meticulously for CoNS, but it is thought that they account for about 40% of bacteraemia cases in neutropenic patients (Becker *et al.*, 2014).

One of the risks of surgery is a post-operative infection, due to the exposure to opportunistic organisms and the natural depletion of the immune system. It has been estimated that skin and soft-tissue infections attributable to staphylococci occur after up to 5% of operations (Tong *et al.*, 2015). Amongst patients who have undergone major surgery, including SOTs, the contribution of CoNS to this statistic is significant. For example, Sommerstein *et al.* (2015) found that 4% of patients developed mediastinitis after major cardiac operations and that CoNS were implicated in over 40% cases (Sommerstein *et al.* 2015). Similarly, Yavuz *et al.* (2013) found that just over 2.5% of their patients developed post-operative sternal infection. Most of these were caused by *Staphlyococcus* spp. but more isolates were CoNS than *Staph. aureus*.

7.6.4   Laboratory Diagnosis

Samples collected from appropriate sites (blood, wound swab, sputum, urine, faeces, nasal swab) depending on the clinical situation, can be cultured on conventional agar plates. The organisms should grow readily on nutrient and blood agar, as small (~2 mm diameter) yellow, cream or white colonies after incubation at 37 °C for 24 to 48 hours. Although Figure 7.1 shows a *Staph. aureus* isolate with golden yellow colonies (Figure 7.1a) and a clearly much paler *Staph. epidermidis* (Figure 7.1b), the distinction in colonial morphology not always quite so obvious for every clinical isolate and further tests are required. All species would be Gram positive cocci and the presence of catalase and production of deoxyribonuclease (DNase) are characteristic of *Staphylococcus* spp. (SMI ID7, 2014). The coagulase test is used to identify isolates as *Staph. aureus*, since this detects Protein A (Humphreys, 2012). While this can be applied reliably to most isolates there are a number of possible complications which should be borne in mind.

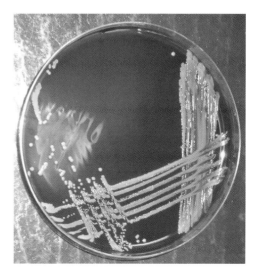

Figure 7.1a *Staphylococcus aureus* growing on blood agar as golden yellow colonies. (*See insert for colour representation of the figure.*)

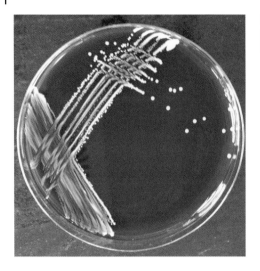

Figure 7.1b *Staphylococcus epidermidis* growing on blood agar as white colonies. (*See insert for colour representation of the figure.*)

One is that *Staph. aureus* is currently classified as two subspecies – *Staph. aureus aureus* and *Staph. aureus anaerobius*. The latter is not commonly found, but it often gives a negative result in the coagulase test and may even be catalase negative (SMI ID7, 2014). In such as case, where other evidence indicates the cause of the patients symptoms might be *Staph. aureus*, however, further investigation is warranted. Another diagnostic problem is posed by the fact that there are some other coagulase-positive *Staphylococcus* species. Again, disease associated with these organisms is only rarely encountered (SMI ID7, 2014), but the clinical laboratory scientist should be aware of this possibility.

Screening of nasal and skin swabs from patients and healthcare workers is used as an infection control measure in certain situations (ARHAI, 2014). Plating out the samples on Mannitol salt agar can also be useful as it selects for tolerance of high NaCl concentrations (*Staphylococcus* spp.) and differentiates between *Staph. aureus* (mannitol fermenting) and CoNS (unable to ferment mannitol). This medium has also been used in screening individuals for type of *Staph. aureus*; the addition of oxacillin allows distinction between MSSA and MRSA on the basis that only the former would be sensitive to this antibiotic (SMI B29, 2014). When screening, the use of an enrichment step (e.g., TSB broth with 7% NaCl) before plating out is recommended, to enhance sensitivity (SMI B29, 2014). A number of chromogenic agars are available and these allow discrimination between colonies of MSSA and MRSA on the basis of their effect on the medium leading to growth of a particular colour (SMI B29, 2014; Morris *et al.* 2012). In Figure 7.2, the growth of MRSA is clearly visible as blackish blue colonies. The medium is designed to select for all *Staph. aureus* meaning that the slightly larger, pale colonies are identified as MSSA. An evaluation of chromogenic agars by Morris *et al.* (2012) indicated that extension of the incubation time above that recommended by the manufacturers improved the MRSA isolation rate. However the optimal time for the best-performing medium was 48 hours (Morris *et al.*, 2012), which might be too long in some clinical situations. Also, some isolates may need confirmatory tests. These would normally be carried out on colonies from a non-selective medium such as nutrient or blood agar; indeed, it is reported that using bacteria grown on chromogenic agar for

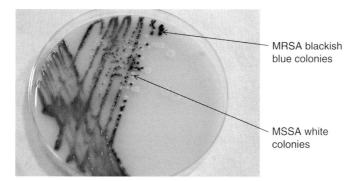

MRSA blackish
blue colonies

MSSA white
colonies

Figure 7.2 Growth of methicillin-resistant *Staphylococcus aureus* (MRSA) and methicillin-sensitive *Staphylococcus aureus* (MSSA) on chromogenic agar. (*See insert for colour representation of the figure.*)

coagulase and DNase test results gives variable and inconsistent results (SMI B29, 2014; Morris *et al.*, 2012). Identification of species by MALDI-TOF is an alternative to be considered (SMI B29, 2014; Zhu *et al.*, 2015). This technique has the potential to distinguish not only between MRSA and MSSA but also to identify separate CoNS species (Zhu *et al.*, 2015). There are also commercial kits available which are intended to detect the PBP2a protein from blood plate culture colonies, although some tests are reported to perform better than others (e.g., Tasse *et al.*, 2016). Real-time PCR to detect the SSC*mec* in screening samples directly has also been evaluated. Danial *et al.* (2011) tested over 3000 swabs collected from over 1200 patients by culture on chromogenic agar and PCR during the course of one month. Approximately 12% of the samples tested positive for MRSA by one or both methods, with PCR giving a false-negative result on only two specimens (Danial *et al.*, 2011). The authors calculated that while the cost per test for the PCR was about five times higher than for the culture method, the mean turnaround times were between five and nine times faster for the molecular assay (Danial *et al.* 2011). This was an 'in house test', but there are commercial kits available. The possibility that the assay may not detect all variants in the SSC*mec* sequence should be borne in mind (SMI B29, 2014). Benchtop PCR analysers intended for point-of-care testing/screening for MRSA are available. Studies of their effectiveness have often involved them being operated by scientific staff in the laboratory and evaluation of their use on wards and clinics is somewhat limited. Nevertheless, findings do suggest that they could be clinically beneficial and have an impact on the laboratory workload. Parcell and Phillips (2014) trained healthcare assistants to collect swabs and put them through the bench-top PCR analyser for patients on an orthopaedic and a vascular ward. During the eight-month trial, a total of 1206 samples were collected and tested (on the POCT device and by conventional culture in the laboratory). MRSA was not detected by either method on the majority of swabs (over 90%). There were some discrepant results and some cases where the PCR result was invalid, but overall the bench-top analyser is reported to have performed well. Parcell and Phillips (2014) suggest that all swabs which test positive for MRSA by bench-top PCR and all uncertain results should be followed up by tests in the main laboratory. This would represent a significant decrease in the number of samples sent to the laboratory, which may be cost effective in terms of staff time and infection control procedures – even though the bench-top PCR would be relatively expensive per test. While the specificity of the POCT PCR assay was good (over 97%), the reported sensitivity was less than 80%, so in a screening situation the merits of this approach would need to be considered carefully.

The above considerations about rapid and accurate detection and identification of *Staphylococcus* spp. also apply in a diagnostic situation. In immunocompromised patients, it is important to bear in mind that any species of *Staphylococcus* could be potentially pathogenic. While accurate identification of MRSA strains is vital – both for the patient and from an infection control perspective – isolates which turn out to be CoNS should not be dismissed without full investigation of the clinical situation. In patients with septicaemia, the laboratory contribution is significant, since identifying the causative agent and determining the antibiotic susceptibility profile as soon as possible helps to improve the outcome (see Chapter 2). Where positive blood culture bottles yield Gram positive, catalase positive bacteria, ascertaining the exact species is important in patient management. If a result of *Staph. aureus* is reported to the clinicians, there is a possibility that they will err on the side of caution and treat with vancomycin (in case it is MRSA) while the antibiotic sensitivity tests are pending (Humphreys, 2012). Similarly, if the isolate is a CoNS, then vancomycin is also suggested as the initial treatment, due to the inconsistent susceptibility patterns found in these species (Humphreys, 2012). The risk with using this antibiotic unnecessarily is that resistant MRSA strains could develop; some cases have already been reported (e.g., Friães *et al.*, 2015). The conventional plate-based antibiotic susceptibility testing adds 24 hours or more to the time taken to achieve the final result once the full identification has been ascertained. The turnaround time can be reduced through creative use of the available methods. For example, it is possible to perform MALDI-TOF on isolates straight from positive blood culture bottles, rather than plating them out and growing them up to pure colonies overnight. However, as components of the growth medium and protein by-products of microbial metabolism need to be removed first, this requires extra reagents and hands on staff time (Kohlmann *et al.*, 2015). Several authors have reported a modification of the protocol to the use of a four-hour plate culture (Fitzgerald *et al.*, 2016; Kohlmann *et al.*, 2015) and reported that results are reliable and acceptable. Kohlmann *et al.* (2015) analysed 598 positive blood cultures obtained from 319 patients across several hospitals over a four-month period. One hundred and sixty one were identified as *Staphylococcus* spp., but over 40% of these results were classed as 'unreliable'. All 32 of the *Staph. aureus* isolates were correctly recognised by the MALDI-TOF and the authors note that the four-hour plate culture provided enough material for follow-up testing for SSC*mec* by PCR. This format (MALDI-TOF plus PCR to distinguish between MRSA and MSSA) is also advocated by other authors, for example, Romero-Gomez *et al.* (2013). Yossepowitch *et al.* (2014) suggest that use of an immunochromatographic assay which is intended to detect *Staph. aureus*-specific proteins in positive blood culture bottles could be used instead of MALDI-TOF with a follow-up MRSA/MSSA PCR assay (Yossepowitch *et al.*, 2014). MALDI-TOF results tend to be less favourable for CoNS species (e.g., Kohlmann *et al.*, 2015), so this might not be acceptable for an immunocompromised patient. An overnight incubation of the subculture plates prior to further testing would appear to be warranted for sensitive and specific CoNS isolation. Also, despite this shortened time to identification for MRSA, the antibiotic susceptibilities would still need to done. Fitzgerald *et al.* (2016) have trialled a protocol which involves using the 4 hour culture for plate antimicrobial sensitivity testing as well as for MALDI-TOF. They reported 100% reliability in identifying *Staph. aureus* and satisfactory sensitivity results (Fitzgerald *et al.*, 2016). In samples from sites other than blood, full species identification of *Staphylococcus* spp. is invaluable wherever possible. The distinction between MRSA, MSSA and CoNS might be thought of as sufficient, but when the patient is immunocompromised and the infection might be iatrogenic, full bacteriological details are extremely helpful. Using MALDI-TOF on isolates from conventional culture, after overnight incubation can provide identification to species level (e.g., Argemi *et al.*, 2015) and discriminate between a range of CoNS species.

Point to consider 7.6: Think about the relative merits of each of the available laboratory detection and identification methods. Would any single assay be sufficient for diagnosis of *Staphylococcus* infection in immunocompromised patients (if so, which one)

7.6.5 Treatment and Management

Screening of patients for MRSA carriage before routine admission to hospital wards is common practice in the UK, as part of infection control procedures. Swabs are collected from the nose and body and (usually) plated out onto chromogenic agar. Which patients are tested, and how often those who are regularly attend the hospital (e.g., post-transplant) are monitored, tends to vary locally (SMI B29, 2014). Those who are found to be MRSA carriers would be treated with antibacterial preparations (mupirocin cream applied to the nasal area plus a chlorhexidine-based body wash or shampoo) before further medical intervention. Clearly this is not possible in every clinical situation (such as for an emergency admission), but it can be useful in reducing the risk of MRSA to immunocompromised patients in an oncology or transplant ward.

Systemic staphylococcal infection is treated with appropriate antibiotics, usually for at least two weeks (Tong *et al.*, 2015). This is a situation where determination of antimicrobial susceptibilities as soon as possible is a valuable contribution to patient outcome. In general, *Staphylococcus* species would be expected to be susceptible to penicillin and therefore a β-lactam antibiotic such as flucloxacillin would be recommended as the first-line treatment. If the isolate was identified as MRSA, then vancomycin would be used instead. Interestingly, there is increasing evidence for geographical variation in resistance patterns for MRSA and MSSA both between countries (e.g., den Heijer *et al.*, 2013) and within regions (e.g., Cuny *et al.*, 2015). Also resistance to standard treatments, such as methicillin and linezolid, amongst nosocomial CoNS isolates appears to be growing (e.g., Becker *et al.*, 2014; Gu *et al.* 2013). Therefore in the case of severely ill, immunocompromised patients it would be particularly important to ascertain the antimicrobial sensitivity patterns of individual *Staphylococcus* spp. isolates.

7.7 *Aspergillus* spp.

Classification
Kingdom: Fungi
Division: Ascomycota

Class: Eurotiomycetes
Order: Eurotiales
Family: Trichocomaceae
Genus: *Aspergillus*
Species *Aspergillus fumigatus, A. flavus A. niger, A. nidulans, A. terreus*

7.7.1 Introduction

Over 200 species of *Aspergillus* have been characterised, but only around 20 are known to be human pathogens (Warnock, 2012) and the five species listed above are the most important (Murray *et al.*, 2013). These spore-forming, filamentous fungi are common soil organisms and are also widely found growing as moulds in organic detritus – including holes in walls and corners of buildings; their conidia, containing spores, are transmitted through air and water (Murray *et al.*, 2013; Warnock, 2012) and therefore people are continually exposed to this fungus. In healthy adults, inhalation of conidia can sometimes result in localised, superficial fungal growth (although it is unusual) and some people develop an allergy to the spores, but generally clinical disease is rarely seen. In contrast, immunocompromised patients can experience disseminated, invasive infection, which is difficult to treat and is often fatal (Murray *et al.*, 2013).

7.7.2 Pathogenesis and Clinical Symptoms

Aspergillus spp. can establish in people who are not immunocompromised. There is a form of sinusitis, usually associated with invasive growth of *A. flavus* or *A. fumigatus* (Warnock, 2012). Up to 20% of asthmatics develop 'allergic aspergillosis' which is hypersensitivity to *A. fumigatus*, resulting in bronchial blockage and fluid collection in the lungs. The fungal mycelia grow in this mucous and may be seen on examination of respiratory specimens (Warnock, 2012). Laboratory investigations would also show raised levels of serum IgE, along with eosinophilia in the bronchial tissue (Murray *et al.*, 2013). This allergic condition can develop into chronic pulmonary aspergillosis (CPA). This chronic form of infection is also found in people who have lung cavities caused by previous damage (e.g., due to tuberculosis) or other forms of lung disease (Denning *et al.*, 2015). An aspergilloma can develop, which is a ball of tangled mycelial growth lodged in the space (Warnock, 2012). Patients generally only harbour one 'fungus ball', often experience either no symptoms or a mild (albeit productive) cough and intervention is not usually deemed necessary (Murray *et al.*, 2013). Nevertheless, multiple lesions do sometimes occur and also some patients with CPA have more extended mycelial growth, which leads to lung fibrosis (Denning *et al.*, 2015). Fungal activity may extend into the small blood vessels, causing haemoptysis and sometimes haemorrhage (Warnock, 2012).

  A deep-seated, disseminated *Aspergillus* spp. infection can occur in patients who are immunocompromised, particularly those who are neutropenic (Murray *et al.*, 2013; Warnock, 2012). The lung is the main site of this invasive aspergillosis (IA), but due to the propensity of the fungus to grow invasively into blood vessels it can spread to other sites including large organs and the CNS (Murray *et al.*, 2013); this occurs in approximately 30% of cases (Warnock, 2012). Patients present with a range of symptoms, depending on the nature and profundity of the compromise to their immune system

(Enoch *et al.*, 2006). Some cases are asymptomatic and in others prolonged pyrexia may be the only indication of a possible infection – which can make diagnosis a challenge, particularly since the fungus often does not grow in blood cultures (Warnock, 2012). The other clinical features are cough, chest pain and haemoptysis (Enoch *et al.*, 2006); fluid may be seen on the lungs (Murray *et al.*, 2013). Early diagnosis is the key to a successful outcome, but the mortality rate for IA is high even after anti-fungal treatment (SMI B57, 2015) and may be up to 70% (Murray *et al.*, 2013).

The pathogenic mechanisms involved in aspergillosis are not well understood, but it is clear that while the fungus is adapted to grow in the human body, the host response also plays a key part in the outcome of the exposure. Inhalation of conidia takes them into the lungs, where macrophages engulf and destroy them, while any spores which germinate into mycelia are attacked by neutrophils (Murray *et al.*, 2013). The fungal cell wall contains a number of polysaccharides, including galactomannan. The activity of germinating and growing exposes these to the host's innate immune response (Cramer *et al.*, 2011), including Toll Like receptors (TLRs) and cytokines. The fungus counters with a variety of proteases and the ability to bind to fibrinogen and laminin on the alveolar membranes, thus establishing growth (Murray *et al.*, 2013). An *Aspergillus*-specific CD4+ T-cell response has been found to be important in restricting its effects (Cramer *et al.*, 2011).

CPA mostly occurs in people who have underlying conditions which provide a favourable physical environment for the fungus, as well as preoccupying the cellular immune response, thus facilitating mycelial growth (Denning *et al.*, 2015). Similarly, those at highest risk of IA are patients who have been neutropenic for more than three weeks, which includes those where immunosuppression has been induced post-transplant (Enoch *et al.*, 2006). This is consistent with a lack of ability to mount either the innate cellular mechanisms or the specific CD4+ T cell response.

Point to consider 7.7: Do you think immunocompromised patients more at risk of exposure to *Aspergillus* spores? If so, how and why might this happen?

7.7.3 Epidemiology

Aspergillus spp. spores are ubiquitous in the everyday environment, but clinical disease is rare and incidence is not recorded systematically. The global prevalence of CPA has been estimated at over 2 million, compromising 411,000 cases of complications of allergic aspergillosis, 1.74 million cases in people with damage from previous/current tuberculosis and an additional 72,000 in people with pulmonary sarcoidosis (Denning *et al.*, 2015). *Aspergillus* spp. are one of the most common causes of invasive fungal infection in immunocompromised patients (Enoch *et al.*, 2006). Although treatments for immunological disorders are improving and the numbers of people undergoing transplants and cancer therapy are increasing, the population at risk for aspergillosis is still relatively small, which means that epidemiological data is limited. It has been estimated that 13% of bone marrow transplant recipients experience IA, as well as 14% to 18% of heart-lung transplant patients, up to 7% of those who have a liver transplant and less than 1% of kidney transplant patients (Enoch *et al.*, 2006). Reported studies tend to look at the frequency of involvement of particular organisms in recognised disease rather than incidence *per se*. For example, a European study of fungal infections which developed after bone marrow transplant analysed reports from 378 patients during 9 years. Fifty-three individuals developed invasive fungal infections and *Aspergillus* spp. was involved in the majority (50%) of these (Corzo-Leon *et al.*, 2015). One U.S. study of organ transplant recipients found that 19% developed IA (Pappas *et al.*, 2010), while another estimated the rate of infection to be 25% (Neofytos *et al.*, 2010) which also illustrates the difficulty with interpreting epidemiological data in this situation.

7.7.4   Laboratory Diagnosis

Diagnosis of aspergillosis is difficult and it requires close working between the clinicians on the wards and the laboratory staff. Investigations will usually involve abdominal X-ray and/or CT scan (depending on the presentation) which would indicate that infection with this organism was a possibility. The laboratory contribution is therefore vital in confirmation and ensuring that the patient receives appropriate treatment in good time. Suitable specimens to collect depend on the nature and site of the *Aspergillus* spp. infection. Sputum or bronchoalveolar lavage would probably be taken for general investigation of respiratory infection (SMI B57, 2015). Thus, where aspergillosis infection is a possibility, microscopic examination would be an appropriate starting point (SMI B57, 2015; Warnock, 2012). Preparation of the respiratory secretion in KOH should reveal the morphology of any hyphae and the characteristic conidiophores. Figure 7.3a is a sample of *A. niger* taken from a plate culture and prepared in lactophenol cotton blue stain. It shows the features of the conidiophore and conidia characteristic of the species clearly highlighted by the blue stain. The fluorescent stain calcofluor – which binds to cellulose and chitin – is recommended to improve sensitivity (SMI B57, 2015; Warnock, 2012). In general, microscopy is useful for diagnosis of allergic aspergillosis, less helpful for aspergilloma where fungal mycelia might be scanty and of almost no value for invasive aspergillosis (Warnock, 2012). Thus a failure to detect *Aspergillus* by this method does not exclude this infection.

After appropriate preparation, respiratory samples should also be cultured on Sabouraud dextrose agar without cycloheximide at 25–37 °C for 24 to 48 hours (SMI B57, 2015). It is important to note that filamentous moulds are common laboratory contaminants, so aseptic technique and scientific judgement are required for best results. Figure 7.3b shows the characteristic features of *Aspergillus niger* seen after several days growth on Sabouraud dextrose agar. Growth starts from the centre of the plate, so that the white fluffy outer layer is younger mycelia. This is a common feature of all *Aspergillus* spp., while the colour of the centre varies according to species. In *A. niger*, this middle section is a greenish/black (Figure 7.3b), while in *A. fumigatus* it is a blueish/green and in *A. flavus* it is a yellowish or olive green. The darker layer is caused by the coloured conidia (asexual spores), which the fungus produces as it matures. When microscopy does not yield detectable fungus (thus indicating that cultures should be set up), then a lung biopsy would be taken. Histological sections should be stained with periodic-acid Schiff (PAS) or Grocott's methenamine silver (Warnock, 2012) and again a suitably prepared homogenate from the biopsy can be cultured (SMI B17, 2015).

Molecular detection methods such as PCR for the 18s RNA ITS region (SMI B57, 2015) or the 28s rRNA gene (Lewis White *et al.*, 2013) reportedly have good sensitivity and the DNA can be analysed further for species identification (and strain variation). In whole blood and serum from patients with IA, PCR for the fungal genome is more sensitive than galactomannan (GM) serology (Lewis White *et al.*, 2013), although the

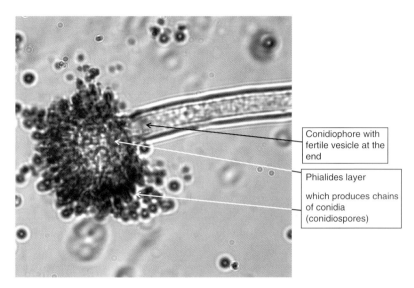

Figure 7.3a Conidiophore *Aspergillus niger* prepared in lactophenol cotton blue. (*See insert for colour representation of the figure.*)

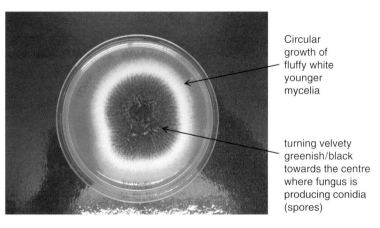

Figure 7.3b *Aspergillus niger* grown on Sabouraud dextrose agar. (*See insert for colour representation of the figure.*)

latter may be more feasible in most situations and provide a faster result (see below). Lewis White *et al.* (2013) evaluated a point-of-care immunochromatographic test intended to detect an extracellular fungal glycoprotein. They compared it with PCR for the *Aspergillus* 28s rRNA gene and an EIA to detect fungal GM using over 600 prospective and stored whole blood and serum samples. Taking the clinical definitions (proven IA, possible IA and no invasive fungal disease) as the gold standard, they reported overall specificities of 91.5%, 96.6% and 98.0% for GM EIA, PCR and the POCT respectively (Lewis White *et al.*, 2013). The PCR method showed the best sensitivity (95.5%), while this parameter was surprisingly low for the GM EIA at 77.3%. The POCT gave a sensitivity of 81.8% (Lewis White *et al.*, 2013), which would make it unsuitable for use as a standalone assay, but suggests that it might offer an advantage over GM serology.

The authors suggest that the POCT could be useful as part of a diagnostic algorithm including PCR (Lewis White *et al.*, 2013).

When the only symptom of invasive infection is pyrexia, the most likely course of action diagnostically would be to take blood cultures to determine the cause of the fever. However, *Aspergillus* spp. are rarely isolated in such cases (Warnock, 2012). In contrast with other types of systemic infection, serology is actually quite useful in aspergillosis diagnosis. For CPA patients with a single, defined aspergilloma (but no production of secretions (or haemoptysis) in which the fungal mycelial could be identified), serology for specific IgG or *Aspergillus* precipitins is usually enough to confirm the diagnosis (Denning *et al.*, 2015). Commercially available kits for detection of *Aspergillus* IgG generally have acceptable performance (e.g., Baxter *et al.*, 2013). GM released from the fungal cell wall during growth is detectable in serum and respiratory fluids of patients with IA. The method to detect this antigen is quantitative and it has therefore proved useful in both diagnosis and monitoring of the infection (e.g., Chai *et al.*, 2012). This assay should be considered when IA is a possibility, since early diagnosis equates with a better outcome (Warnock, 2012).

Point to consider 7.8: When *Aspergillus* is isolated from a respiratory sample from a patient with invasive aspergillosis (IA) is it important to determine the exact species? Explain your answer.

7.7.5   Treatment and Management

Surgery is required to remove aspergillomas, although each case is assessed on merit, as it is not always considered necessary or safe to do so. Where there is a risk of haemorrhage it would be recommended (Warnock, 2012). IA infection needs intravenous antifungal treatment. Amphotericin B is generally effective but it is quite toxic and does not work against *A. terreus* (Murray *et al.*, 2013); in this case voriconazole is a suitable alternative (Warnock, 2012). Studies comparing the two drugs have been limited by the selection of 'cases' according to clinical assessment rather than mycological confirmation of aspergillus infection (Herbrecht *et al.*, 2015). This highlights the importance of accurate and timely laboratory diagnosis and illustrates the role laboratory scientists can play in enhancing care through developing improved techniques. However, trials do indicate that voriconazole therapy is tolerated more easily and leads to better outcomes than amphotericin (e.g., Herbrecht *et al.*, 2015).

7.8   *Cryptococcus* spp.

Kingdom: Fungi
Phylum: Basidiomycota
Class: Tremellomycetes
Order: Tremellales
Family: Tremelaceae
Genus: *Cryptococcus*
Species: *Cryptococcus neoformans, C. gattii*

7.8.1   Introduction

Cryptococcus spp. are yeasts and they are classified as basidiomycetes. This is an eclectic phylum which also includes edible mushrooms, smuts and rusts. It should be noted that other yeasts, such as *Candida* spp. and *Saccharomyces* spp. are in the phylum Ascomycota. *Cryptococcus* spp. are important opportunistic pathogens, particularly in patients with T-lymphocyte disorders. A marked increase in the incidence of cryptococcal infections since the late 1980s has been associated with the emergence of human immunodeficiency virus (HIV), and a rise in the number of people on T-cell suppressive chemotherapy (Warnock, 2012).

Two species of *Cryptococcus* are known to be important pathogens in humans, *C. neoformans* and *C. gattii* and molecular analysis has led to further divisions into subspecies (Perfect and Bicanic, 2015). The number of genetic groups and nomenclature of species is subject to regular revisions due to the application of ever more discriminatory laboratory methods. The key point is that isolates in genetic groups VNI-VNIV are classified as variants of *C. neoformans*, while a different set of genetic groups (VGI, VGII, VGIII and VGIV) account for all *C. gattii* sub species (Gullo *et al.*, 2013) – which some authors argue should be separate species (e.g., Nyazika *et al.*, 2016). The laboratory scientist should also be aware that there are sporadic reports of human disease associated with other *Cryptococcus* species, notably *C. laurentii* (Gullo *et al.*, 2013).

7.8.2   Pathogenesis and Clinical Symptoms

Cryptococcus spp. are commonly found in the environment and therefore people are continually exposed to them, usually through inhalation. Healthy adults rarely develop overt disease, but may experience a mild lung infection, which resolves uneventfully (Warnock, 2012). Pulmonary infection does occasionally occur, with fungal lesions and this sometimes develops into a form of pneumonia (Warnock, 2012). In patients with T-lymphocyte deficiency, there is a risk of developing cryptococcal meningitis, which is potentially fatal (Murray *et al.*, 2013). The symptoms appear relatively slowly over a period of months, starting with headaches and low-level pyrexia, and then progressing to behavioural changes and visual impairment suggestive of brain damage (Warnock, 2012). Disseminated infection can also occur, mainly affecting the skin and bones. However, it should be noted that cases have been reported where *Cryptococcus*-associated skin and bone lesions were found in the absence of any detectable CNS or pulmonary cryptococcal infection (Warnock, 2012).

Point to consider 7.9: Why are patients with T-lymphocyte disorders particularly vulnerable to cryptococcal-associated disease?

In a healthy adult host, the *Cryptococcus* spp. cells would normally be ingested by macrophages, which then induce inflammatory cytokines. The macrophages also take the function of antigen presenting cells, thus triggering specific T- and B-cell responses (Murray *et al.*, 2013). The yeasts can produce a polysaccharide capsule, which affords protection against the phagocytosis and can disrupt the antigen presentation process. Unless a very high dose of yeast cells is inhaled, immunocompetent people can usually deal with the infection. Patients with impaired cellular immunity are less likely to respond effectively to clear the *Cryptococcus* and since it can grow at 37 °C, it is adapted to proliferate inside the human body (Murray *et al.*, 2013). A number of virulence factors which contribute to pathogenicity been identified (e.g., Alspaugh, 2015) and there is evidence that the expression of these may vary with genotype (e.g., Campbell *et al.*, 2015). The most prominent factor is the production of the polysaccharide capsule, which appears to be induced by the yeast metabolism when it is within the mammalian host, in response to the environment (Murray *et al.*, 2013; Alspaugh, 2015). The biochemical structure of the cell wall can also be changed (e.g., increased chitin content or production of melanin) as a result of cell stress (Murray *et al.*, 2013; Campbell *et al.*, 2015). In common with other fungi, *Cryptococcus* spp. secrete proteins in extracellular vesicles (Alspaugh, 2015) and these include enzymes which digest host substances, thus causing damage (Campbell *et al.*, 2015). Both innate and adaptive cellular responses are important in protection against cryptococcal-mediated pathology and the Th-1 inflammatory response appears to be crucial (Rohatgi and Pirofski, 2015). Clearly, an immunocompromised host, particularly someone with T-cell deficiency would not be able to mount the necessary defence. Healthy adults also produce protective antibodies against the capsule and the cell wall (Rohatgi and Pirofski, 2015). It appears that *C. gattii* can occasionally evade or overcome the response even in an immunocompetent host, although the circumstances which facilitate this and the mechanisms which the fungus uses are not fully understood.

7.8.3 Epidemiology

Both species of this yeast are common environmental organisms, but places that they are usually found are quite distinct. Yeasts of *C. neoformans* are commonly found in the faeces of the feral (town/city) pigeon (*Columbia livia*) and other birds, while *C. gattii* comes from *Eucalyptus* spp. (eucalyptus) trees (Gullo *et al.*, 2013; Warnock, 2012). This means that *C. neoformans* is transmitted throughout the world in urban and suburban areas. As highlighted in Figure 7.4a, pigeon faeces is an unavoidable feature of life in built-up areas, meaning that people are regularly exposed to the yeast particles. In contrast, *C. gattii* has a rather more restricted distribution since *Eucalyptus* spp. are native to Australia (Figure 7.4b) and Southeast Asia (though they are cultivated in other geographical areas). *C. gatti* is therefore considered to be a tropical and sub-tropical species. Although the incidence of the latter species has been increasing, the majority of cases of serious cryptococcal disease in humans are attributable to the former (Perfect and Bicanic, 2015; Gullo *et al.*, 2013).

In the mid-2000s, a large study estimated that there were about a million cases of cryptococcal meningitis each year around the world, with over 600,000 deaths (Park *et al.*, 2009). The majority of cases were reported from sub-Saharan Africa and the data showed a clear relationship with HIV prevalence. The calculated rate of cases in Western Europe at that time was less than 1000 (Park *et al.*, 2009). Reliable data is lacking, but it is thought that the incidence pattern is changing, partly due to effective and more widely available anti-retroviral therapy, coupled with greater awareness of the risk from *Cryptococcus* spp. amongst healthcare workers, leading to more routine screening for cryptococcal antigen (CRAG) in HIV positive patients (e.g., Meya *et al.*, 2010). Patel *et al.* (2013) screened 157 people who had attended a London hospital following a recent a new diagnosis of HIV. They found that eight of them had serological evidence of CRAG and in seven of these cases it was the presentation with cryptococcal

Figure 7.4a Pigeons (*Columba livia*) in an urban setting in the United Kingdom. Source: Courtesy of Dr A Gunn. (*See insert for colour representation of the figure.*)

Figure 7.4b Eucalyptus tree (Eucalyptus spp.) in a suburban setting in Australia.
Source: Courtesy of Mrs S. Weeks. (*See insert for colour representation of the figure.*)

meningitis which led to the HIV diagnosis (Patel *et al.*, 2013). In Canada and the United States, *C. gattii* is considered to be an emerging infection (Gullo *et al.*, 2013). A few cases have been reported in Europe (e.g., McCormick Smith *et al.*, 2015), although molecular typing investigations tend to indicate that the infections are usually acquired outside of the continent (McCormick Smith *et al.*, 2015; Gullo *et al.* 2013). Cryptococcal disease is also a known risk for immunocompromised patients who do not have HIV; for example, it occurs in up to 8% of people after SOT (Sloan and Parris, 2014). However, it is notable that there is also significant mortality and morbidity in apparently immunocompetent people (Perfect and Bicanic, 2015). This may be due to late diagnosis in cases where cryptococcosis is not initially considered as a possible causative agent; in this scenario *C. gattii* is more likely to be the culprit (Sloan and Parris, 2014).

Point to consider 7.10: What are the issues which make collecting reliable data on *Cryptococcus* spp. infection rates difficult?

7.8.4 Laboratory Diagnosis

Where cryptococcal meningitis is suspected or a strong possibility, an accurate and timely diagnosis is needed, in order to avoid a fatality. A number of options are available in the laboratory and it is likely that a combination would be employed. The yeast cells are often present at sufficient concentration to be detected by microscopy in cerebrospinal fluid (CSF) samples. They are quite small (diameter of 4–10 μm), so Indian ink (or other suitable stain) should be used to highlight the polysaccharide capsule and internal morphology of the yeast cell (SMI B27, 2015), although this does not always work very well (Warnock, 2012). In around 40% of cases, the yeast is not seen by microscopy and therefore the latex agglutination test for the cryptococcal capsular antigen is recommended (SMI B27, 2015), along with the serological test for the cryptococcal antigen (CRAG) to contribute to the diagnosis (SMI B27, 2015). It has been suggested that routine screening of AIDS patients for CRAG can be effective in identifying patients at risk of developing cryptococcal meningitis and starting anti-fungal treatment appropriately before they become symptomatic (Perfect and Bicanic, 2015; Meya *et al.*, 2010). Studies involving patients with HIV and CD4+ cell counts of <100 cells/μL have found this approach to be life-saving and cost-effective (e.g., Meya *et al.*, 2010; Jarvis *et al.*, 2013).

Immunochromatographic lateral flow assays to detect CRAG have been developed. They are reported to have good sensitivity and evaluation studies show favourable comparisons with other laboratory methods (e.g., Perfect and Bicanic, 2015; McMullan *et al.*, 2012). They are reportedly able to detect *C. gattii* as well as *C. neoformans* in a range of samples, including serum, CSF and urine (McMullan *et al.*, 2012). They have been found to be rapid, easy to use and cost-effective (including staff time in the cost per test calculation), thus making them suitable options for diagnosis in laboratories with limited experience of *Cryptococcus* spp. diagnosis (McMullan *et al.*, 2012) and as point of care tests (Perfect and Bicanic, 2015). Some formats are semi-quantitative which means they can also be useful in monitoring patients after treatment (Perfect and Bicanic, 2015), although they may not yet be as robust as required (McMullan *et al.*, 2012; Pfaller, 2015). Williams *et al.* (2015) tested the use of blood collected from finger pricks in a point-of-care CRAG kit on HIV patients in Uganda. They compared results obtained from venous blood and CSF using this kit with the finger-prick samples from 207 patients with possible cryptococcal meningitis. Overall, CRAG was detected in the finger-prick blood of 149 patients and the results were in full agreement with the tests on whole blood. The antigen was also detected in CSF from 138 of these CRAG positive patients, using the same point of care kit and confirmed by culture (Williams *et al.*, 2015). The test took 10 minutes to perform and given that it found patients who had cryptococcal fungaemia without meningitis (Williams *et al.*, 2015), it shows promise for early diagnosis, potentially allowing pre-emptive treatment to be started.

If sufficient quantities of the CSF sample remain after the microscopy and/or CRAG testing, then it should be plated out onto Sabouraud dextrose agar without cyclohex-imide, regardless of the initial results. If the yeast has not been detected so far, then culture is a chance to check; if *Cryptococcus* has been found, then an isolate is needed to confirm the species and determine anti-fungal susceptibility. Culturing blood can also be useful in severely immunocompromised patients, who might have fungaemia. Sets of plates should be incubated at both the usual, relatively low temperature, used for fungal cultures (25–30 °C) and at 37 °C for three weeks – although the organism usually grows within a few days (Warnock, 2012). Confirmation of species can be carried out in a number of ways, including PCR and MALDI-TOF (Pfaller, 2015).

In lung infections, sputum can be examined microscopically after treatment with a chemical such as potassium hydroxide to digest the surrounding material and release the yeast cells (Warnock, 2012). In a disseminated infection, biopsies can be taken and histological sections prepared; these are stained with periodic acid–Schiff to detect the fungus and a capsular stain such as Alcian blue (Warnock, 2012). Serological tests for IgG, IgM and IgA antibodies raised against cryptococcal capsular antigen are also available (Pfaller, 2015), but their diagnostic usefulness is fairly limited.

7.8.5 Treatment

In serious CNS and disseminated infections, treatment is a combination of ampho-tericin B and flucytosine (5-fluorocytosine), given intravenously, followed by a course of oral fluconazole to prevent a recurrence. This prophylactic-type phase may be required for 6 to 12 months (Perfect and Bicanic, 2015) or longer in AIDS patients (Warnock, 2012). Several different preparations of the drugs are available and although there are limitations with all anti-fungal agents (see Chapter 1), this schedule for treating crypto-coccal infection has been extensively trialled and is recommended by the WHO (Perfect and Bicanic, 2015). Resistant strains of *Cryptococcus* spp. are not common (Gullo *et al.*, 2013), but these drugs are still used cautiously because they are expensive and patients need careful monitoring due to their toxicity. Some clinicians recommend reduction of any immunosuppressive therapy during the course of antifungal treatment to improve the patient's outcome (e.g., Sloan and Parris, 2014). Unfortunately, most of those who need antifungal therapy for cryptococcosis are HIV-positive people in sub-Saharan Africa (Jarvis and Harrison, 2016; Meya *et al.*, 2010) and many do not receive adequate medical intervention (Perfect and Bicanic, 2015). Different doses and preparations of amphotericin B and flucytosine, as well new anti-fungal drugs are being trialled, with a view to creating regimens more suitable in low resource settings (Jarvis and Harrison, 2016; Perfect and Bicanic, 2015).

7.9 Exercises

7.1 **A** Think about the infectious agents which transplant donors and recipients need to be screened for prior to HSCT and SOT transplant procedures. Make a list for each.

    B Research protocols for pre-transplant screening, such as national guidelines and/or your laboratory's policy and compare these with your lists.

C Are there any notable differences between policies for the two types of transplant? Discuss reasons for the similarities/differences.

D Consider whether any significant pathogens have been omitted from the screening policy (e.g., newly recognised organisms).

7.2 A Investigate national guidelines and/or your laboratory's practice for monitoring HIV positive patients (which markers are tested for, which infections are screened for, how often).

B Discuss whether the available evidence supports the usual approach taken to monitoring.

C What role does monitoring play in determining long-term outcomes for the patient?

7.3 A Investigate published protocols and/or your laboratory's practice for screening of patients for MRSA carriage in advance of hospital admission. If you find variation in the types of patients or frequency of screening, discuss possible reasons for this.

B A number of point-of-care tests are available for MRSA testing. Carry out literature-based research on these and decide whether they should have a place in routine pre-procedure screening.

7.10   Case Studies

7.1 Mrs CM, who is now 53, was diagnosed with heart failure about four years ago. After waiting for a suitable donor for some time, she had a cardiac transplant four weeks ago. Two days ago, she developed a cough and a high temperature, which have not resolved.

a) Given her clinical situation, which infections would you need to exclude?

b) Which investigations would be recommended? Which sample(s) should be collected for laboratory testing and which tests should be undertaken?

c) After two days, the initial results are available. The chest X-ray indicates possible filamentous fungal infection, but all the microbiology laboratory tests reported so far have failed to detect any organisms. How should the laboratory proceed?

d) The follow-up tests indicate *Aspergillus* spp. infection. How should this be treated and managed? What is the prognosis for Mrs CM?

7.2 Mr AD is a 45-year-old man who was diagnosed with HIV infection nine years ago. He has attended all hospital and general practitioner appointments as required, so that his HIV viral load, CD4+ count and antiretroviral treatment regimen have been regularly monitored. He has now presented with a three-week history of an unproductive (dry) cough, shortness of breath and feeling unusually tired.

a) Which investigations would be undertaken first?

b) Clinically Mr AD is diagnosed with pneumonitis. Which infectious cause(s) would be suspected? Which samples would be collected for laboratory investigation and which tests should be done?

c) The virology results are:

CMV viral load: 3890 genomes mL^{-1} blood
HIV viral load: 380 copies mL^{-1} blood
CD4$^+$ count: 150 cells μL^{-1} of blood

What do these indicate?

d) How should Mr AD's pneumonitis be treated?

References

Aguilar C *et al.* (2014). Prevention of infections during primary immunodeficiency. *Clinical Infectious Diseases*, **59**: 1462–1470. doi: 10.1093/cid/ciu646

Alspaugh JA (2015). Virulence mechanisms and Cryptococcus neoformans pathogenesis. *Fungal Genetics and Biology*, **78**: 55–58. doi: 10.1016/j.fgb.2014.09.004

ARHAI (2014). Implementation of Modified Admission MRSA Screening Guidance for NHS. Leeds: Department of Health.

Argemi X *et al.* (2015). Implementation of MALDI-TOF MS in routine clinical laboratories improves identification of coagulase negative staphylococci and reveals the pathogenic role of *Staphylococcus lugdunensis*. *Journal of Clinical Microbiology*, **53**: 2030–2036. doi: 10.1128/JCM.00177-15

Atkinson C and Emery VC (2011). Cytomegalovirus quantification: Where to next in optimising patient management? *Journal of Clinical Virology*, **51**: 223–228. doi: 10.1016/j.jcv.2011.04.007

Baxter CG *et al.* (2013). Performance of two *Aspergillus* IgG EIA assays compared with the precipitin test in chronic and allergic aspergillosis. *Clinical Microbiology and Infection*, **19**: E197–E204. doi: 10.1111/1469-0691.12133

Beam E and Razonable RR (2012). Cytomegalovirus in solid organ transplantation: Epidemiology, prevention, and treatment. *Current Infectious Disease Reports*, **14**: 633–641. doi: 10.1007/s11908-012-0292-2

Becker K, Heilmann C and Peters G (2014). Coagulase-negative staphylococci. *Clinical Microbiology Reviews*, **27**: 870–926. doi: 10.1128/CMR.00109-13

Blann A and Ahmed N (2014). *Blood Science: Principles and Pathology*. Chichester: Wiley-Blackwell.

Campbell LT *et al.* (2015). Cryptococcus strains with different pathogenic potentials have diverse protein secretomes. *Eukaryotic Cell*, **14**: 554–563. doi: 10.1128/EC.00052-15

Cannas G *et al.* (2012). Infectious complications in adult acute myeloid leukemia: Analysis of the Acute Leukemia French Association-9802 prospective multicenter clinical trial. *Leukemia & Lymphoma*, **53**: 1068–1076. doi: 10.3109/10428194.2011.636812.

Chai LYA *et al.* (2012). Early serum galactomannan trend as a predictor of outcome of invasive Aspergillosis. *Journal of Clinical Microbiology*, **50**: 2330–2336. doi: 10.1128/JCM.06513-11

Chinen J and Buckley RH (2010). Transplantation immunology: Solid organ and bone marrow. *Journal of Allergy and Clinical Immunology*, **125**: S324–S325. doi: 10.1016/j.jaci.2009.11.014.

Cohen S, Tyrrell DAJ, Smith AP (1991). Psychological stress and susceptibility to the common cold. *New England Journal of Medicine*, **325**: 606–612.

Corzo-León DE *et al.* (2015). Epidemiology and outcomes of invasive fungal infections in allogeneic haematopoietic stem cell transplant recipients in the era of antifungal prophylaxis: A single-centre study with focus on emerging pathogens. *Mycoses*, **58**: 325–336. doi: 10.1111/myc.12318

Cramer RA, Rivera A and Hohl TM (2011). Immune responses against *Aspergillus fumigatus*: What have we learned? *Current Opinion in Infectious Disease*, **24**: 315–322. doi: 10.1097/QCO.0b013e328348b159

Cuny C *et al.* (2015). State-wide surveillance of antibiotic resistance patterns and spa types of methicillin-resistant Staphylococcus aureus from blood cultures in North Rhine-Westphalia, 2011–2013. *Clinical Microbiology and Infection*, **21**: 750–757. doi: 10.1016/j.cmi.2015.02.013

Danial J *et al.* (2011). Real-time evaluation of an optimized real-time PCR assay versus brilliance chromogenic MRSA agar for the detection of methicillin-resistant *Staphylococcus aureus* from clinical specimens. *Journal of Medical Microbiology*, **60**: 323–328. doi: 10.1099/jmm.0.025288-0

Delforge ML, Desomberg L and Montesinos I (2015). Evaluation of the new LIAISON® CMV IgG, IgM and IgG Avidity II assays. *Journal of Clinical Virology*, **72**: 42–45. doi: 10.1016/j.jcv.2015.09.002

Delves PJ, Martin SJ, Burton DR and Roitt IM (2011). *Roitt's Essential Immunology, 12th edn*. Chichester: Wiley-Blackwell.

Denning DW *et al.* (2015). Chronic pulmonary aspergillosis: Rationale and clinical guidelines for diagnosis and management. *European Respiratory Journal*, **47**: 45–68. doi: 10.1183/13993003.00583-2015

Enoch DA, Ludlam HA and Brown NM (2006). Invasive fungal infections: A review of epidemiology and management options. *Journal of Medical Microbiology*, **55**: 809–818. doi: 10.1099/jmm.0.46548-0

Fitzgerald C *et al.* (2016). Rapid identification and antimicrobial susceptibility testing of positive blood cultures using MALDI-TOF MS and a modification of the standardised disc diffusion test: a pilot study. *Journal of Clinical Pathology*, **69**: 1025–1034. doi: 10.1136/jclinpath-2015-203436

Friães A *et al.* (2015). Epidemiological survey of the first case of vancomycin-resistant *Staphylococcus aureus* infection in Europe. *Epidemiology and Infection*, **143**: 745–748. doi: 10.1017/S0950268814001423.

Fryer JF *et al.* (2010) Collaborative Study to Evaluate the Proposed 1st WHO International Standard for Human Cytomegalovirus (HCMV) for Nucleic Acid Amplification (NAT)-Based Assays. http://www.who.int/biologicals/expert_committee/BS_2138_HCMV_ECBS_Report.pdf

Gavaldà J, Aguado JM, Manuel O, Grossi P & Hirsch HH. (2014). A special issue on infections in solid organ transplant recipients. *Clinical Microbiology and Infection*, **20(s7)**: 1–3. doi: 10.1111/1469-0691.12658

George B *et al.* (2010). Pre-transplant cytomegalovirus (CMV) serostatus remains the most important determinant of CMV reactivation after allogeneic hematopoietic stem cell transplantation in the era of surveillance and preemptive therapy. *Transplant Infectious Disease*, **12**: 322–329. doi: 10.1111/j.1399-3062.2010.00504.x

Griffiths PD (2009). Cytomegalovirus. In: *Principles and Practice of Clinical Virology, 6th edn*. Zuckerman AJ, Banatvala JE, Schoub BD, Griffiths PD and Mortimer P, eds. Chichester: Wiley-Blackwell.

Griffiths P, Baraniak I and Reeves M (2015). The pathogenesis of human cytomegalovirus. *Journal of Pathology*, **235**: 288–297. doi: 10.1002/path.4437

Griffiths P *et al.* (2013). Desirability and feasibility of a vaccine against cytomegalovirus. *Vaccine*, **31**: B197–B203. doi: 10.1016/j.vaccine.2012.10.074

Griffiths P and Lumley S (2014). Cytomegalovirus. *Current Opinion in Infectious Diseases*, **27**: 554–559. doi: 10.1097/QCO.0000000000000107

Gu B *et al.* (2013). The emerging problem of linezolid-resistant Staphylococcus. *Journal of Antimicrobial Chemotherapy*, **68**: 4–11. doi: 10.1093/jac/dks354

Gudiol C *et al.* (2013). Changing aetiology, clinical features, antimicrobial resistance, and outcomes of bloodstream infection in neutropenic cancer patients. *Clinical Microbiology and Infection*, **19**: 474–479. doi: 10.1111/j.1469-0691.2012.03879.x

Gullo FP *et al.* (2013). Cryptococcosis: Epidemiology, fungal resistance and new alternatives for treatment. *European Journal of Microbiology and Infectious Diseases* **32**: 1377–1391. doi: 10.1007/s10096-013-1915-8

Halliday E, Winkelstein J and Webster A DB (2003). Enteroviral infections in primary immunodeficiency (PID): A survey of morbidity and mortality. *Journal of Infection*, **46**: 1–8. doi: 10.1053/jinf.2002.1066

den Heijer CD *et al.* and APRES Study Team. (2013). Prevalence and resistance of commensal Staphylococcus aureus, including meticillin-resistant S aureus, in nine European countries: A cross-sectional study. *The Lancet Infectious Diseases*, **13**: 409–415. doi: 10.1016/S1463-3099(13)70036-7

Herbrecht R *et al.* (2015). Application of the 2008 definitions for invasive fungal diseases to the trial comparing voriconazole versus Amphotericin B for therapy of invasive aspergillosis: A collaborative study of the Mycoses Study Group (MSG 05) and the European Organization for Research and Treatment of Cancer Infectious Diseases Group. *Clinical Infectious Diseases*, **60**: 713–720. doi: 10.1093/cid/ciu911

Hoffbrand AV and Moss PAH (2016). *Essential Haematology, 7th edn.* Chichester: Wiley-Blackwell.

Humphreys H (2012). Staphylococcus. In: *Medical Microbiology: A Guide to Microbial Infections, 18th edn.* Greenwood D, Barer M, Slack R and Irving W, eds. London: Churchill Livingston – Elsevier.

Jarvis JN *et al.* (2013). Cost effectiveness of cryptococcal antigen screening as a strategy to prevent HIV-associated cryptococcal meningitis in South Africa. *PloS One*, **8 (7):** e69288. doi: 10.1371/journal.pone.0069288

Jarvis JN and Harrison TS (2016). Forgotten but not gone: HIV-associated cryptococcal meningitis. *Lancet Infectious Diseases*, **16**: 756–758. doi: 10.1016/S1473-3099(16)00128-6

Jenq RR and van den Brink MRM (2010). Allogenic haematopoietic stem cell transplantation: individualized stem cells and immune therapy of cancer. *Nature Reviews Cancer*, **10**: 213–221. doi: 10.1038/nrc2804

Johannessen I and Ogilvie MM (2012). Herpesviruses. In: *Medical Microbiology: A Guide to Microbial Infections, 18th edn.* Greenwood D, Barer M, Slack R and Irving W, eds. London: Churchill Livingston – Elsevier.

Kerr JR (2015). A review of blood diseases and cytopenias associated with human parvovirus B19 infection. *Reviews in Medical Virology*, **25**: 224–240. doi: 10.1002/rmv.1839

Kohlmann R *et al.* (2015). MALDI-TOF mass spectrometry following short incubation on a solid medium is a valuable tool for rapid pathogen identification from positive blood cultures. *International Journal of Medical Microbiology*, **305**: 469–479. doi: 10.1016/j. ijmm.2015.04.004

Kotton CN (2010). Management of cytomegalovirus infection in solid organ transplantation. *Nature Reviews Nephrology*, **6**: 711–721. doi: 10.1038/nrneph.2010.141

Lewis White P, Parr C, Thornton C and Barnes RA (2013). Evaluation of real-time PCR, galactomannan enzyme-linked immunosorbent assay (ELISA), and a novel lateral-flow device for diagnosis of invasive aspergillosis. *Journal of Clinical Microbiology*, **51**: 1510–1516. doi: 10.1128/JCM.03189-12

Lisboa LF *et al.* (2012). Analysis and clinical correlation of genetic variation in cytomegalovirus. *Transplant Infectious Disease*, **14**: 132–140. doi: 10.1111/j.1399-3062.2011.00685.x

Ljungman P *et al.* (2014). Donor cytomegalovirus status influences the outcome of allogeneic stem cell transplant: A study by the European group for blood and marrow transplantation. *Clinical Infectious Diseases*, **59**: 473–481. doi: 10.1093/cid/ciu364

McCormick Smith I *et al.* (2015). Cryptococcosis due to *Cryptococcus gattii* in Germany from 2003-2013. *International Journal of Medical Microbiology*, **305**: 719–723. doi: 10.1016/j.ijmm.2015.08.023

McIntyre CL, Knowles NJ and Simmonds P (2013). Proposals for the classification of human rhinovirus species A, B and C into genotypically assigned types. *Journal of General Virology*, **94**: 1791–1806. doi: 10.1099/vir.0.053686-0

McMullan BJ *et al.* (2012). Clinical utility of the cryptococcal antigen lateral flow assay in a diagnostic mycology laboratory. *PLoS ONE* **7**(11)**:** e49541. doi: 10.1371/journal. pone.0049541

Manicklal S, Emery VC, Lazzarotto T, Boppana SB and Gupta RK (2013). The "silent" global burden of congenital cytomegalovirus. *Clinical Microbiology Reviews*, **26**: 86–102. doi: 10.1128/CMR.00062-12

Manuel O *et al.* (2009). Impact of genetic polymorphisms in cytomegalovirus glycoprotein B on outcomes in solid-organ transplant recipients with cytomegalovirus disease. *Clinical Infectious Diseases*, **49**: 1160–1166. doi: 10.1086/605633

de Martel C *et al.* (2012). Global burden of cancers attributable to infections in 2008: a review and synthetic analysis. *Lancet Oncology*, **13**: 607–615. doi: 10.1016/S1470-2045(12)70137-7

Meya DB *et al.* (2010). Cost-effectiveness of serum cryptococcal antigen screening to prevent deaths among HIV-infected persons with a CD4+ cell count ≤ 100 cells/µL who start HIV therapy in resource-limited settings. *Clinical Infectious Diseases*, **51**: 448–455. doi: 10.1086/655143

Morris K, Wilson C and Wilcox MH (2012). Evaluation of chromogenic meticillin-resistant Staphylococcus aureus media: sensitivity versus turnaround time. *Journal of Hospital Infection*, **81**: 20–24. doi: 10.1016/j.jhin.2012.02.003

Mortensen LH *et al.* (2012). Early-life environment influencing susceptibility to cytomegalovirus infection: Evidence from the Leiden Longevity Study and the Longitudinal Study of Aging Danish Twins. *Epidemiology and Infection*, **140**: 835–841. doi: 10.1017/S0950268811001397

Murphy KM (2011). *Janeway's Immunobiology, 8th edn.* New York: Garland Science.

Murray PR, Rosenthal KS and Pfaller MA (2013). *Medical Microbiology, 7th edn.* Philadelphia: Elsevier-Saunders.

Neofytos D *et al.* (2010). Epidemiology and outcome of invasive fungal infections in solid organ transplant recipients. *Transplant Infectious Disease*, **12**: 220–229. doi: 10.1111/j.1399-3062.2010.00492.x

Nyazika TK *et al.* (2016). *Cryptococcus tetragattii* as a major cause of cryptococcal meningitis among HIV-infected individuals in Harare, Zimbabwe. *Journal of Infection*, **72**: 745–752. doi: 10.1016/j.jinf.2016.02.018

Pappas PG, *et al.* (2010). Invasive fungal infections among organ transplant recipients: results of the Transplant-Associated Infection Surveillance Network (TRANSNET). *Clinical Infectious Diseases*, **50**: 1101–1111. doi: 10.1086/651262

Parcell BP and Phillips G (2014). Use of Xpert® MRSA PCR point-of-care testing beyond the laboratory. *Journal of Hospital Infection*, **87**: 119–121. doi: 10.1016/j.jhin.2014.04.002

Park BJ *et al.* (2009). Estimation of the current global burden of cryptococcal meningitis among persons living with HIV/AIDS. *AIDS*, **23**: 525–530. doi: 10.1097/QAD.0b013e32822ffac

Patel S *et al.* (2013). The prevalence of cryptococcal antigenemia in newly diagnosed HIV patients in a Southwest London cohort. *Journal of Infection*, **66**: 75–79. doi: 10.1016/j.jinf.2012.09.014

Perfect JR and Bicanic T (2015). Cryptococcosis diagnosis and treatment: What do we know now. *Fungal Genetics and Biology*, **78**: 49–54. doi: 10.1016/j.fgb.2014.10.003.

Pfaller MA (2015). Application of culture-independent rapid diagnostic tests in the management of invasive candidiasis and cryptococcosis. *Journal of Fungi*, **1**: 217–251. doi: 10.3390/jof1020217

PHE. (2015). Annual Epidemiological Commentary: Mandatory MRSA, MSSA and *E. coli* bacteraemia and *C. difficile* infection data, 2014/15. London: Public Health England.

Pignatelli S, Monte PD, Rossini G and Landini MP. (2004). Genetic polymorphisms among human cytomegalovirus (HCMV) wild-type strains. *Reviews in Medical Virology*, **14**: 383–410. doi: 10.1002/rmv.438V

Pillay D, Geretti AM and Weiss RA (2009). Human immunodeficiency viruses. In: *Principles and Practice of Clinical Virology*, *6th edn.* Zuckerman AJ, Banatvala JE, Schoub BD, Griffiths PD and Mortimer P, eds. Chichester: Wiley-Blackwell.

Razonable RR and Hayden RT (2013). Clinical utility of viral load in management of cytomegalovirus infection after solid organ transplantation. *Clinical Microbiology Reviews*, **26**: 703–727. doi: 10.1128/CMR.00015-13

Rohatgi S and Pirofski LA. (2015). Host immunity to *Cryptococcus neoformans*. *Future Microbiology*, **10**: 565–581. doi: 10.2217/FMB.14.132

Romero-Gómez MP *et al.* (2013). Evaluation of combined use of MALDI-TOF and Xpert (®) MRSA/SA BC assay for the direct detection of methicillin resistance in *Staphylococcus aureus* from positive blood culture bottles. *The Journal of Infection*, **67**: 91–92. doi: 10.1016/j/jinf.2013.03.014

Schneider-Schaulies S and ter Meulen V (2009). Measles virus. In: *Principles and Practice of Clinical Virology*, *6th edn.* Zuckerman AJ, Banatvala JE, Schoub BD, Griffiths PD and Mortimer P, eds. Chichester: Wiley – Blackwell.

Shallcross LJ *et al.* (2013). The role of the Panton-Valentine leucocidin toxin in staphylococcal disease: A systematic review and meta-analysis. *Lancet Infectious Diseases*, **13**: 43–54. doi: 10.1016/S1473-3099(12)70238-4

Shore AC and Coleman DC (2013). Staphylococcal cassette chromosome *mec*: Recent advances and new insights. *International Journal of Medical Microbiology*, **303**: 350–359. doi: 10.1016/j.ijmm.2013.02.002

Sijmons S, Van Ranst M and Maes P. (2014). Genomic and functional characteristics of human cytomegalovirus revealed by next-generation sequencing. *Viruses*, **6**: 1049–1072. doi: 10.3390/v6031049.

Sloan DJ and Parris V (2014). Cryptococcal meningitis: epidemiology and therapeutic options. *Clinical Epidemiology*, **6**: 169–182. doi: 10.2147/CLEP.S38850

SMI B17: Public Health England. (2014). Investigation of Tissues and Biopsies. UK Standards for Microbiology Investigations. B17 Issue 5.3. http://www.hpa.org.uk/SMI/pdf

SMI B27: Public Health England. (2015). Investigation of Cerebrospinal Fluid. UK Standards for Microbiology Investigations. B27 Issue 6. https://www.gov.uk/uk-standards-for-microbiology-investigations-smi-quality-and-consistency-in-clinical-laboratories

SMI B29: Public Health England. (2014). Investigation of Specimens for Screening for MRSA. UK Standards for Microbiology Investigations. B29 Issue 6. http://www.hpa.org.uk/SMI/pdf

SMI B57: Public Health England. (2015). Investigation of bronchoalveolar lavage, sputum and associated specimens. UK Standards for Microbiology Investigations. B57 Issue 3.1. https://www.gov.uk/uk-standards-for-microbiology-investigations-smi-quality-and-consistency-in-clinical-laboratories

SMI ID7: Public Health England. (2014). Identification of *Staphylococcus* species, *Micrococcus* species and *Rothia* species. UK Standards for Microbiology Investigations. ID7 Issue 2.3. http://www.hpa.org.uk/SMI/pdf

SMI V28: Public Health England. (2015). Cytomegalovirus Serology. UK Standards for Microbiology Investigations. V28 Issue 3. https://www.gov.uk/uk-standards-for-microbiology-investigations-smi-quality-and-consistency-in-clinical-laboratories

Sommerstein R *et al.* (2015). Factors associated with methicillin-resistant coagulase-negative staphylococci as causing organisms in deep sternal wound infections after cardiac surgery. *New Microbes and New Infections*, **6**: 15–21. doi: 10.1016/j.nmni.2015.04.003

Tasse J *et al.* (2016). Rapid bench identification of methicillin-sensitive and methicillin-resistant Staphylococcus aureus: A multicenter comparative evaluation of Alere PBP2a Culture Colony Test (Alere) Versus Slidex MRSA detection (bioMérieux). *Diagnostic Microbiology and Infectious Disease.* **85**: 419–421. doi: 10.1016/j.diagmicrobio.2016.04.008

Taylor GP (2009). Human T-lymphotropic viruses. In: *Principles and Practice of Clinical Virology*, 6th edn. Zuckerman AJ, Banatvala JE, Schoub BD, Griffiths PD and Mortimer P, eds. Chichester: Wiley-Blackwell.

Thorley-Lawson D *et al.* (2016). The link between *Plasmodium falciparum* malaria and endemic Burkitt's lymphoma—New insight into a 50-year-old enigma. *PLOS Pathogens*, **12**(1): e1005331. doi: 10.1371/journal. ppat.1005331

Tong SYC *et al.* (2015). *Staphylococcus aureus* infections: epidemiology, pathophysiology, clinical manifestations, and management. *Clinical Microbiology Reviews*, **28**: 603–661. doi: 10.1128/CMR.00134-14

Warnock DW (2012). Fungi. In: *Medical Microbiology: A Guide to Microbial Infections, 18th edn.* Greenwood D, Barer M, Slack R and Irving W, eds. London: Churchill Livingston – Elsevier.

Williams D *et al.* (2015). Evaluation of fingerstick cryptococcal antigen lateral flow assay in HIV-infected persons: A diagnostic accuracy study. *Clinical Infectious Diseases*, **61**: 464–467. doi: 10.1093/cid/civ263

Yavuz SS *et al.* (2013). Incidence, aetiology, and control of sternal surgical site infections. *Journal of Hospital Infection*, **85**: 206–212. doi: 10.1016/j.jhin.2013.07.010

Yossepowitch O *et al.* (2014). A cost-saving algorithm for rapid diagnosis of Staphylococcus aureus and susceptibility to oxacillin directly from positive blood culture bottles by combined testing with BinaxNOW® S. aureus and Xpert MRSA/SA Assay. *Diagnostic Microbiology and Infectious Disease*, **78**: 352–355. doi: 10.1016/j.diagmicrobio.2014.01.001

Zhu W *et al.* (2015). Evaluation of the Biotyper MALDI-TOF MS system for identification of *Staphylococcus* species. *Journal of Microbiological Methods*, **117**: 14–17. doi: 10.1016/j.mimet.2015.07.014.

Appendix

Answers for Case Studies

Chapter 2

Case study 1

A Would you report this result to the haematology/oncology clinicians at this stage?

Although coagulase negative staphylococci (CoNS) are a common contaminant of IV lines, it is important not to dismiss this result as not significant without more information about the collection of the samples and the condition of the patient. She is showing signs of sepsis and no other microorganism was found in the blood culture. CoNS can cause sepsis and they are particularly associated with nosocomial infections. The results of other tests (API or MALDI-TOF) to confirm the exact species are awaited, but would be prudent to let the clinicians treating Mrs AB know about this isolate while additional results are pending.

B What should be taken into account when deciding on appropriate antibiotics?

Local hospital/trust policy and or national guidelines about antibiotic prescription. It is important to bear in mind that there are anti-microbial resistant strains of S. epidermidis *which are resistant to, for example, methicillin and vancomycin.*

Case study 2

A What did these results suggest

The HIV results indicate that Mr FG has met the criteria to start anti-retroviral treatment. His HCV viral load is high. He still has no evidence of infection with HBV, but because he is in a high-risk group, monitoring should continue. It is not clear from the information available, but he may have been offered the HBV vaccine.

B How do you explain these results?

The anti-retroviral treatment appears to be effective in reducing the HIV activity and thus increasing the CD4+ count. The general improvement in FG's condition with respect to the HIV seems to be having a positive impact on his response to the HCV as well; the viral load has also decreased for HCV. While there might be a temptation to doubt either the previous or current HCV viral load results, this is likely to be a genuine effect even though no specific treatment for the HCV has been given.

Clinical Microbiology for Diagnostic Laboratory Scientists, First Edition. Sarah J. Pitt.
© 2018 John Wiley & Sons Ltd. Published 2018 by John Wiley & Sons Ltd.

Chapter 3

Case study 1

A   Which pathogens do you think would be the mostly causes of Mr AJ's symptoms?

Symptoms of possible 'atypical pneumonia': Possible pathogens include Mycoplasma pneumoniae, Legionella pneumophila *and* Chlamydophila pneumoniae.
*   However, the patient history suggests CMV pneumonitis could be a risk (no information about CMV serological status provided). Also respiratory viruses could cause these symptoms in an immunocompromised patient. Pneumocystis pneumonia should also be considered.*

B   Discuss these findings and suitable treatment and management for Mr AJ.

P. jirovecii pneumonia is usually considered in HIV-positive patients with atypical pneumonia, but it is also a risk for patients post bone marrow transplant. Mr AJ should be managed for his symptoms and also started on a course of treatment with co-trimoxazole (unless contraindicated).

Case study 2

A   Which viruses would you suspect might be causing BK's symptoms?

Possible pathogens include respiratory syncytial virus, adenoviruses, parainfluenza viruses, and influenza viruses. Co-infection with two or more of these is also a possibility, especially if this has occurred in the winter season.

B   Which laboratory test would you suggest in this case and why?

Multiplex PCR for respiratory viruses, which should include the most common and locally most prevalent viruses. RSV is likely and information about possible outbreak in the local area should be sought. However, other causes cannot be discounted, particularly as the wheezing is not (yet) severe.

C   Under which circumstances would you suggest treatment for BK?

If her condition deteriorates and the wheezing becomes severe respiratory distress, then oxygen may be used to relieve symptoms. If RSV is confirmed, then ribavirin might be considered.

Chapter 4

Case study 1

A   What do these symptoms suggest and which investigations should be carried out (not only microbiology laboratory tests)?

Symptoms suggest stomach ulcer, which could be due to Helicobacter pylori *infection. Investigations include gastroscopy, collection of stomach tissue biopsy, urease test on biopsy material, urea breath test, and H. pylori stool antigen test.*

B   Would this be enough to confirm the clinical diagnosis? If not, which further tests could be performed on the tissue sample?

This should be sufficient since histological appearance is distinctive. However, if there is any uncertainty, the presence of bacterial antigens can be verified using immunocytochemistry.

C Assuming that the diagnosis is confirmed, which treatment would be recommended for Mr DE?

Management of symptoms and a proton pump inhibitor plus antibiotic treatment such as clarithromycin plus either amoxicillin or metronidazole to eliminate the H. pylori.

Case study 2

A What do these results indicate?

i) *Patient has received HBV vaccine (anti-HBS but no* **core** *antibody which would be indicative of past infection)*
ii) *Patient may be in acute stage of HCV infection or equivocal result may be an anomaly*
iii) *Patient appears to be experiencing acute HAV infection, though it should be ascertained whether he was recently given anti-HAV vaccine*

B Which follow-up tests should be carried out and what time interval would be appropriate?

Suggest follow-up in one month. Repeat HCV and HAV antibody tests. If the 'equivocal' HCV antibody result was due to early stages of acute infection, the test should now give a definitive result. The HAV IgG should also be detectable by this time, although IgM will probably not have cleared by then.

Chapter 5

Case study 1

A Which infectious causes of these symptoms would you consider testing for and why?

Given the 'flu-like' symptoms it would be sensible to test for influenza virus since this has been reported to be dangerous to the mother in pregnancy (particularly pandemic Influenza A H1 N1).
The swollen glands could be indicative of primary CMV which could cause congenital infection in the baby.
The symptoms also suggest primary toxoplasmosis which can also be a source of congenital infection.

B Which samples should be collected and which investigations should be performed?

Respiratory samples (e.g., throat swab) for influenza; PCR for Influenza A and B.
Blood samples for serology; IgM and IgG for both Cytomegalovirus and Toxoplasma gondii.

C What do these results indicate?

They indicate that Mrs GH has acute primary toxoplasmosis.

D How should Mrs GH be treated and counselled?

Amniotic fluid should be collected if possible for PCR for T. gondii *DNA.*
She should be prescribed antibiotics – if confirmation of foetal infection, then sulfadiazine and pyrimethamine; if not, then spiromycin.
Due to her overt symptoms, it is likely that the toxoplasmosis infection has been detected and treated in good time to prevent transmission to the foetus. The chances of vertical transmission early in the second trimester are estimated to be relatively low (30–40%).
Both these factors indicate a low risk of serious adverse consequences to the baby, but Mrs GH must be advised that her baby may be affected and he/she must be monitored carefully post-partum.

Case study 2

A Provide an interpretation for each of these results

JK is immune to rubella and there is no evidence of infection with HIV or syphilis. However, she does have active infection with HBV.

B Which of these results should be followed up and which further tests should be undertaken?

Request a second blood sample; re-test for HBV to confirm the result.
Test for other markers – core total and IgM to see whether acute or chronic infection
e- antigen/ e-antibody to determine risk of perinatal infection
HBV viral load

C The follow-up tests indicate a pattern of markers indicating a chronic HBV infection and a high risk of transmission of the virus to her baby perinatally. Outline the likely results and what plans should be put in place for the birth.

Would expect results to be: Core total antibody DETECTED

e- antigen DETECTED
HBV viral load: > 10^6 HBV DNA copies mL^{-1}

Plan: Order HBV vaccine and HBIG; consider use of antivirals; consider caesarean section (though need to consider mother's needs here).

Chapter 6

Case study 1

A Which laboratory investigations would you recommend and why?

Since the infection might not be gonorrhoea, Gram stain of a smear prepared from the swab plus culture (using a selection of specialized media) could be a possible way to proceed. However, given the presumptive diagnosis and the relatively low sensitivity of GC culture, NAAT for Neisseria gonorrhoeae *would be recommended if available.*

B What further laboratory tests might be required in order to treat the individual patient effectively and from a public health point of view?

In this situation,there is no bacterial isolate for determination of antibiotic sensitivities for the N. gonorrhoeae *so treatment of EP would be started without that information. However, follow-up swabs should be collected and tested once the course of antibiotics has been completed. Depending on Ms EP's history of possible contact with STIs, a full screen for a range of STIs might be considered. This would mean a request to the GP to collect further samples. Contact tracing of all EP's possible sexual contacts, starting with FT would be very important for public health considerations.*

A What are the implications for EP and FT of this result?

There is a psychological aspect of being diagnosed with an STI which must also be taken into account. Has EP acquired the infection from FT or someone else? She might be angry with FT and embarrassed by the diagnosis.

Case study 2

A What treatment and management would be recommended for the patient?

The patient should be treated with a course of metronidazole and be advised that it is common infection but usually asymptomatic. He should take precautions to avoid re-infection.

B The GUM clinic decide to try and contact as many of GT's recent sexual contacts as possible and test them for infection with the parasite. Which test method/assay would you recommend?

Examination of swabs is the usual routine method, but there are issues with sensitivity, particularly for screening of asymptomatic patients. If available, the Transcription Mediated Amplification (TMA) would lend itself well to this type of situation as it would be less labour intensive for screening a number of samples. Also in case of any collection and transport issues, the TMA is a good option because it does not rely on viable parasites for optimal detection.

Chapter 7

Case study 1

A Given her clinical situation, which infections would you need to exclude?

Since she is on immunosuppressive therapy and only a few weeks post a major operation, she is at risk from a range of opportunistic infections. The symptoms are fairly non-specific, but could suggest the early stages of pneumonia. Nosocomial infections are a possibility. Possible bacterial causes of bacterial pneumonia include Staphylococcus aureus *(MRSA),* Pseudomonas aeruginosa, Klebsiella pneumoniae, Escherichia coli. *Mrs CM's Cytomegalovirus status should be confirmed and if she was seropositive pre-transplant, a reactivation of the virus is possible. Also,* Pneumocystis jirovecii *pneumonia and fungal infections should be considered.*

B   Which investigations would be recommended? Which sample(s) should be collected for laboratory testing and which tests should be undertaken?

Chest x-ray and microbiological testing of bronchoalveolar lavage (BAL). Laboratory tests should include microscopy for fungi, bacterial/mycological culture, and PCR for viruses.

C   How should the laboratory proceed?

Continue with the Sabouraud agar incubation since filamentous fungi may take longer than 48 hours to grow; suggest collection of lung biopsy for histological examination; consider PCR for fungal species.

D   The follow-up tests indicate *Aspergillus* spp. infection. How should this be treated and managed? What is the prognosis for Mrs CM?

Treatment with antifungal agent – amphotericin B or voriconazole. Since the infection has been detected before the symptoms were severe, it is to be hoped that it early in the infection. However, invasive aspergillosis in immunocompromised patients carries a high mortality rate.

Case study 2

A   Which investigations would be undertaken first?

As Mr AD's cough is not productive, clinical pulmonary investigations such as lung function tests and x-ray would be indicated at this stage.

B   Clinically Mr AD is diagnoses with pneumonitis. Which infectious cause(s) would be suspected? Which samples would be collected for laboratory investigation and which tests should be done?

Cytomegalovirus pneumonitis should be suspected in an HIV-positive patient with these symptoms. Bronchalvelolar lavage should be collected for CMV PCR. Whole blood should also be tested for CMV and HIV viral loads, along with CD4+ count.

C   What do these indicate?

As the HIV viral load is $> 10^2$ copies per mL and the $CD4^+$ count is < 200 cells per mL*, this indicates that the anti-retroviral therapy is sub-optimal. This is possibly due to non-compliance (unlikely since Mr AD seems to be taking care of his health, attending regular medical appointments and managing his condition well) or failure of one of the drugs in his treatment regimen (e.g., due to HIV strain developing resistance). It is the resulting immunosuppression which has allowed the CMV to cause symptomatic disease.

D   How should Mr AD's pneumonitis be treated?

Anti-retroviral therapy should be reviewed. The aim would be to manage Mr AD's HIV infection optimally, to boost the patient's own immune response to the CMV, meaning that anti-CMV therapy would not be required in order for the pneumonitis to resolve. Ganciclovir treatment might be considered if this did not happen.

*see Chapter 2

Index

Clinical Microbiology for Diagnostic Laboratory Scientists, First Edition. Sarah J. Pitt.
© 2018 John Wiley & Sons Ltd. Published 2018 by John Wiley & Sons Ltd.